Carbonate Platform Systems: components and interactions

Special Publication reviewing procedures

The Society makes every effort to ensure that the scientific and production quality of its books matches that of its journals. Since 1997, all book proposals have been refereed by specialist reviewers as well as by the Society's Publications Committee. If the referees identify weaknesses in the proposal, these must be addressed before the proposal is accepted.

Once the book is accepted, the Society has a team of series editors (listed above) who ensure that the volume editors follow strict guidelines on refereeing and quality control. We insist that individual papers can only be accepted after satisfactory review by two independent referees. The questions on the review forms are similar to those for *Journal of the Geological Society*. The referees' forms and comments must be available to the Society's series editors on request.

Although many of the books result from meetings, the editors are expected to commission papers that were not presented at the meeting to ensure that the book provides a balanced coverage of the subject. Being accepted for presentation at the meeting does not guarantee inclusion in the book.

Geological Society Special Publications are included in the ISI Science Citation Index, but they do not have an impact factor, the latter being applicable only to journals.

More information about submitting a proposal and producing a Special Publication can be found on the Society's web site: www.geolsoc.org.uk.

It is recommended that reference to all or part of this book should be made in one of the following ways.

INSALACO, E., SKELTON, P. W. & PALMER, T. J. (eds) 2000. *Carbonate Platform Systems: components and interactions*. Geological Society, London, Special Publications, **178**.

WRIGHT, D. T. & ALTERMANN, W. 2000. Microfacies development in Late Archaean stromatolites and oolites of the Ghaap Group of South Africa. *In*: INSALACO, E., SKELTON, P. W. & PALMER, T. J. (eds) 2000. *Carbonate Platform Systems: components and interactions*. Geological Society, London, Special Publications, **178**, 51–70.

GEOLOGICAL SOCIETY SPECIAL PUBLICATION NO. 178

Carbonate Platform Systems: components and interactions

EDITED BY

E. INSALACO
TotalFinaElf Exploration UK PLC, Geoscience Research Centre

P. W. SKELTON
The Open University, UK

and

T. J. PALMER
University of Wales, Aberystwyth, UK

2000
Published by
The Geological Society
London

THE GEOLOGICAL SOCIETY

The Geological Society of London was founded in 1807 and is the oldest geological society in the world. It received its Royal Charter in 1825 for the purpose of 'investigating the mineral structure of the Earth' and is now Britain's national society for geology.

Both a learned society and a professional body, the Geological Society is recognized by the Department of Trade and Industry (DTI) as the chartering authority for geoscience, able to award Chartered Geologist status upon appropriately qualified Fellows. The Society has a memberhip of 9099, of whom about 1500 live outside the UK.

Fellowship of the Society is open to persons holding a recognized honours degree in geology or a cognate subject, or not less than six years' relevant experience in geology or a cognate subject. A Fellow with a minimum of five years' relevant postgraduate experience in the practice of geology may apply for chartered status. Successful applicants are entitled to use the designatory postnominal CGeol (Chartered Geologist). Fellows of the Society may use the letters FGS. Other grades of membership are available to members not yet qualifying for Fellowship.

The Society has its own Publishing House based in Bath, UK. It produces the Society's international journals, books and maps, and is the European distributor for publications of the American Association of Petroleum Geologists (AAPG), the Society for Sedimentary Geology (SEPM) and the Geological Society of America (GSA). Members of the Society can buy books at considerable discounts. The Publishing House has an online bookshop (http://bookshop.geolsoc.org.uk).

Further information on Society membership may be obtained from the Membership Services Manager, The Geological Society, Burlington House, Piccadilly, London W1V 0JU (E-mail: enquiries@geolsoc.org.uk; tel: +44 (0) 207 434 9944).

The Society's Web Site can be found at http://www.geolsoc.org.uk/. The Society is a Registered Charity, number 210161.

Published by The Geological Society from:
The Geological Society Publishing House
Unit 7, Brassmill Enterprise Centre
Brassmill Lane
Bath BA1 3JN
UK

(*Orders*: Tel. +44 (0)1225 445046
Fax +44 (0)1225 442836)
Online bookshop: http://bookshop.geolsoc.org.uk

British Library Cataloguing in Publication Data
A catalogue record for this book is available from the British Library.

ISBN 1–86239–074–6

Typeset by Type Study, Scarborough, UK
Printed by Bell & Bain, UK

Distributors

USA
AAPG Bookstore
PO Box 979
Tulsa
OK 74101–0979
USA
Orders: Tel. +1 918 584-2555
Fax +1 918 560-2652
E-mail *bookstore@aapg.org*

Australia
Australian Mineral Foundation Bookshop
63 Conyngham Street
Glenside
South Australia 5065
Australia
Orders: Tel. +61 88 379-0444
Fax +61 88 379-4634
E-mail *bookshop@amf.com.au*

India
Affiliated East–West Press PVT Ltd
G-1/16 Ansari Road, Daryaganj,
New Delhi 110 002
India
Orders: Tel. +91 11 327-9113
Fax +91 11 326-0538
E-mail *affiliat@nda.vsnl.net.in*

Japan
Kanda Book Trading Co.
Cityhouse Tama 204
Tsurumaki 1-3-10
Tama-shi
Tokyo 206–0034
Japan
Orders: Tel. +81 (0)423 57-7650
Fax +81 (0)423 57-7651

Contents

INSALACO, E., SKELTON, P. W. & PALMER, T. J. Carbonate platform systems: components and interactions – an introduction 1

NAYLOR. L. A. & VILES, H. A. A temperate reef builder: an evaluation of the growth, morphology and composition of *Sabellaria alveolata* (L.) colonies on carbonate platforms in South Wales 9

STEUBER, T. Skeletal growth rates of Upper Cretaceous rudist bivalves: implications for carbonate production and organism – environment feedbacks 21

PERRY, C. T. & BERTLING, M. Spatial and temporal patterns of macroboring within Mesozoic and Cenozoic coral reef systems 33

WRIGHT, D. T. & ALTERMANN, W. Microfacies development in Late Archaean stromatolites and oolites of the Ghaap Group of South Africa 51

RIEGL, B. & PILLER, W. E. Reefs and coral carpets in the northern Red Sea as models for organism – environment feedback in coral communities and its reflection in growth fabrics 71

NEBELSICK, J. H. & BASSI, D. Diversity, growth-forms and taphonomy: key factors controlling the fabric of coralline algal dominated shelf carbonates 89

GILI, E. & SKELTON, P. W. Factors regulating the development of elevator rudist congregations 109

GLYNN, P. W. El Niño-Southern Oscillation mass mortalities of reef corals: a model of high temperature marine extinctions? 117

GISCHLER, E. & LOMANDO, A. J. Isolated carbonate platforms of Belize, Central America: sedimentary facies, late Quaternary history and controlling factors 135

HOUSE, M. R., MENNER, V. V., BECKER, R. T., KLAPPER, G., OVNATANOVA, N. S. & KUZ'MIN, V. Reef episodes, anoxia and sea-level changes in the Frasnian of the southern Timan (NE Russian platform) 147

STÖSSEL, I. & BERNOULLI, D. Rudist lithosome development on the Maiella Carbonate Platform margin 177

KIESSLING, W., FLÜGEL, E. & GOLONKA, J. Fluctuations in the carbonate production of Phanerozoic reefs 191

SCHLAGER, W. Sedimentation rates and growth potential of tropical, cool-water and mud-mound carbonate systems 217

Index 229

Preface

This volume arises from the 1999 Lyell Meeting on '*Organism-environment feedbacks in carbonate platforms and reefs*', which was held at the Geological Society, London, on 1–2 March, 1999. It was convened by Enzo Insalaco (Elf, now TotalFinaElf Exploration UK), Peter Skelton (Open University) and Tim Palmer (University of Wales, Aberystwyth). The aim of the meeting was to explore examples of how interactions between organisms and environments have generated the variety of carbonate platform facies and geometries seen in the ancient and modern record. All hierarchical levels of interaction were considered, ranging from that between organismal growth and ambient conditions, via the development of facies mosaics according to climate and the provision of accommodation space, to the interplay of global and evolutionary change. Such an approach was felt to be timely in view of the rise to prominence of 'Earth Systems Science' over the last few years, with its focus upon how complex systems of influences and feedbacks between components of the Earth operate. Thirty-nine presentations (oral and poster) were made at the meeting, addressing aspects ranging from the growth of individual carbonate skeletons to global changes in the burial of carbonate carbon through time. Thirteen of these presentations are expanded here as papers.

The meeting was commissioned by the Joint Committee for Palaeontology, on behalf of the Geological Society, the Palaeontological Association, the British Micropalaeontological Society and the Palaeontographical Society. Generous financial support for the meeting was received from Amerada Hess Ltd, BP Exploration Operating Co. Ltd., and the Geological Society, to all of whom we express our gratitude. We would also like to thank the staff of the Geological Society, and Janet Dryden at the Open University, for their friendly and efficient help with the organisation of the meeting. Likewise, we thank the staff of the Society's Publishing House, as well as Series Editor Martyn Stoker, and our army of referees, for their assistance in bringing this publication to fruition.

Enzo Insalaco, Peter Skelton & Tim Palmer

Carbonate platform systems: components and interactions – an introduction

ENZO INSALACO[1], PETER SKELTON[2] & TIM J. PALMER[3]

[1] *TotalFinaElf Exploration UK PLC, Geoscience Research Centre, 30 Buckingham Gate, London, SW1E 6NN, UK*

[2] *Department of Earth Sciences, Open University, Milton Keynes MK7 6AA, UK*

[3] *Institute of Geography and Earth Sciences, University of Wales Aberystwyth, Aberystwyth, Ceredigion SY23 3BD, UK*

Carbonate platforms are open systems with natural boundaries in space and time. Across their spatial boundaries there are fluxes of energy (e.g. light, chemical energy in compounds, and kinetic energy in currents and mass flows) and matter (e.g. nutrients, dissolved gases such as CO_2, and sediment – especially, of course, carbonates). Internally, these fluxes are regulated by myriads of interactions and feedbacks (Masse 1995), and the residue is consigned to the geological record. The most distinctive aspect of carbonate platforms is the predominant role of organisms in producing, processing and/or trapping carbonate sediment, even in Precambrian examples.

Because of evolutionary changes in this strong biotic input, it is harder to generalize about carbonate platforms than about most other sedimentary systems. Evolution has altered both the constructive and destructive effects of platform-dwelling organisms on carbonate fabrics, with profound consequences for facies development. Moreover, changing patterns in the provision of accommodation space (e.g. between greenhouse and icehouse climatic regimes) have also left their stamp on facies geometries, in turn feeding back to the evolution of the platform biotas. Hence simplistic analogies between modern and ancient platforms may give rise to misleading interpretations of what the latter were like and how they formed. Although a number of carbonate platform and reef specialists have warned of the dangers of such misplaced uniformitarianism (e.g. Braithwaite 1973; Gili *et al.* 1995; Wood 1999), it remains depressingly commonplace in the literature on ancient carbonate platforms. The endless quest in the literature for an all-purpose definition of 'reefs' in the fossil record is symptomatic of this delusion. Like the Holy Grail of Medieval legend, the object of the search remains cloaked in the vagueness of myth. The aim of this volume is more pragmatic – to present case studies that describe the components of some ancient and modern examples, analyse their interactions, detect significant differences as well as similarities between them, and so explore the possible causes and effects of changes through time. In short, our quest is for the reality of variety, not imagined unity.

The studies presented here concentrate on shallow-water platform systems, with an occasional nod towards fossil mudmounds of deeper water origin, largely because these are the most accessible and best-understood examples of substantial carbonate bodies. Nevertheless, we should also note in passing the existence of significant tracts of deep-water coral mounds today, around the North Atlantic (Mortensen *et al.* 1995), for example. Ongoing work on these should provide interesting comparisons with their shallow-water counterparts, especially with respect to the effects on fabrics and facies of their very different circumstances of development.

The geological outcomes of platform development depend upon a hierarchy of interactions. At the lowest level are those that structure communities, often involving taphonomic feedbacks. Over larger scales of time and space, tectonics, eustacy, climate and oceanographic factors set their own imprints on the physiography and facies anatomy of platforms, including the determination of their beginnings and ends – the temporal boundaries of the platform systems. We have accordingly arranged the thirteen papers presented in this volume in two parts, to reflect this hierarchical scaling. Part 1 concerns community level aspects, from organisms and sediment production, to growth fabrics. Themes include:

(a) the ecology and palaeoecology of benthic biotas, particularly factors influencing growth fabric genesis;
(b) processes and rates of skeletal growth, bioerosion and sediment production; and
(c) taphonomic and diagenetic influences.

Part 2 concerns larger scale aspects, from influences on the growth and demise of individual

From: INSALACO, E., SKELTON, P. W. & PALMER, T. J. (eds) 2000. *Carbonate Platform Systems: components and interactions*. Geological Society, London, Special Publications, **178,** 1–8. 0305–8719/00/$15.00

platforms, to global patterns of change. Themes here include:

(a) factors influencing the establishment and demise of carbonate platform systems;
(b) the sequence stratigraphy of carbonate platforms and its bearing on the development of biogenic lithosomes within them; and
(c) global estimates of carbonate production by various carbonate platform systems through the Phanerozoic.

Part 1: Community level processes and products

Growth fabrics (*sensu* Insalaco 1998) emerge from tightly integrated systems of abiotic and biotic components that display a variety of organism – environment feedbacks. Processes that have a bearing on what type of fabric is preserved include: recruitment, growth and reproduction of skeletalized organisms; bioerosion; secondary encrustation; sediment autoproduction by mechanical and biological erosion *in situ*; sediment import (both siliciclastic and carbonate) and export (including dissolution); and early marine cementation. Figure 1 presents a conceptual dynamic model that integrates these processes. When viewed in this fashion it is clear that changes in one component can result in positive or negative feedbacks from others.

The rates at which these component processes operate, over very short temporal scales (seasons to decades), and their effects and linkages with other factors are variously examined in this first part of the book. Their geological implications and the degree to which they can be inferred from the geological record are also considered. There is a huge literature on reef science spread across the biological, palaeontological and sedimentological domains, and a comprehensive review is beyond the scope of this book. Recent reviews of the general subject area are given in Fagerstrom (1987), Birkeland (1997) and Wood (1999).

These themes are discussed with respect to corals (**Riegl & Piller**), coralline algae (**Nebelsick & Bassi**), cyanobacteria (**Wright & Altermann**), rudist bivalves (**Steuber, Gili & Skelton**), polychaetes (**Naylor & Viles**) and macroborers (**Perry & Bertling**), and with reference to different time periods from the Precambrian to the present day. The different temporal scales of processes are also addressed, ranging from those

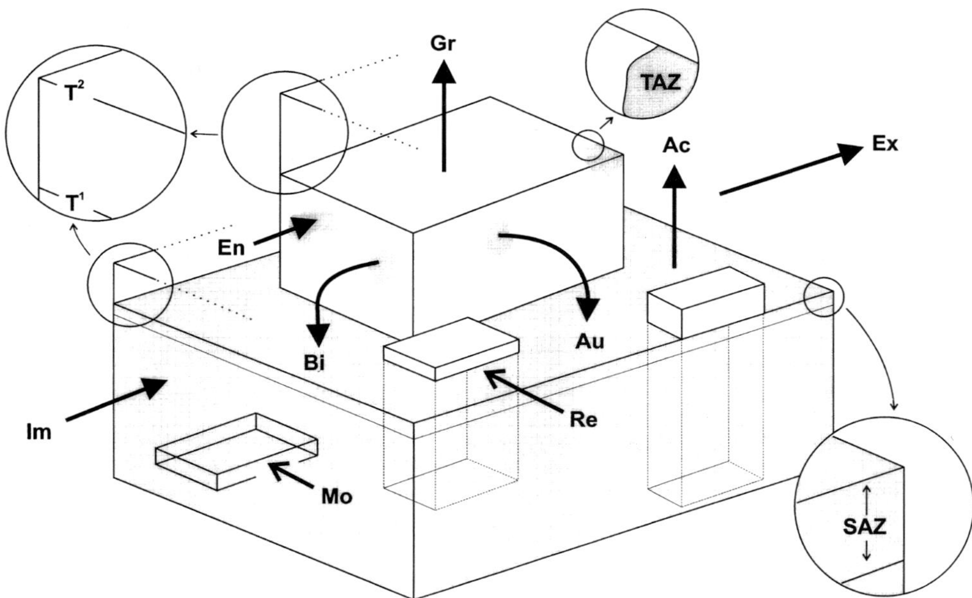

Fig. 1. Conceptual diagram illustrating the biological and sedimentological components and interactions involved in growth fabric production by shelly benthos on carbonate platforms. (Ac) sediment accumulation rate; (Au) Autoproduction (mechanical and biological erosion); (Bi) bioerosion; (En) encrustation; (Ex) sediment export (including dissolution); (Gr) skeletal growth rate; (Im) sediment import; (Mo) mortality; (Re) recruitment; (SAZ) sedimentologically active zone; (T^n) successive time slices; (TAZ) taphonomically active zone.

operating within the lifetime of an individual (**Naylor & Viles, Steuber, Riegl & Piller**) to the time involved in the formation of entire biogenic lithosomes (**Gili & Skelton**). The studies also concern a variety of depositional contexts (ramps, shelves, mixed carbonate-siliciclastic systems, and both tropical and temperate regimes).

The first two papers look at the role of skeletal accretion (either by sand-trapping or carbonate shell growth) and environmental consequences.

Though set in a siliciclastic context, the study of **Naylor & Viles** illustrates the environmental feedbacks of bioconstructions, with reference to *Sabellaria alveolata* (Linné), a sedentary polychaete that builds significant, but poorly documented, wave-resistant reefs from sand particles. Reefs can form quite rapidly in temperate waters with a large, continuous supply of sand-grade sediment and turbulent water in the lower eulittoral zone. They have three main environmental effects. First, they increase topographical variation in the lower eulittoral zone, such as enhancing pool forms, thereby increasing the amount of sheltered habitat for other marine fauna. Secondly, they increase surface roughness, which possibly reduces wave energy as swash dissipates more quickly over rougher surfaces ('bioprotection'). Thirdly, the reefs appear to protect the underlying limestone from direct wave attack and abrasion by physically covering the surface and by storing sediment that might otherwise be available to abrade the shore.

In a more classically carbonate context, **Steuber** has deployed geochemical sclerochronology to measure skeletal growth rates of Upper Cretaceous rudist bivalves, in order to evaluate their implications for carbonate production and organism-environment feedbacks. Estimated values of annual $CaCO_3$ production in dense congregations range from 4.6 to 28.5 kg m^{-2} – comparable to those reported from modern coral reefs. Study of a rudist-bearing sedimentary sequence revealed a decrease both in shell size and carbonate production of a single species in a transect from turbulent outer platform, to lagoonal inner platform deposits. Accumulating bioclastic sediment and fouling by faeces and pseudofaeces may have inhibited growth and excluded many other taxa in the restricted settings where current-flushing was infrequent, while nutrient flux and frequent flushing allowed optimal growth on the outer platform. The rapid vertical growth of the rudists would have mitigated against the effects of fouling, but only where sedimentary accumulation was itself sufficient to maintain shell stability. Thus there appears to have been a complex system of feedbacks between shell growth, packing density and current activity, which yielded a highly efficient, if ephemeral, kind of carbonate factory in propitious conditions.

In modern platform environments, organisms also play a major role in skeletal breakdown (Bromley 1994). Such bioerosional processes are considered in the paper by **Perry & Bertling**, who have assessed the impact of boring by macroscopic organisms on Mesozoic and Cenozoic coral reefs. Macroboring of coral reefs has varied significantly through time, with the modern intensity and producer composition a relatively recent phenomenon. Given the impact of bioerosion on framework morphology, community composition and sediment production today, Neogene reef systems make poor analogues for understanding the structure and dynamics of older examples. Sponges (dominant now) appear to have played a subordinate role in Mesozoic coral-dominated buildups. Worms and barnacles dominated in the early Mesozoic (Triassic and early Jurassic), with a progressive increase in bivalve borers through the Jurassic. Macroborers may have radiated to colonize new ecological niches during the early stages of coral reef diversification. The trend was nonetheless influenced by other biotic changes in the marine realm, especially switches in nutrient status and the origins or diversification of reef grazers.

Besides the constructive and destructive processes discussed above, diagenesis also takes a hand in fabric development. **Wright & Altermann** have looked at growth fabric taphonomy from a diagenetic angle in Late Archaean age stromatolites and ooids of the Ghaap Group, South Africa. This case study illustrates the importance of microbial communities as process-drivers for carbonate precipitation (see also Webb 1996). Detailed microfabric analysis illustrates the close relationship between organic decay processes, carbonate mineralogy and fabric development. It appears that the outcomes were in part controlled by diagenesis under reducing conditions. Anoxic microbial decay, by modifying ambient water chemistry, set the context for carbonate precipitation. Moreover, the degree of organic degradation was a significant control on the mineralogy (e.g. dolomite formation). Hence, even from Archaean times, carbonate platform development bore the imprint of biotic activity.

The influence of topographical context is considered next, in the paper by **Riegl & Piller**, who discuss Recent coral carpets and reefs in the northern Red Sea and the gulfs of Suez and Aqaba. They show the dependence of growth fabric types on a combination of sea-floor topography, hydrodynamic aspect and the ecology of constituent coral species. Reef frameworks show a clear ecological zonation along depth and hydrodynamic exposure gradients. Coral carpets, by contrast, build a framework lacking a distinct internal zonation since they grow only in areas without pronounced gradients. The initiation of the differing framework types was governed by bottom topography. They differ, moreover, in their patterns of sediment retention – the carpet tends to retain its bioclastic production, while the reef exports it. Synchronously with framework growth, the environment itself is modified, which in turn modifies the coral communities. Thus an environment-organism-environment feedback loop exists.

To discover what kinds of formative processes and conditions can be interpreted from the fossil record, **Nebelsick & Bassi** have analysed growth fabric variation in coralline algal-dominated Lower Oligocene shelf carbonates from northern Slovenia. They demonstrate the importance of assemblage diversity, growth-forms and taphonomy in controlling the type of fabric preservation. A wide range of taphonomic features, including disarticulation, encrustation, fragmentation and abrasion, can be observed. It is noted that the taphonomy of red algae is highly dependent on initial growth-form and specific environment. The determination of diversity is dependent on taxonomic identification, necessarily based only on preserved diagnostic characters. Uncertainties can thus arise through conflicts between palaeontological and botanical systematics. Growth-form determination in thin section is influenced by orientation and sectioning affects. Despite such problems, it is suggested that these aspects of limestone fabrics, when carefully analysed, can form a sound basis for microfacies differentiation, and hence palaeoenvironmental interpretation.

The final paper in this part of the volume, by **Gili & Skelton**, examines the feedbacks that appear to have regulated the development of elevator rudist lithosomes in the Upper Cretaceous of the southern Central Pyrenees. This study complements that of **Steuber**, viewing these distinctive biosedimentary systems in a broader perspective in order to discern the factors involved in their initiation, consolidation and termination. Initiation was associated with pauses in allochthonous sediment influx, themselves linked with minor increments of accommodation space. Two positive feedback loops to establishment are apparent: first, pioneer settlers provided more hard substrates for subsequent recruitment; and secondly, bioerosion of shells fuelled the *in situ* formation of bioclastic sediment, leading to the embedding and consolidation of congregations. Thereafter, an inferred correlation between rudist density and sediment destabilization at the benthic boundary layer is postulated to have affected rudist recruitment. Thus, successful rudist larval settlement declined with increasing numerical density of individuals – a crucial negative feedback mechanism. The rudist congregations could then have been maintained at about the same density through time by this stabilizing process, until shoaling and/or swamping by renewed allochthonous influxes terminated them.

Of this set of seven papers, only one (**Riegl & Piller**) concerns itself directly with growth fabric development in modern coral reefs. This may seem surprising in a book about interactions and feedbacks in carbonate platform ecosystems, of which coral reefs are probably the most widely investigated examples. Yet the concerns of the other papers – shell growth and carbonate production, bioerosion, taphonomy (including diagenesis) and factors controlling recruitment and mortality – are relevant across the entire spectrum of carbonate platform ecosystems, including modern coral reefs. This point brings us back to consideration of the dynamic model proposed in Fig. 1: the very diversity of the examples discussed here illustrates the chaotic tendency of such a complex system to yield contrasting results according to differing initial conditions, both biotic and environmental.

Part 2: Larger scale aspects

Beyond community interactions and local taphonomic feedbacks, a host of larger-scale and/or longer-term factors modulate the dynamic relationship between biogenic growth fabrics and associated facies, and ultimately control the establishment and demise of the platforms themselves. A serious hindrance to understanding these aspects of platform development is the gulf between ecological and geological scales of observation. We know all too little about the long-term geological repercussions of short temporal scale environmental events, and how to recognize their geological signatures. The scaling issue is addressed by **Glynn**, who looks at the effects of episodic El Niño-Southern Oscillation (ENSO) events on coral reef systems. These are effectively geologically instantaneous events, which may nevertheless have long term consequences for evolution and extinction, and **Glynn** asks how such short-lived ecological impacts may be recognized in the geological record. This paper thus effectively bridges the artificial gap between the two parts of the volume.

Extreme environmental conditions related to ENSO activity are (1) high sea surface temperatures (SSTs) and low photic zone nutrient availability during El Niño events, and (2) low SSTs and high nutrient availability during La Niña events. Extreme ENSO events (e.g. 1982–1983 and 1997–1998) stressed eastern tropical Pacific reef-building corals, with severe mortality promoted by 'bleaching' during prolonged sea warming, and restoration inhibited by plankton blooms and overgrowth by benthic algae during periods of elevated nutrient concentrations. Recruitment of corals has been nil to slow in many disturbed areas. Intense external and internal bioerosion by echinoids and endolithic bivalves, respectively, has already occurred on reefs affected by the 1982–1983 El Niño event, and reef frameworks in the Galápagos Islands that had been established for between 1000 and 5000 years have been reduced to cobbles and sand. The depauperate coral fauna and meagre reef development in the eastern tropical Pacific could thus be due largely to the severe and episodic conditions resulting from ENSO perturbations and subsequent low coral recruitment and intense bioerosion, which limits reef substrates. But how may such events be recognized in the geological record? Possibilities proposed by **Glynn** include:

(a) temperature-related oxygen isotope signatures,
(b) skeletal stress bands and growth discontinuities,
(c) increases in coral debris in beach storm deposits,
(d) increase in coral clastics resulting from intensified bioerosion, and
(e) increased preservation of bioeroded skeletal material.

This timely study shows that taphonomy offers one key to bridging the gap between short temporal scale interactions and longer-term consequences. It is an area that still has much potential to offer (see also Scoffin 1992).

The next two papers consider factors involved in the establishment of platforms. The effect of topography on growth fabrics was noted earlier, in the paper by **Riegl & Piller**. Over a longer time-scale, the positive feedback of reef framework growth on prominences during sea-level rise may amplify the relationship, so affecting the overall structure of a platform and its reef tracts. This relationship was explored in a classic paper on the role of antecedent karst topography by Purdy (1974), and the theme is pursued here by **Gischler & Lomando**, in a paper on the late Quaternary development of the Belize reef tract. They find that differences in physiography and facies between the various reefs relate to a combination of antecedent topography, due to differential subsidence and latitudinal variation in karstification, and hydrodynamic aspect.

House *et al.*, consider events over a longer time-span – the comings and goings of a prograding series of carbonate platforms in the southern Timan and Pechora region of northern European Russia during late Devonian (Frasnian) times. With the benefit of recent refinements in correlation based on conodonts and ammonoids, they distinguish globally effective influences (especially eustacy) from regional effects. Episodes of reef development in the region co-incided with transgressive phases, some evidently eustatic, but they appear to have ended in shallowing events, unlike many examples in western Europe where termination involved eutrophication and drowning. Moreover, reef developments continued into the late Frasnian, whilst in other European regions they mostly ended earlier in the Frasnian. The global incidence of anoxic facies is a striking feature, postulated to reflect the periodic flooding of cold, nutrient-rich waters over epicontinental areas in connection with upwelling systems associated with the distinctive climatic regime of the time. This substantial study illustrates the vital importance of detailed stratigraphical analysis for any attempt to make sense of global versus regional controls on the development of ancient carbonate platforms.

Stössel & Bernoulli focus more closely on the internal sedimentary dynamics of carbonate platform development, in their study of the Upper Cretaceous succession of the Maiella Platform in the central Apennines, Italy. Like **Steuber**, and **Gili & Skelton**, they too note the prodigious bioclastic productivity of rudist-dominated associations, particularly in the outer platform zone (see also Carannante *et al.* 1993, 1997, 1999; Gili *et al.* 1995; Ruberti 1997). This production regularly outpaced the provision of accommodation space, resulting in redistribution of the bioclastic sediment both on and off platform, yielding the distinctively tabular facies geometry so typical of rudist platforms. Sheet-like rudist lithosomes with individuals in life position make up only some 20% of the stratigraphic thickness of the Maiella outer platform zone, and are preferentially preserved in the thicker sedimentary cycles indicative of relatively larger increments of accommodation space. The latter observation is particularly interesting, pointing to an important longer-term taphonomic feedback to eventual facies architecture. In this context, another long-term feedback to the development of platform facies that should be mentioned is that of differential compaction (Hunt *et al.* 1996), a frequently overlooked factor. The three studies of rudist lithosomes in this volume together paint a remarkably consistent picture of this distinctive type of non-reefal biosedimentary system, which contrasts markedly with that of modern coral reefs, despite more than matching the latter in terms of carbonate production. Once again, we are confronted with the contrasts – as much as the similarities – between platform ecosystems from different times and places, referred to earlier. In the particular case of the Cretaceous, this has also been an emergent theme in the publications of the Global Sedimentary Geology Programme (GSGP) CRER-Working Group 4 on Cretaceous Carbonate Platforms (Schlager & Philip 1990; Simó *et al.* 1993; Philip & Skelton 1995).

The final two papers of the volume consider patterns on a global scale, through geological time. In recent years, documentation of the stratigraphical distribution of Phanerozoic carbonate platforms has become sufficiently comprehensive to allow broad semi-quantitative reviews of their history to be attempted. Dercourt *et al.* (1993) published a widely referred-to series of palaeogeographical maps showing platform development during 14 selected time intervals from the late Murgabian (Permian) to the Tortonian (Tertiary). More recently, Kiessling *et al.* (1999) have assembled a 'comprehensive database on Phanerozoic reefs'. Here, these same authors use their database to calculate relative carbonate production rates by reefs through the Phanerozoic, using certain assumptions concerning, for example, inferred potentials for debris production and export, and progressive loss of record through time. Four maxima for gross production are noted (Wenlock/Ludlow, Givetian/Frasnian, late Jurassic and Neogene). Of these, the calculated Givetian/Frasnian peak is the highest, matching the large numbers and extents of platform margin reef systems known from that interval – one major example of which was described earlier in this volume by **House *et al.*** **Kiessling *et al.*** compare their plot with inferred variations in such extrinsic factors as eustatic sea-level, ocean crust production, atmospheric CO_2 levels, palaeoclimate and nutrient availability, but find few correlations. They conclude that either the controls on reefal carbonate production are too complex to allow reliable predictions, or biotic factors represent more important controls than physico-chemical parameters. The constructed curve of Phanerozoic reefal carbonate production is also poorly correlated with proposed curves of global carbonate platform areas, suggesting that reefs *per se* rarely made a significant contribution to the global carbonate budget. Such grand syntheses are bound to prove controversial – as this one did at the conference – not least because of the assumptions used in the calculations. Whether or not the reader regards the present results as valid, however, they do serve to draw attention to the issues that need to be tackled if we are to arrive at a quantitative account of the chequered and variegated history of carbonate platforms and reefs.

The volume closes with **Schlager's** global review, which covers an even broader canvas, though employing a different approach. He recognizes three major kinds of benthic production factory for marine carbonates – tropical, cool-water, and mud-mound carbonate systems – and compares data on their sedimentation rates and growth potentials. An important consideration stressed here, in relation to all the systems, is the decrease in measured sedimentation rates according to the time span of observation. The importance of normalizing for time span when making comparisons cannot therefore be over-emphasized. 'Growth potential' is estimated from maximum observed rates of aggradation for each system. Highest rates overall are shown by the tropical system. Although values for the cool-water system can match these for short time spans (up to 200 000 years), they fall back more markedly for time spans of over 1 Ma. Reworking and local trapping of the frequently more mobile cool-water carbonates is regarded as probably responsible for their high short-term values. The mud-mound system, surprisingly, shows rates of aggradation that rival those of the tropical system, though since they export considerably less sediment than the latter, their gross production rate is less. Again, the assumptions involved in such a broad comparison lay it open to controversial discussion. A question that particularly needs to be addressed is the extent to which **Schlager's** tripartite classification of carbonate factories, based essentially on modern carbonate sediments, can be applied legitimately to the past. Carannante *et al.* (1997, 1999), for example, have cogently argued that 'foramol-type' Upper Cretaceous rudist limestones, though geographically tropical, were more like 'cool-water' carbonates in terms of their sedimentological attributes.

Prospect

As with any thriving area of scientific enquiry, the studies here beg as many questions as they offer answers. They invoke models that require further testing and assumptions that may be open to debate, and so sustain the need for more primary data. Hence, in addition to what they report, they also serve to identify where new initiatives might be directed. Each highlights particular needs, but two general issues are worth commenting upon in conclusion.

As noted at the beginning of this introduction, differing biotic and other environmental inputs to the growth of carbonate platforms have led to some significant differences in their forms and distributions. The variability of the examples discussed in this book illustrate the point, so reinforcing the need to generate synthetic interpretative models of ancient platform systems from primary observational data. Simplistic imposition of generalized models based on (certain) modern reef systems, for example, can be misleading. At issue is the appropriate level at which to apply uniformitarian analogy. At the level of simple physical, chemical and to some extent biological relationships, it is a good friend. At the level of whole, complex depositional systems, however, it can deceive.

A second general point is that at all hierarchical levels, from the genesis of growth fabrics to global patterns of carbonate platform development, the rates of processes are seen to be as important as their nature in moulding the variety of outcomes in the record. In many instances we have a fair idea of the main processes concerned, but our knowledge of their rates is often based on limited data. Hence there is a need for more quantitative estimation of rates, so that conceptual models (like Fig. 1) can be converted to predictive numerical models amenable to rigorous testing. For example, how have relative rates of skeletal carbonate production and sediment autoproduction through bioerosion varied in time and space? And how have these variations, in tandem with diagenetic influences, affected the potentials of platform systems for *in situ* carbonate aggradation and export? At around what (short-term) rates did given types of growth fabric accumulate, and how long did it take for lithosomes comprised of them to form, relative to other deposits? How have the frequency and distribution of ecological perturbations such as ENSO events, hurricanes and – especially further back in time – oceanic anoxic events varied, and what (if any) correlation do they show with the changing fortunes of platforms? How have relative rates of carbonate production and redistribution varied along with changing patterns in the provision of accommodation space? Allied with such questions is the crucial issue, raised several times in the book, of the temporal scaling of rates – the non-linear translation of short-term effects (e.g. sedimentation and disturbance) into long-term outcomes (platform sequences). These and other matters concerning the dynamics of carbonate platform systems have been tackled in a preliminary fashion in this book. The many questions that remain should help to set the agenda for future work in this rapidly growing field of study.

References

BIRKELAND, C. (ed.) 1997. *Life and Death of Coral Reefs*. Chapman & Hall, New York.

BRAITHWAITE, C. 1973. Reefs: just a problem of semantics? *American Association of Petroleum Geologists Bulletin*, **57**, 1100–1116.

BROMLEY, R. G. 1994. The palaeoecology of bioerosion. *In*: DONOVAN, S. K. (ed.) *The palaeobiology of trace fossils*. John Wiley, Chichester, 134–154.

CARANNANTE, G., RUBERTI, D. & SIMONE, L. 1993. Rudists and related sediments in late Cretaceous open shelf settings. A case history from Matese area (central southern Apennines, Italy). *Giornale di Geologia*, (3a), **55**, 21–36.

CARANNANTE, G., GRAZIANO, R., RUBERTI, D. & SIMONE, L. 1997. Upper Cretaceous temperate-type open shelves from northern (Sardinia) and southern (Apennines-Apulia) Mesozoic Tethyan margins. *In*: JAMES, N. P. & CLARKE, J. A. D. (eds) *Cool-water carbonates*. Society of Economic Paleontologists and Mineralogists, Tulsa, Oklahoma, Special Publication, **56**, 309–325.

CARANNANTE, G., GRAZIANO, R., PAPPONE, G., RUBERTI, D. & SIMONE, L. 1999. Depositional system and response to sea level oscillations of the Senonian rudist-bearing carbonate shelves. Examples from central Mediterranean areas. *Facies*, **40**, 1–24.

DERCOURT, J., RICOU, L. E. & VRIELYNCK, B. (eds) 1993. *Atlas Tethys palaeoenvironmental maps*. Gauthier-Villars, Paris, 1–307.

FAGERSTROM, J. A. 1987. *The evolution of reef communities*. John Wiley and Sons, New York.

GILI, E., MASSE, J.-P. & SKELTON, P. W. 1995. Rudists as gregarious sediment-dwellers, not reef-builders, on Cretaceous carbonate platforms. *Palaeogeography, Palaeoclimatology, Palaeoecology*, **118**, 245–267.

HUNT, D., ALLSOP, T. & SWARBRICK, R. E. 1996. Compaction as a primary control on the architecture and development of depositional sequences: conceptual framework, applications and implications. *In*: HOWELL, J. A. & AITKEN, J. F. (eds) *High Resolution Sequence Stratigraphy: Innovations and Applications*. Geological Society, London, Special Publications, **104**, 321–345.

INSALACO, E. 1998. The descriptive nomenclature and classification of growth fabrics in fossil scleractinian reefs. *Sedimentary Geology*, **118**, 159–186.

KIESSLING, W., FLÜGEL, E. & GOLONKA, J. 1999. Paleo Reef Maps: Evaluation of a comprehensive database on Phanerozoic reefs. *American Association of Petroleum Geologists Bulletin*, **83**, 1552–1587.

MASSE, J.-P. 1995. Carbonate platforms as systems. *Géologie Méditerranéenne*, **21**, 125–126.

MORTENSEN, P. B., HOVLUND, M., BRATTEGARD, T. & FARESTVEIT R. 1995. Deep water bioherms of the scleractinian coral *Lophelia pertusa* (L.) at 64° N on the Norwegian shelf: structure and associated megafauna. *Sarsia*, **80**, 145–158.

PHILIP, J. & SKELTON, P. W. (eds) 1995. Palaeoenvironmental models for the benthic associations of Cretaceous carbonate platforms in the Tethyan realm. *Palaeogeography, Palaeoclimatology, Palaeoecology*, **119**, 1–199.

PURDY, E. G. 1974. Karst-determined facies patterns in British Honduras: Holocene carbonate sedimentation model. *American Association of Petroleum Geologists Bulletin*, **58**, 825–855.

RUBERTI, D. 1997. Facies analysis of an Upper Cretaceous high-energy rudist-dominated carbonate ramp (Matese Mountains, central-southern Italy): subtidal and peritidal cycles. *Sedimentary Geology*, **113**, 81–110.

SCHLAGER, W. & PHILIP, J. 1990. Cretaceous carbonate platforms. *In*: GINSBURG, R. N. & BEAUDOIN, B. (eds) *Cretaceous resources, events and rhythms* (NATO ASC. Series, **304**), Kluwer, Dordrecht, 173–195.

SCOFFIN, T. P. 1992. Taphonomy of coral reefs: a review. *Coral Reefs*, **11**, 57–77.

SIMÓ, J. A. T., SCOTT, R. W. & MASSE, J.-P. (eds) 1993. *Cretaceous carbonate platforms*. American Association of Petroleum Geologists, Memoir, **56**, 1–479.

WEBB, G. E. 1996. Was Phanerozoic reef history controlled by the distribution of non-enzymatically secreted reef carbonates (microbial carbonate and biologically induced cements)? *Sedimentology*, **43**, 947–971.

WOOD, R. 1999. *Reef Evolution*. Oxford University Press, Oxford.

A temperate reef builder: an evaluation of the growth, morphology and composition of *Sabellaria alveolata* (L.) colonies on carbonate platforms in South Wales

LARISSA A. NAYLOR* & HEATHER A. VILES
School of Geography, University of Oxford, Mansfield Road Oxford, OX1 3TB, UK (e-mail: larissa.naylor@geog.ox.ac.uk)
**Name changed from Motiuk to Naylor*

Abstract: *Sabellaria alveolata* (Linné) (Polychaeta: sabellariidae) is a sedentary polychaete that builds wave-resistant reefs from sand-sized particles. Reefs are formed in areas with a large, continuous supply of sand-sized sediment and turbulent water, such as the Bristol Channel, UK. Although several studies have documented the extent, growth, form and geological importance of Sabellariidae, their bioconstructive role has not been adequately assessed. *S. alveolata* occurs rarely in the UK and is classified as a distinct ecological unit by the Marine Nature Conservation Review, yet it has been little studied. Thus, there is a need for greater understanding of the species' role in UK coastal ecology and geomorphology. An evaluation of reef development in terms of composition, growth and extent of *S. alveolata* constructions on carbonate shore platforms in South Wales is presented here. Results indicate that *S. alveolata* is a fast-growing species capable of juvenile settlement, tube formation and growth of 2.5–5.0 cm in as little as two months after installation of exposure blocks of artificial substrata. Particle size analysis of 24 randomly selected reefs and ten adjacent sand samples shows a significant difference in mean grain size, with worms consolidating particles of a coarser size distribution than the mean particle size of surrounding sand. Preliminary scanning electron microscope observations indicate preferential use of flat, platy and elongate particles in the worm tubes. This research forms part of a larger study concerned with quantifying the bioconstructive capability of *S. alveolata*.

A biogeomorphological approach has been used to evaluate the role of *Sabellaria alveolata* (Linné) (Polychaeta: Sabellariidae) colonies in current shore platform processes. Biogeomorphology is a type of geomorphological research that examines the two-way interplay between ecological and geomorphological processes (Viles 1988). The distribution of species is often related to the underlying geomorphological forms while the surface morphology may be altered by organisms. Some species chemically or physically alter rocky substrates and are generally referred to as bioeroders or more specifically as agents of biocorrosion or bioabrasion, respectively (Spencer 1988, 1992). In contrast, some species have been found to perform a bioconstructional role as they build features that change the morphology of the landscape (for example, see Kelletat 1989; Dalongeville *et al.* 1994; Gektidis 1997). Some researchers (see, for instance, Trudgill 1988; Kelletat 1989; Dalongeville 1995) have postulated that bioconstructions may also protect the substrate from erosion by physically covering the surface, but, as yet, little effort has been made to quantify their protective role (Kelletat 1989). For this reason, we feel it is important to classify organisms that accrete material and/or build biogenic forms as bioconstructors and those that provide a protective cover of the surface as bioprotectors. Existing research has primarily been focused on quantifying bioerosion and the production of biokarst (Schneider & Torunski 1983) rather than measuring the amount of bioconstruction or degree of bioprotection provided by organisms.

When designing biogeomorphological field studies it is necessary to understand and incorporate ecological and geomorphological theories, methodologies and techniques (Trudgill 1988). The task of integrating ecological and geomorphological field methods and laboratory techniques is often a challenging one and typically, existing methods are modified to effectively evaluate a given subject from a biogeomorphological perspective.

Aims

This paper presents preliminary findings of a multidisciplinary evaluation of the bioconstructive role of *S. alveolata* colonies on Blue Lias limestone shore platforms in South Wales. A series of surveys has been conducted to determine the extent, composition and growth rates of *S. alveolata* colonies as a means of assessing

From: INSALACO, E., SKELTON, P. W. & PALMER, T. J. (eds) 2000. *Carbonate Platform Systems: components and interactions*. Geological Society, London, Special Publications, **178,** 9–19. 0305–8719/00/$15.00

the bioconstructive capability of the species. This work is part of a larger project to quantify the effects of multiple processes (ecological, geomorphological and geological) on shore platform process rates (Naylor in prep).

Sabellaria alveolata (Linné): species and habitat characteristics

Species characteristics

Sabellaria alveolata (Linné) (Polychaeta: Sabellariidae) is a filter-feeding sedentary polychaete that builds wave-resistant reefs from sand-sized (between 63 μm and 2 mm on the Wentworth Scale) mineral grains and shell debris. Individual worms are 2–5 cm long and they build tubes up to 15 cm in length from sand-sized mineral grains with diameters ranging from 0.5 to 5.0 mm (Wilson 1971; Gruet 1992; see Fig. 1 for greater detail). Individuals generally live for four to five years with growth quickest during the first year and steadily tapering in subsequent years, although they have been known to live for up to ten years in rare cases (Wilson 1971; Gruet 1992).

Gruet (1984) found the diameter of individual tubes to be positively correlated with the size of the organism and Vovelle (1965) determined that the particle size used by the organism is directly related to size of their building organ (see Fig. 1). However, there are no concentric rings or growth bands to accurately determine growth rates or age distributions of a given colony and as a consequence, there has been little effort to evaluate seasonal growth rates. Moreover, the tube length is much longer than the animal itself (up to 15 cm) and older, unused portions of worm tubes have been found to break off or to be covered by younger individuals as juveniles preferentially settle on existing reefs (Wilson 1971). *S. alveolata* typically develops large colonies of reefs and at the site studied, they cover approximately 5 ha (see Fig. 2a) (Rupert & Barnes 1994). Individual reefs have an irregular three-dimensional form and exhibit gaps and/or patchy growth (for example, see Fig. 2b, c). As such, an estimate of the population and age dynamics of, or volume of sediment stored by, a particular colony of worms is extremely difficult. Although the density of worm tubes has been measured in the field (for example Herdman (1920) found colonies on Hilbre Island, UK, with tube densities of approximately 3120 per square foot), it is not necessarily a representative indication of the number of living individuals.

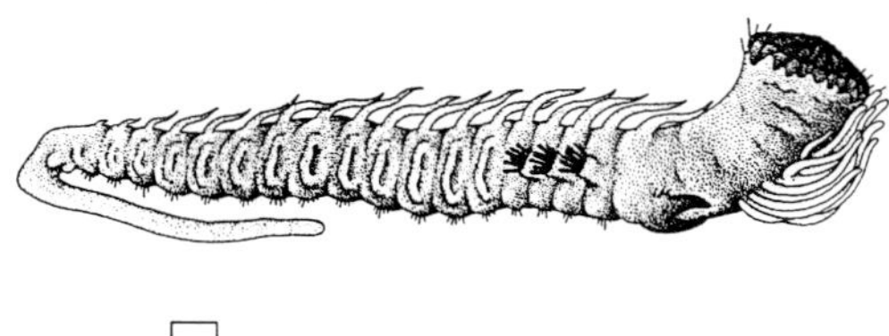

Fig. 1. Right ventral view of *Sabellaria alveolata* removed from its tube (after Gruet 1992, drawn by Ben Naylor).

Habitat characteristics

Sabellariidae are adapted to and require turbulent, sediment-laden waters with a continual supply of sand-sized particles (Kirtley & Tanner 1968). They typically form large reef colonies (up to 1.5 m in height) or encrusting reefs (up to 30 cm in height) which cover extensive areas of the mid-eulittoral, lower eulittoral and sublittoral zones and are particularly exposed during low water neap and low water spring tides (LWN-LWS) (Wilson 1971). In addition to the two main reef forms, *S. alveolata* can also occur singularly (Mendelssohn 1976). At the site studied, *S. alveolata* is primarily found in the latter two forms. Most reefs form encrusting reefs and are located in the lower littoral and upper eulittoral zones, while several individual worms are found growing in the base of rock pools or along crevices and occur much higher up the shore, in the mid-eulittoral zone. Sabellariidae typically inhabit hard rocky substrates in or adjacent to sandy beaches but they have also been found growing on peat (Multer & Milliman 1967) or oyster beds (Galaine & Houlbert 1916) adjacent to a sandy beach. However, at the site studied, the reefs are found on a rocky substrate without an adjacent sand supply.

Species persistence and resilience

Species analogous to *S. alveolata* have existed since the Carboniferous period (Dawson 1980; Howell 1962) and researchers have postulated that they may have played a considerable role in shaping ancient coastlines and influencing sedimentary cycles (see, for instance, Herdman 1920; Multer & Milliman 1967; Kirtley & Tanner 1968). The reef structures have been observed to be quite persistent over decadal to century temporal scales in recent history (see, for instance, Audouin & Edwards (1832), Galaine & Houlbert (1916), Renaud (1917), Lucas (1959), Dolfus (1960), Mathieu (1967), Gruet (1986) who have all studied *S. alveolata* reefs in

Fig. 2. Photographs showing the extent, habitat and characteristics of *S. alveolata* tubes and reefs. (**a**) The areal extent of *S. alveolata* colonies in the lower eulittoral zone at the study site (the metal frame in the foreground is 1 × 1 m and colony relief is up to 30 cm); (**b**) the orientation, size and density of individual tubes from a typical adult reef; (**c**) the pool-enhancing properties of the reefs (the pool rim in the foreground is heightened by approximately 15 cm by the presence of reefs, which improves the amount of shelter for other organisms); (**d**) cross-section of an individual worm tube showing the imbricate structure and the preferential use of platy grains for the inner tube diameter.

Mont-Saint-Michel bay, France). However, it is likely that populations fluctuate over time and as Gruet (1986) determined, colonies in Mont-Saint-Michel bay exhibit different life history characteristics with some colonies growing rapidly while others appear more senescent. These dynamics have yet to be related to environmental parameters such as position on the shore or fluctuations in larval settlement.

There is limited information on the global geographical extent of the Sabellariidae with the exception of the study of Kirtley & Tanner (1968) who noted that the family has been recorded between 72° north and 53° south. *S. alveolata* are typically found in exposed areas prone to severe storms and abrasion rather than sheltered areas with low turbulence and high sedimentation. In Florida, Sabellariidae reefs have been found to withstand large storm events and researchers have noted the rapid ability of individuals to rebuild their tubes (Kirtley 1992). However, colonies have been found to go into decline if there is too much fine sediment (i.e. mud) that will plug individual tubes, or if there is an algal bloom or competition by other filter feeders such as encrusting barnacles and mussels (Multer & Milliman 1967). Such conditions can lead to the weakening of reef structures which renders them more susceptible to erosion by storm waves (Wilson 1971). Thus, the presence and persistence of *S. alveolata* is dependent upon exposed environmental conditions where abrasion is dominant, waters are turbulent and there is a fixed, stable substrate together with a continuous supply of coarse (i.e. sand-sized) rather than fine (i.e. silt- and clay-sized) mineral grains.

Existing research on Sabellariidae

Numerous German, French, English and American researchers have discussed the presence of Sabellariidae reefs during the past 150 years, although studies have typically been limited to colonies in selected areas of France (see, for instance, Dollfus 1960; Vovelle 1963, 1965, 1971; Gruet 1984, 1986), the United Kingdom (Herdman 1920; Wilson 1929, 1968*a*,*b*, 1971, 1974, 1976; Mendelssohn 1976), America (Fager 1964; Multer & Milliman 1967; Gram 1968; Kirtley & Tanner 1968; Posey *et al.* 1984; Kirtley 1992) and the North Sea (Richter 1928; Linke 1951). Since the 1950s research on the Sabellariidae has become increasingly quantitative and has trended along disciplinary lines. One body of research has examined the biological characteristics of the genera through detailed studies evaluating the anatomy and/or physiology of individual worms and the processes of settlement, tube formation and reproduction (see, for instance, Wilson 1929, 1968*a*,*b*; Vovelle 1965; Orrhage 1978; Gruet *et al.* 1987). The other body of research has evaluated the amount of sediment selection and heavy mineral storage and the effects on hydrodynamics, in an effort to determine the geological significance of the genera (for example, see Renaud 1917; Fager 1964; Multer & Milliman 1967; Gram 1968; Kirtley & Tanner 1968). Although several authors have commented on the protection and sediment storage afforded by these large reef structures and/or barriers (Galaine & Houlbert 1916; Renaud 1917; Multer & Milliman 1967; Kirtley 1992), limited work has quantified the large-scale morphology of the reefs (Gruet 1986) and previous studies have not evaluated the bioconstructive role of the species.

UK research and distribution of S. alveolata

In the UK, pioneering research on *S. alveolata* was carried out by Wilson (1929, 1968*a*,*b*, 1970, 1971, 1974, 1976). Wilson concentrated his research efforts at Duckpool, Cornwall, where he observed temporal population changes, documented patterns of settlement and evaluated the stages of larval development, the latter through laboratory research. Since Wilson's work, little research has specifically evaluated the species in the UK. However, Crisp (1964) documented the reduction of *S. alveolata* populations in the UK after an extremely cold winter. Shortly thereafter, an investigation into the distribution and morphology of polychaete worms in the UK documented the existing geographical extent of the species as well as adding several new locations where *S. alveolata* was present (Mendessohn 1976). Recently, the Marine Nature Conservation Review (MNCR) developed a program to map marine ecosystems by combining habitat and ecology into a 'biotope' classification system (Conner *et al.* 1997).

S. alveolata was designated a unique biotope in its littoral and sublittoral classification program and the MNCR considers the *S. alveolata* biotope to be rare in the eulittoral zone (Conner *et al.* 1997) as the population covers less than 0.5% of the total UK coastline (Sanderson 1996). Species records of *S. alveolata* in the UK have been compiled from three key sources (Crisp 1964; Wilson 1971; Mendelssohn 1976) to produce a map indicating species' presence (Fig. 3). These observations are limited in scope as they primarily document the intertidal occurrences of the species, solely in accessible areas. There are, however, accounts of *S. alveolata* and

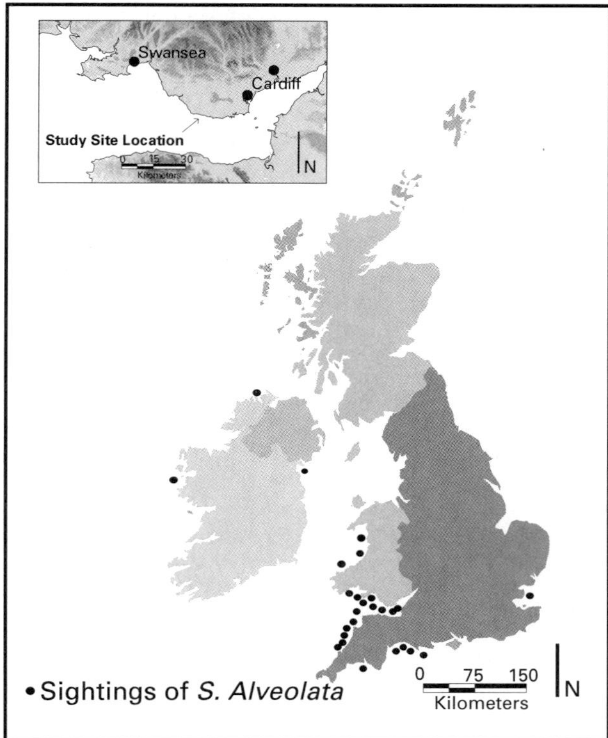

Fig. 3. United Kingdom map outlining the occurrence of *S. alveolata* and detailing the study site.

S. spinulosa subtidally from dredging companies and over time there will be more information on the species' occurrence as the biotope mapping is completed for the UK.

Study site location

The study site is located within the Glamorgan Heritage Coast which is a protected area along the shores of the Bristol Channel between Cardiff and Swansea (Fig. 3). The Bristol Channel has the second largest tidal range in the world, with large storm events occurring throughout the winter months. The shore platforms of the area are composed of Blue Lias limestone, grade seawards at a low angle, extend for up to 300 m in width and span approximately 20 km of coastline (Trenhaile 1972). The platforms are subject to a morphogenic gradient from high energy in the west to low energy in the east. The study area for detailed surveys, located just east of Nash Point (51°23′00″ N, 003°33′02″ W), was selected to minimize geological variability as it has only one band of Blue Lias limestone, extending from high to low water, which is structurally oriented parallel to the wave front.

Field and laboratory methods

Field methods

Mapping the extent of S. alveolata *colonies*. The entire littoral zone of the Heritage Coast was mapped during the spring low tides of September 1997 using the guidelines for phase one biotope mapping produced by the Countryside Council for Wales (Richards *et al.* 1995), which is an earlier version of the MNCR's biotopes mapping program (Conner *et al.* 1997). The maps were hand drawn by Heritage Coast personnel at 1: 12 000 scale using ordnance survey base maps, with information subsequently verified using aerial photographs. Their data were used to determine the geographical extent of the *S. alveolata* biotope at the study site.

Exposure block trials and growth measurements. Exposure block trials were established in the mid-upper eulittoral zones to measure the amount of colonization, bioerosion and weathering over an 18 month period, as part of a larger project to evaluate biogenic processes. A systematic random sampling design was used to lay out a network of sampling points along a series of transects from the upper littoral to lower eulittoral zones (see Fig. 4). In the mid-upper eulittoral zones, three transects were established perpendicular to the shore with the first transect selected randomly and each subsequent transect offset by 30 m. Eight stations were selected from the sampling area using a random number table and 24 exposure blocks were emplaced during 8–11 October 1998. At each station three 50 × 50 × 20 mm blocks of Bath Stone, were affixed to the limestone substrate in the base of rock pools using marine epoxy. Bath stone was selected for experimental trials because it is a softer limestone than Blue Lias and in short-term experimental studies, such as this one, a softer stone generally yields more results. Exposure block trials were established in mid-upper eulittoral zone rock pools to evaluate pool-forming processes and colonization rates. Several environmental constraints, such as short periods of exposure, the ability to drain rock pools and the time required for the marine epoxy to set, prevented the establishment of exposure blocks in the lower eulittoral zone.

S. alveolata growth rates have been mentioned by other researchers but are often based on semi-quantitative information (for example, see Kirtley & Tanner 1968) or measurements of decadal changes in morphological form (Gruet 1986), and Wilson (1976) comments on the difficulties faced when measuring *S. alveolata* colony growth. Given that *S. alveolata* preferentially colonizes on pre-existing tubes (Wilson 1971), conventional methods of removing organisms, from known areas, to observe colonization rates were not possible. As such, a variety of techniques used to measure growth of coral reef structures (such as Alizarin red dyeing of tube cement) were pilot tested from May to September 1998 to measure *S. alveolata* colony growth rates in the lower eulittoral zone, but the techniques were either not feasible or unsuccessful. Consequently, growth measurements of *S. alveolata* were made in rock pools containing exposure blocks on a bimonthly basis for a six month period. Pools containing exposure blocks were used for growth measurements as they provided a clear dateline from which subsequent settlement patterns and growth rates could be measured.

Granulometric surveys. In the lower eulittoral zone, where the dominant *S. alveolata* colonies are found, a 60 × 40 m sampling area was established as an extension of the upper shore transects (Fig. 4). The area was subdivided into a grid of stations and 24 sampling points were selected using a random number generator. At each point, a portion (approximately 10 × 10 × 10 cm) of the nearest reef sample was removed with a knife and within the sampling area ten loose sand samples were randomly collected. In total, 24 *S. alveolata* tube clusters, ten individual *S. alveolata* tubes and ten adjacent sand samples were collected on 10 October 1998 for particle size and scanning electron microscope (SEM) analyses.

Laboratory techniques

Sample preparation for particle size analysis. Each *S. alveolata* reef sample was gently disaggregated manually to remove worms, treated with 10% KOH solution to remove proteins followed by removal of organic material using 6% H_2O_2 solution. The samples were subsequently dried to a constant weight (total dry weights were in the range of 130–475 g), rehydrated in distilled water and wet sieved. The fraction greater than 63 μm was dried, weighed and a 40–100 g subsample, obtained using a sample splitter, was used in all subsequent analyses.

Ten sand samples were rinsed in distilled water, dried at 50°C and subsampled using a sample splitter. Samples were wet sieved to 63 μm and 50–100 g of coarse fraction was dried for subsequent sieve analysis. For both sediment types, the coarse fraction was mechanically sieved for ten minutes using British Standard Sieves at 0.25 ϕ intervals ranging from –2.0 to 4.0 ϕ.

Statistical analysis of particle size data. Histograms were produced for each sample to compare distributions between samples. Method of moments analysis was used to compute descriptive statistics and a Student's t-test was used to evaluate the significance of patterns observed.

SEM sample preparation. The ten individual worm tube samples were dried at 40°C, cut transversely into small segments, mounted and gold coated for SEM analyses. Both disaggregated and intact *S. alveolata* tube samples were analysed to evaluate sediment characteristics in their natural and disaggregated forms.

Results and discussion

Extent

The Sabellarian biotope was found scattered along the entire Heritage Coast, covering 44 ha of the lower eulittoral zone. The form and extent of reefs varies along the Heritage Coast with large reef structures (between 50 and 150 cm in

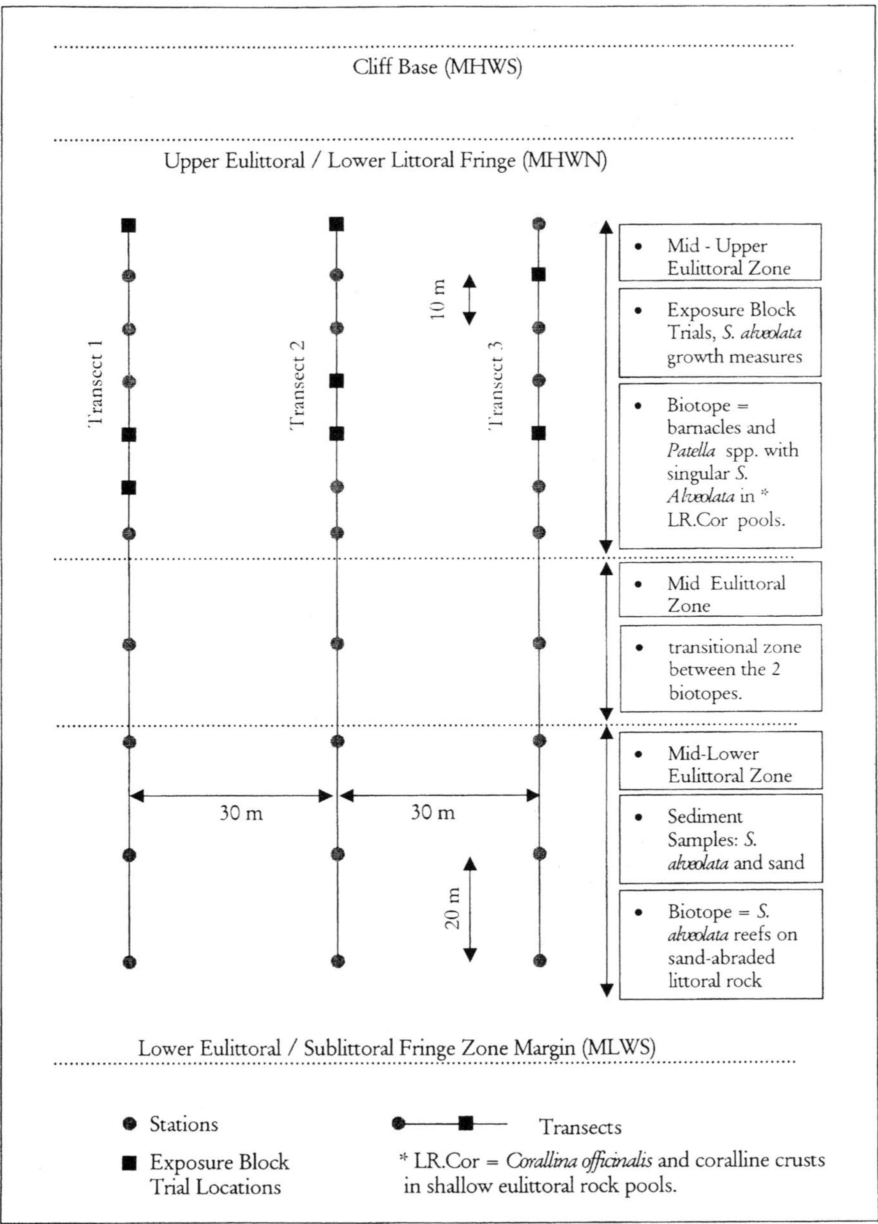

Fig. 4. Schematic diagram of sampling design, plot locations, biotopes and shore positions.

height) of varying morphological form and limited extent found at the western end of the coast near Southerndown and Temple Bays. Further east near Nash Point, where the study site is located, the reefs are more extensive and contiguous and form a thin crust of reef material (up to 30 cm in height) over a much larger surface area (approximately 5 ha).

Growth measurements

Growth measurements are reported from rock pools in the mid-upper eulittoral zone where exposure blocks were established in October 1998. The worms settled, affixed themselves and grew to 2.5 and 5.0 cm in length in less than 52 days. These growth rates are significantly faster than those measured in laboratory settlement trials conducted by Wilson (1971). The rock pools have been monitored bimonthly since December 1998 and by May 1999, 36% of pools had *S. alveolata* settlement and some individual tubes had grown to 9.8 cm in length. Although these measurements do indicate considerable growth in a short time period, they are not indicative of the general population as the individuals measured are near the uppermost extent of their range on the shore rather than from areas with large assemblages in the lower eulittoral zone (see Fig. 4). Thus, growth rates lower down the coastal profile may be greater and these results represent a minimum rate for this area. However, the results do indicate that the species can settle, establish itself and grow quite quickly in a relatively short period of time.

Composition

Results are reported for the 63 μm to 2 mm fraction, including carbonates, of 24 worm tube and ten sand samples. The fine fraction (i.e. less than 63 mm) differed considerably between *S. alveolata* and sand samples with considerably more fine material found in the worm tube samples. This variation is possibly a result of the ability of reef structures to store a 'passive' fraction which Vovelle (1965) identified as fine material that was trapped in reef crevices without forming part of the 'active' matrix of individual worm tubes. This material is less likely to be cemented into the reef structure than larger clasts which are 'trapped' by worm tubes growing over them and as such the storage time of fine fraction is likely to be considerably less than that of larger clasts or material forming the tube matrix. Thus, the fine fraction was considered a short-term component of *S. alveolata* reefs likely to fluctuate considerably both spatially and temporally and as such is not discussed further.

Particle size distributions

Both sediment types had very distinctive sediment distribution with minimal variation between samples within each sediment type. Analysis of 24 *S. alveolata* tube samples revealed an almost normal distribution (see Fig. 5) with the median and mean grain size of *S. alveolata* samples being almost equal (median and mean size were 1.59 and 1.58ϕ, respectively). The standard deviation of the mean sediment size was extremely small (0.07). Meanwhile, the ten sand samples were also similar to each other as the mean and median grain size were both 1.81ϕ and the mean standard deviation was 0.09. Thus both sediment types have very little variation between individual samples but the mean value differed significantly between sediment type with *S. alveolata* samples containing a larger proportion of coarser grains. As the variance between samples within each sediment type is minimal, one sample from each sediment type has been selected to illustrate preliminary trends. The difference in the two sediment distributions is clearly evident in Fig. 5. The *S. alveolata* tube

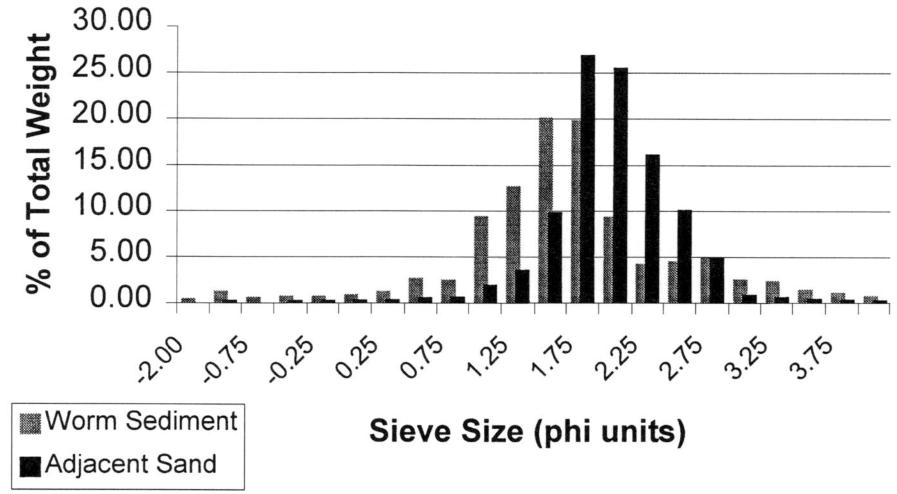

Fig. 5. Frequency histogram plot showing the difference in particle size distribution between sand and worm samples (samples 6 and 18, respectively).

sample contains more sediment from a wider range of grain sizes and the largest sample weights are from grain sizes slighter coarser than the median size of adjacent sand samples. Figure 5 clearly illustrates that the worms are preferentially selecting coarser grain sizes from the adjacent sand. An independent, two-sample Student's t-test was conducted for all samples and the results confirmed that there was a significant difference between sample means at the 99.99% confidence interval.

As the selection of individual worms is related to the size of the individual and its building organ, the trend of selecting coarser or finer grains will vary depending on the age structure of the population and the particle size distribution of the surrounding sand. In this case, the mean particle size of surrounding sand is at the fine end of medium sand (1.81ϕ) and the largest proportion of sand is contained in the size range 1.75–2.25ϕ, while the worms are preferentially selecting slightly coarser sand (mean = 1.59ϕ) with the largest proportion of sand from size range 1.25–1.75ϕ (see Fig. 5 for more details). This tendency for the Sabellariidae to preferentially select grains has been quantified by other researchers (Vovelle 1958, 1963, 1965, 1971; Fager 1964; Multer & Milliman 1967; Gruet 1984) and appears to be a defining characteristic of the species; however, it is important to recognize that the tendency to select grains that are either coarser or finer depends on the biological and sedimentological conditions of each site. Sand samples are now being collected at the study site on a seasonal basis to determine if the composition of loose sand changes seasonally and to relate seasonal sand composition to reef sand samples, as the reefs are likely to contain particles trapped over several months and/or years.

Scanning electron microscope (SEM) analysis

Individual worm tubes were examined with a Cambridge 90 Stereoscan SEM and an initial survey was conducted to document tube structure, clast orientation, particle size distribution and organic cement characteristics. Investigation of tube sections clearly identified the structure of individual tubes and the range of particle sizes used in tube construction (Fig. 2d). From the image (Fig. 2d) it is obvious that the particles surrounding the organic tube lining are flat, platy or elongate in shape and display an imbricate structure. The tube lining and the role of organic cement in binding the sediments were also observed (Fig. 2d). It is clear that the worms are selective in their choice of grains and that individual tubes have a distinctive structure.

Both SEM and particle size analyses (natural and disaggregated forms, respectively) of tube sediments were conducted to observe the structural orientation of *S. alveolata* colonies. The preferential selection of grains is clearly evident in both the granulometric and SEM analyses of *S. alveolata* tubes presented here.

Biogeomorphological significance

At the study site

The presence of *S. alveolata* bioconstructions in the lower eulittoral zone (Fig. 2a-c) has three main effects on environmental conditions. First, the colonies increase morphological variation in the lower eulittoral zone, such as enhancing pool forms, thereby increasing the amount of sheltered habitat for other marine fauna (Fig. 2c). This has been found to greatly improve secondary diversity in Florida and Oregon (Kirtley & Tanner 1968; Posey *et al.* 1984). Second, the colonies increase surface roughness which possibly reduces wave energy as swash dissipates more quickly over rougher surfaces (Trenhaile pers. comm. 1998). A reduction in wave energy would reduce the pressure that waves exert on the shore platform, thereby affecting the severity of weathering and erosion caused by wave action. Third, the reef structures appear to protect the underlying limestone from wave attack and abrasion by physically covering the surface and by storing sediment that would otherwise be available to abrade the shore. For example, the worm tubes have been found growing over the surface of loose cobbles and this gradually cements fragments into the reef structure thereby storing them as long as the reef persists. This could potentially reduce surface lowering and/or abrasion potential and needs to be investigated further. However, the effect of worm colonies on the chemical breakdown of the underlying rock has yet to be evaluated.

Larger-scale significance

Worm colonies have the ability to store considerable quantities of shell debris, rock fragments and sand grains for several to tens of years (Fager 1964; Multer & Milliman 1967; Gram 1968; Kirtley & Tanner 1968). This is sediment that would otherwise form part of the 'loose' material in sediment cycling and is being temporarily stored on a decadal scale in the reef deposits. Given the large extent of reefs in the Bristol Channel (see Fig. 3) and their geomorphological and ecological importance at individual sites, the value of these organisms as sediment reservoirs, 'natural' defence structures and biodiversity facilitators, requires greater consideration. Moreover, the sediment requirements of the species needs to be further researched to ensure that sediment stores which supply the reefs are not too heavily depleted (by processes such as aggregate dredging). It is hoped that further research into the sediment requirements of, and sediment use by, the species can improve our understanding of the sedimentological importance of the species and its relationship with available sediment supplies in the Bristol Channel. Moreover it may prove useful to evaluate the sensitivity of *S. alveolata* to varying sediment intensities in laboratory and field trials, in order to determine the species' resilience to changing sediment conditions thereby evaluating its role as a 'sensitivity indicator' to dredging operations in the Bristol Channel.

From the work carried out thus far, the species appears to be of biogeomorphological importance in the Bristol Channel at both small and large spatial scales. More importantly, this research has identified several new research questions and stressed the importance of this species as a rare temperate reef builder (i.e. bioconstructor) that appears to play a considerable bioprotective role on shore platforms in South Wales. The amount of bioprotection provided by *S. alveolata*, both in terms of sediment storage (i.e. volume of sediment and duration of storage) and substrate protection, is an important subject that requires further research.

Financial support for field work expenses was provided by a PGS 'A' award from the Natural Sciences and Engineering Research Council of Canada (NSERC), the School of Geography, University of Oxford and Worcester College, Oxford. Logistical assistance was provided by volunteers at Atlantic College, students at the University of Oxford and staff at the Glamorgan Heritage Coast Centre. The authors greatly appreciate the assistance of Ben Naylor, for the illustration.

References

AUDOUIN, M. & EDWARDS, M. 1832. *Reserches pour servir a l'histoire naturelle du littoral de la France ou recueil des mémoires sur l'anatomie, la physiologie, la classification et les moeurs des animaux de nos côtes*, Volume **1**. Crochard, Paris.

CONNER, D. W., BRAZIER, D. P., HILL, T. O. & NORTHEN, K. O. 1997. *Marine Nature Conservation Review: marine biotope classification for Britain and Ireland: Volume 1. Littoral Biotopes, version 97.06*. Joint Nature Conservation Committee Report **229**.

CRISP, D. J. 1964. The effects of the winter of 1962/1963 on the British marine fauna. *Helgolaender Wissenschaftliche Meeresuntersuchungen,* **10**, 313–327.

DALONGEVILLE, R., LE CAMPION, T. & FONTAINE, M. F. 1994. Bilan bioconstruction – biodestruction dans les roches carbonatées en mer Méditerraneé: étudie expérimentale et implications géomorphologiques. *Zeitschrift für Geomorphologie N. F.,* **38**(4), 457–474.

—— 1995. Le rôle des organismes constructeurs dans la morphologie des littoraux de la mer méditerranée: algues calcaires et vermetidés. *Norois,* **42**, 73–88.

DAWSON, J. W. 1890. *On* burrows *and* tracks *of* invertebrate animals *in* palaeozoic rocks, *and other* markings. *Quarterly Journal of the Geological Society,* **46**, 595–618.

DOLLFUS, R. P. 1960. Sur un récif actuel: le banc des Hermelles de la baie du Mont-Saint-Michel. Quelques renseignements documentaires. *Bulletin de la Société Géologique de France,* **11, series 7e**, 133–140.

FAGER, E. W. 1964. Marine sediments: Effects of a tube-building polychaete. *Science,* **143**, 356–359.

GALAINE, C. & HOULBERT, C. 1916. Les récifs d'Hermelles et l'assèchment de la baie du Mont-Saint-Michel. *Comptes Rendus Hebdomadaires des séances de l'académie des sciences*, **163**(21), 613–616.

GEKTIDIS, M. 1997. Microbioerosion in Bahamian ooids. *Cour Forschungsinstitut Senckenberg,* **201**, 109–121.

GRAM, R. 1968. A Florida Sabellaridae reef and its effect on sediment distribution. *Journal of Sedimentary Petrology,* **38**, 863–868.

GRUET, Y. 1984. Granulometric evolution of the sand tube in relation to growth of the polychaete annelid *Sabellaria alveolata* (Linné) (Sabellariidae). *Ophelia,* **23**, 181–193.

—— 1986. Spatio-temporal changes of Sabellarian reefs built by the sedentary polychaete *Sabellaria alveolata* (Linné). *Marine Ecology*, **7**, 303–319.

—— 1992. Biology and Ecology of the polychaete annelid *Sabellaria alveolata* (Linnaeus). *In*: Kirtley, D. W. (ed.) *The Sabellariid reefs in the Bay of Mont Saint-Michel, France. Ecology, Geomoprhology, Sedimentology, and Geologic Implications*. English Translation of *Les récifs à Annélides (hermelles) en Baie du Mont Saint-Michel. Écologie, Géomorphologie, Sédimentologie, et implications Géologiques et Minières*. Florida Oceanographic Institute, Florida.

——, VOVELLE, J. & GRASSET, M. 1987. Composition biomminerale du ciment du tube chez *Sabellaria alveolata* (L.), Annelide Polychete. *Canadian Journal of Zoology,* **65**, 837–842.

HERDMAN, W. A. 1920. Spolia Runiana. – IV. Notes on the abundance of some common marine animals and a preliminary quantitative survey of their occurrence. *Journal of the Linnean Society (Zoology)*, **34**, 247–259.

HOWELL, B. F. 1962. Worms. *In*: Moore, R. C. (ed.) *Treatise on Invertebrate Palaeontology*. Geological Society of America and University of Kansas Press, W144–W177.

KELLETAT, D. 1989. Zonality of rocky shores. *Essener geographische Arbeiten,* Paderborn, **18**, 1–29.

KIRTLEY, D. 1992. Built to last: Worm reefs a feat of natural engineering. *Florida Oceanographic,* **13**, 12–19.

KIRTLEY, D. W. & TANNER, W. F. 1968. Sabellarid worms: builders of a major reef type. *Journal of Sedimentary Petrology,* **38**, 73–78.

LINKE, O. 1951. Neue Beobachtungen uber Sandkorallen Riff in der Norsee. *Natur und Volk*, **81**, 77–84.

LUCAS, G. 1959. Deux examples actuels de 'biolithosores' construits par des Annélides. *Bulletin de la société géologique du France*, **7**(1), 385–389.

MATHIEU, R. 1967. Le banc des Hermelles de la baie du Mont-Saint-Michel, bioherme à Annélides. Sédimentologie, structure et genesè. *Bulletin de la société géologique du France*, **9**(7), 68–78.

MENDELSSOHN, J. M. 1976. *Studies on the taxonomy and biology of some British polychaete worms.* DPhil Thesis, University of Oxford.

MULTER, H. G. & MILLIMAN, J. G. 1967. Geologic aspects of Sabellarian reefs, Southeastern Florida. *Bulletin of Marine Science,* **17**, 257–267.

NAYLOR, L. A. In prep. *Biogeomorphological responses of rocky coasts in South Wales, UK and Western Crete, Greece.* DPhil Thesis, University of Oxford.

ORRHAGE, L. 1978. On the structure and evolution of the anterior end of the Sabellariidae (Polychaeta Sedentaria). With some remarks on the general organisation of the polychaete brain. *Zoologische Jahrbücher Abteilung für Anatomie und Ontogenie der Tiere,* **100**, 343–374.

POSEY, M. H., PREGNALL, A. M. & GRAHAM, R. A. 1984. A brief description of a subtidal Sabellariid (Polychaeta) reef of the Southern Oregon coast. *Pacific Science,* **38**, 28–33.

RENAUD, 1917. De l'influence des Hermelles sur le régime de la baie du Mont Saint-Michel. *Comptes Rendus Hebdomadaires des séances de l'académie des sciences*, **164**, 549–551.

RICHARDS, A., BUNKER, F. & FOSTER-SMITH, R. 1995. *Handbook for Marine Phase 1 Survey and Mapping.* CCW Report **95/6/1**. Countryside Council for Wales, Bangor.

RICHTER, R. 1928. Die fossilen Fährten und Bauten der Würmer, ein Überblick über ihre biologischen Grundformen und deren geologische Bedeutung. *Paläontologische Zeitschrift,* **9**, 195–235.

RUPERT, E. E. & BARNES, R. D. 1994. *Invertebrate Zoology*. 6th edition. Sanders, New York.

SANDERSON, W. G. 1996. Rarity of marine benthic species in Great Britain: development and application of assessment criteria. *Aquatic Conservation: Marine and Freshwater Ecosystems,* **6**, 245–256.

SCHNEIDER, J. & TORUNSKI, H. 1983. Biokarst on limestone coasts, morphogenesis and sediment production. *Marine Ecology,* **4**, 45–63.

SPENCER, T. 1988. Coastal biogeomorphology. *In:* Viles, H. A. (ed.) *Biogeomorphology.* Blackwell, Oxford, 255–318.

—— 1992. Bioerosion and biogeomorphology. *In:* JOHN, D. M., HAWKINS, S. J. & PRICE, J. H. (eds) *Plant–Animal Interactions in the Marine Benthos.* Systematics Association Special Volume, **46**, Clarendon, Oxford, 493–509.

TRENHAILE, A. S. 1972. The shore platforms of the Vale of Glamorgan, Wales. *Transactions of the Institute of British Geographers,* **56**, 127–144.

TRUDGILL, S. 1988. Integrated geomorphological and ecological studies on rocky shores in southern Britain. *Field Studies,* **7**, 239–279.

VILES, H. A. 1988. Introduction. *In:* VILES, H. A. (ed.) *Biogeomorphology.* Blackwell, Oxford, 1–10.

VOVELLE, J. 1958. Remarques sur la structure du tube de *Sabellaria alveolata* (L.) et les formes glandularies impliquées dans son édification. *Archives de zoologie éxperimentale et generale. Notes et Revue*, **95**, 52–68.

—— 1963. Donnes granulométriques sur le tube de quelques annélides polychaètes de la plage de Saint-Efflam. *Cahiers de Biologie Marine* **4**, 315–319.

—— 1965. Le tube de *Sabellaria alveolata* (L.) annelide polychaète *Hermellidae* et son ciment étudie ecologique, experimentale, histologique et histochemique. *Archives de zoologie experimentale et generale,* **106**, 1–187.

—— 1971. Selection des grains du tube chez les amphictenidae (Polychaètes sedentaires). *Cahiers de Biologie Marine,* **12**, 365–380.

WILSON, D. P. 1929. The larvae of the British Sabellarians. *Journal of the Marine Biological Association of the United Kingdom,* **16**, 221–268.

—— 1968*a*. Some aspects of the development of eggs and larvae of *Sabellaria alveolata* (L.), *Journal of the Marine Biological Association of the United Kingdom,* **48**, 367–386.

—— 1968*b*. The settlement behaviour of the larvae of *Sabellaria alveolata* (L.), *Journal of the Marine Biological Association of the United Kingdom,* **48**, 387–435.

—— 1970. Additional observations on larval growth and settlement of *Sabellaria alveolata. Journal of the Marine Biological Association of the United Kingdom*, **50**, 1–31.

—— 1971. *Sabellaria alveolata* (L.) at Duckpool, North Cornwall, 1961–1970. *Journal of the Marine Biological Association of the United Kingdom,* **51**, 509–580.

—— 1974. *Sabellaria alveolata* (L.) at Duckpool, North Cornwall, 1971–1972, with a note for May 1973. *Journal of the Marine Biological Association of the United Kingdom,* **54**, 393–436.

—— 1976. *Sabellaria alveolata* (L.) at Duckpool, North Cornwall, 1975. *Journal of the Marine Biological Association of the United Kingdom,* **56**, 305–310.

Skeletal growth rates of Upper Cretaceous rudist bivalves: implications for carbonate production and organism–environment feedbacks

THOMAS STEUBER

Institut für Geologie, Ruhr-Universität Bochum, Universitätsstr. 150, 44801 Bochum, Germany (e-mail: thomas.steuber@ruhr-uni-bochum.de)

Abstract: The skeletal growth rates of late Cretaceous rudist bivalves have been inferred from cyclic variations of isotopic and chemical compositions which are found in sclerochronological profiles of outer shell layers. Annual shell accretion of 11 studied shells from different environmental settings was in the range of less than 10 to 54 mm. $CaCO_3$ production of individual rudists was calculated to range from 12 to 214 g a^{-1}, and estimates of annual production of rudist communities assuming dense growth fabrics range from 4.6 to 28.5 kg m^{-2}. These production rates are not significantly higher when compared to modern mussel or oyster beds but, in contrast to modern analogues, rudist associations were much more important for the sedimentary budget of low-latitude shallow-water depositional environments.

Controlling factors for the formation of growth fabrics and carbonate production are evaluated in a case study of rudist formations from a single depositional sequence. From turbulent outer platform environments to lagoonal inner platform settings, a decrease both in carbonate production and size of shells was found among specimens of a single species. No differences in carbonate production are evident when carbonate production in siliciclastic and calcareous environments is compared. Potential feedbacks between rudists and their environment induced by vertical growth, large production of bioclastic sediment and ejected biodeposits are discussed.

Carbonate platforms were particularly widespread during the late Cretaceous (Philip *et al.* 1995), and their benthic communities were dominated by rudist bivalves in various depositional environments (Ross & Skelton 1993). Post-Cenomanian rudist associations consisted predominantly of taxonomic groups (Hippuritidae and Radiolitidae; Fig. 1) which were able to actively elevate the commissure above the sediment surface by symmetrical shell accretion along the complete growth margin. Such elevator ecological morphotypes (Skelton & Gili, cited in Skelton 1991) are characterized by thick outer shell layers of low-Mg calcite. The widespread distribution of rudists in Tethyan shallow marine environments suggests that they contributed significantly to the production of calcareous sediments (Ross & Skelton 1993). The elevator growth strategy and an inferred high organic productivity of dense congregations of numerous individuals (Fig. 2) certainly affected sedimentary processes within their environment.

Stable isotope sclerochronology of rudist shells has provided information about skeletal growth rates (Steuber 1996; Steuber *et al.* 1998). Combined with studies on growth fabrics, the carbonate production of rudist associations can be estimated from these data, and potential organism–environment feedbacks in rudist-dominated depositional settings can be delineated. This allows for the quantification of important aspects of the sedimentary budget of Tethyan shelves.

Materials and methods

Numerous shells of late Cretaceous rudists have been analysed for intra-shell variations in isotopic ($\delta^{13}C$, $\delta^{18}O$) and chemical (Mg, Sr, Fe, Mn) compositions by microsampling of sclerochronological profiles. The investigated shells derive from various Turonian–Campanian rudist formations of the eastern Mediterranean and Middle East (Table 1). Bioerosion and diagenetic alteration generally preclude the analysis of fossil shell carbonates in sclerochronological profiles, and specimens which are sufficiently well preserved for such studies are extremely rare. The selection of analysed specimens was, consequently, determined by the availability of suitable shells. Nevertheless, several specimens from two well defined depositional environments in Greece (western margin of Pelagonian continental fragment; Steuber 1999*b*) and Turkey (Pontid magmatic arc; Steuber *et al.* 1998) proved to be sufficiently well preserved for geochemical analyses and provided further clues on factors

From. INSALACO, E., SKELTON, P. W. & PALMER, T. J. (eds) 2000. *Carbonate Platform Systems: components and interactions*. Geological Society, London, Special Publications, **178**, 21–32. 0305–8719/00/$15.00 © The Geological Society of London 2000.

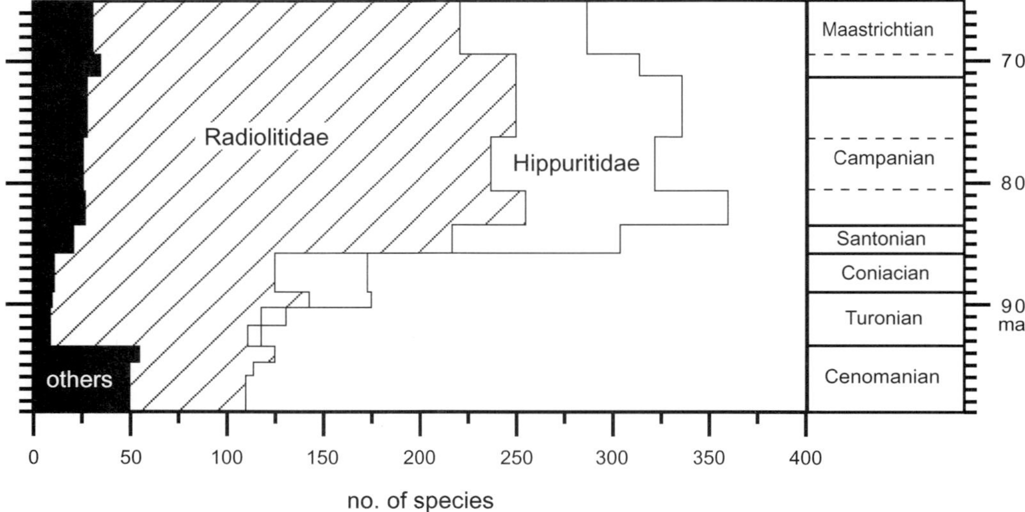

Fig. 1. Species richness of Upper Cretaceous rudist bivalves from the central-eastern Mediterranean and Middle East Tethys (excluding France, Spain and Gosau deposits of northern Alps). Data from Steuber & Löser (in press).

Fig. 2. Example of dense growth fabric of cylindrical shells of hippuritid rudists. Santonian, Montagne des Cornes (Aude), France.

controlling rates of skeletal growth and carbonate production. Detailed descriptions of depositional environments of these rudist formations are given in Steuber *et al.* (1998) and Steuber (1999*b*), and only aspects which pertain to carbonate production and organism–environment feedbacks are discussed here. Methods of geochemical analyses are described elsewhere

Table 1. *Growth rates, individual and potential community production of $CaCO_3$ (see text) in outer shell layers of late Cretaceous rudists.*

Locality/age/ species/morphotype	Shell accretion ($cm\ a^{-1}$)	Individual production ($g\ a^{-1}$)	Individual production ($g\ cm^{-2}\ a^{-1}$)	Community production ($kg\ m^{-2}\ a^{-1}$)	Depositional setting
Central Greece					
(Santonian–Lower Campanian)					
cf. Figs 5a, 6					
V. cornuvaccinum					calcareous lagoon
conical-elongate	1.8	42	4.7	11.7	
geniculate	1.5	25	3.9	9.8	
juvenile	1.7	14	4.4	8.0	
Gorjanovicia* cf. *costata					calc. foreslope
cylindrical	4.4	57	11.4	28.5	
Northern Turkey					
(Lower Campanian)					
cf. Fig. 5b					
V. ultimus					calc.-siliciclastic ramp
conical-elongate	3.0	214	7.8	26.8	
cylindrical	2.7	145	7.8	26.8	
Yvaniella alpani					calc.-siliciclastic ramp
cylindrical	1.0	18	2.6	13.4	
curved	1.0	17	2.6	13.4	
Greece					
(Santonian–Lower Campanian)					
V. inaequicostatus					calc.-siliciclastic ramp
cylindrical	1.4	75	3.6	9.1	
Greece					
(Upper Turonian)					
Vaccinites inferus					calc.-siliciclastic ramp
cylindrical	1.5	12	3.6	21.2	
Oman					
(Lower Campanian)					
Torreites sanchezi					calcareous platform
conical					
adult	0.7	55	1.8	4.6	
juvenile	3.1	69	8.0	20.1	

(Steuber 1999*a*), and results of isotopic and geochemical analyses are presented in Steuber (1996, 1997, 1999*a*) and Steuber *et al.* (1998).

Cyclic variations in $\delta^{18}O$ and Mg concentrations are commonly found within the shells (Fig. 3), and have been shown to reflect predominantly seasonal variations in palaeotemperature (Steuber 1996, 1999*a*). Consequently, cycles in isotopic and chemical compositions delineate annual growth increments.

Rates of carbonate production (P) of individuals were calculated from annual growth rates and the area of the outer shell layer on which $CaCO_3$ was deposited. This area was determined by image analysis of transverse sections (Fig. 3), i.e. sections perpendicular to the growth axis, of analysed specimens:

$$P_{CaCO_3}\ [g] = \text{area of outer shell}\ [cm^2] \times \text{linear shell accretion}\ [cm] \times \text{density}\ [g\ cm^{-3}]$$

The density of precipitated $CaCO_3$ was estimated as 2.6 g cm^{-3}. Density measurements of recent bivalve shells yielded 2.6 to 2.9 g cm^{-3} and matrix content of most shells was less than 1%, although up to 5% were found in prismatic calcite (Taylor & Layman 1972). The acid-insoluble organic residue of fibrous–prismatic rudist shells analysed during the present study was generally less than 100 ppm.

The contribution of the inner rudist shell, which was originally composed of aragonite, to carbonate production is difficult to quantify because the inner shell did not form continuously. Tabulae which sealed the bottom of the

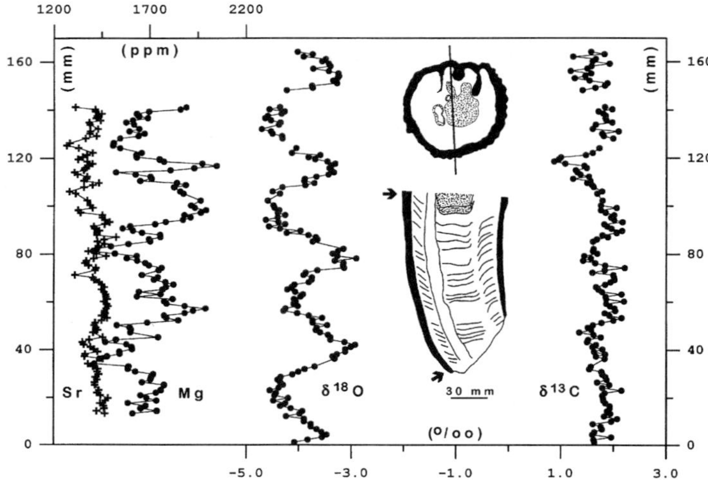

Fig. 3. Intra-shell variations in carbon and oxygen isotopic compositions ($\delta^{13}C$, $\delta^{18}O$), Mg and Sr concentrations of *Vaccinites ultimus* (Lower Campanian, northern Turkey). Inset shows transverse section of shell (top) with indicated position of longitudinal section (bottom); outer shell layer black; visceral cavity, now filled with internal sediments, dotted. Tabulae are indicated in longitudinal section. Sclerochronological profile has been sampled between arrows shown along longitudinal section.

visceral cavity formed occasionally and enclosed large hollow spaces within the shell (Fig. 3). Also, the contribution of the left, opercular valve is difficult to calculate. In cylindrical morphotypes, the size of the left valve remained almost constant after the final diameter of the adult shell was reached so that the contribution to carbonate production by the left valve was only minor. It is estimated that $CaCO_3$ that was precipitated in the inner shell layers and the left valve increases the amount calculated for $CaCO_3$ deposition in the outer shell layer of the right valve by 20%. The $CaCO_3$ production discussed below refers only to the fraction precipitated to form the outer shell of right valves, as this amount can be calculated precisely. These values, therefore, are minimum estimates of total individual and community production.

The carbonate production of rudist communities was estimated from the number of shells that settled on 1 m^2 of the sea-floor (Fig. 4). A potential community production was calculated assuming a dense growth fabric of shells in mutual contact. For the sake of simplicity, 'cubic' packing of shells was assumed (Fig. 4a). Less than 10% matrix between rudist shells was measured on some cross-sections of Santonian rudist lithosomes (Gili 1992). However, such dense growth fabrics (Fig. 2) are exceptional, and rudist congregations were more frequently restricted to clusters of less than ten shells.

Results

Skeletal growth rates

Annual skeletal growth rates of rudist bivalves which have been analysed by stable isotope sclerochronology range from less than 10 mm to 44 mm (Table 1). Even higher rates of 54 mm have been inferred (Steuber 1996), but are not used here in the calculation of $CaCO_3$ production.

Several shells which were sufficiently well preserved for geochemical analyses have been collected from two distinct depositional systems in Greece (western margin of Pelagonian continental fragment, Fig. 5a) and Turkey (Pontid magmatic arc, Fig. 5b). Growth rates of different shells of the same species are similar (Fig. 5a,b), although shell accretion may differ significantly between species from the same environment (Fig. 5b). The life spans of the studied individuals did not exceed 11 years. Measurements of the length of shell lamellae in two specimens of *Sauvagesia* sp. (Radiolitidae) yielded similar rates of shell accretion when compared to those obtained by geochemical analyses of co-occurring species (Fig. 5b). In the Radiolitidae, the original isotopic and geochemical composition of the shells is generally obliterated because of large amounts of diagenetic cements which now fill a boxwork of originally hollow prisms. Prominent shell lamellae and related variations

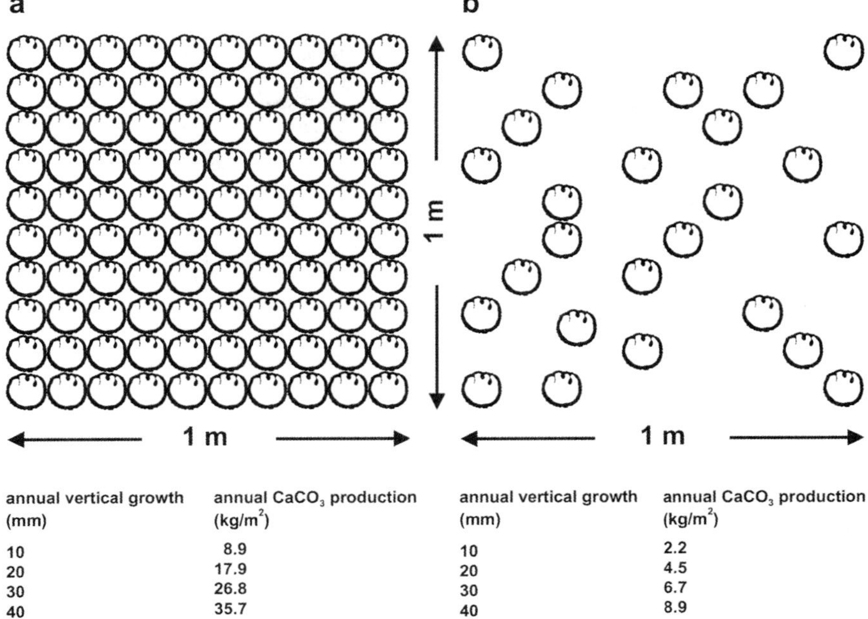

Fig. 4 Model of carbonate production by vertically growing shells of hippuritid rudists (*Vaccinites ultimus*) assuming various vertical growth rates in (**a**) dense growth fabrics and (**b**) more open associations.

in shell structures of the Radiolitidae have been previously interpreted to indicate annual growth increments (Amico 1978), and geochemical analyses of a compact-shelled radiolitid, *Gorjanovicia* cf. *costata*, showed that such shell lamellae indeed formed annually (Steuber 1996).

The apices of the investigated shells were not sufficiently well preserved to obtain information about growth rates in juvenile shells. In the analysed, ontogenetically older parts of the shells, accretion remained constant during life in several of the studied specimens, while it decreased with age in others (Fig. 5). This reflects different growth strategies: unlike other bivalves, the body mass remained most probably constant in individuals that formed cylindrical shells. Only the upper part of the right valve was inhabited, and lower parts of the shells were successively abandoned during growth and sealed by tabulae, as occurs in scleractinian corals. Therefore, the precipitation of similar amounts of $CaCO_3$ by the mantle margin each year resulted in constant vertical growth increments if the diameter of the shells remained constant. In conical morphotypes, precipitated $CaCO_3$ contributed not only to linear shell accretion but also to increasing the diameter of the shells. Consequently, $CaCO_3$ was precipitated on a continuously expanding area of the shell surface. Assuming constant rates of annual $CaCO_3$ precipitation, this explains the decrease in growth rates with ontogenetic age observed in conical shells, which results in growth histories that are similar to other, non-rudist bivalves.

Different growth rates of species which co-occurred within a 2 m thick part of a biostrome investigated in Turkey (Fig. 5b) indicate a variable amount of projection of the shells over the sediment surface (Steuber *et al.* 1998). An upper limit of mean sedimentation rates in this environment is constrained by vertical shell accretion of the slow-growing *Yvaniella alpani*, as sedimentation rates in excess of skeletal growth would have caused burial of the shells. Consequently, cylindrical morphotypes of the fast-growing *Vaccinites ultimus* must have projected with about two-thirds of total shell length over the sediment surface. Similar figures of projection have been derived from field studies of growth fabrics of other hippuritid rudist congregations (Skelton *et al.* 1995).

$CaCO_3$ production

The carbonate production of individual rudists (outer shell layer of right valves) ranges from 12 to 214 g a^{-1} (Table 1). This large range is a

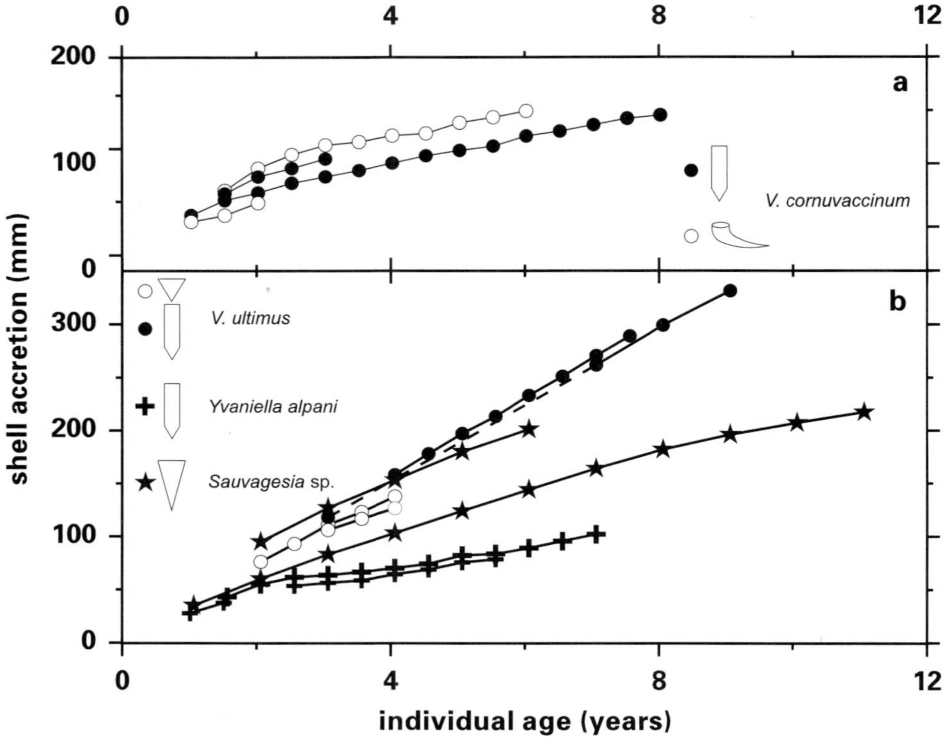

Fig. 5. Growth histories of rudists which shared the same (b) or closely adjacent (a) environments, as reconstructed from cycles in oxygen isotopic compositions (cf. Fig. 3). (**a**) Santonian–Lower Campanian shells from central Greece (see text and Fig. 6). (**b**) Shells of three species (Lower Campanian, northern Turkey; Steuber *et al.* 1998). In geniculate morphotypes, shell accretion was determined along the convex side of shells. Sketches of analysed morphotypes of different species are shown.

reflection predominantly of the different sizes of analysed individuals. To provide a better comparison between rudist specimens, and with modern bivalves, the production is also expressed as the mass of $CaCO_3$ precipitated annually per square centimetre of outer shell layer. These values range from 1.8 to 11.4 g cm^{-2}. Shells of different species that shared the same environment could have had significantly different production rates, although inferred production rates of individuals of one species are always similar. This is evident in the case of two specimens, respectively, of *Vaccinites ultimus* (7.8 g cm^{-2}) and *Yvaniella alpani* (2.6 g cm^{-2}), and in three shells of *V. cornuvaccinum* (3.9–4.7 g cm^{-2}).

Potential community production ranges from 4.6 to 28.5 kg m^{-2} a^{-1} (Table 1). These are, however, hypothetical values, as they do not account for the effects of dense populations on growth rates. Competition for resources and poisoning by faeces and pseudofaeces may have affected skeletal growth rates in crowded populations. Nevertheless, the ranges of $CaCO_3$ production of rudist associations can be constrained assuming variable population density and growth rates. This has been modelled for the case of *Vaccinites ultimus* (Fig. 4). Production rates are 2.2 kg m^{-2} a^{-1} if annual growth rates are only 10 mm and 25% of the sea floor is covered by rudist shells, and may amount to 35.7 kg m^{-2} a^{-1} in dense populations that grow vertically as much as 40 mm a^{-1}.

From the present data, it is difficult to constrain the factors which controlled the rather different rates of $CaCO_3$ production determined from different species. Most analysed specimens have been collected from Upper Turonian–Lower Campanian deposits in Greece and Turkey and, thus, from the former northern margin of Tethys at palaeolatitudes of 20–30°N (van der Voo 1993). Only *Torreites* from Lower Campanian deposits of Oman lived at an equatorial palaeolatitude.

There is no clear relationship between rates of $CaCO_3$ production and environmental setting (Table 1), but it is important that siliciclastic background sedimentation was apparently not detrimental for the establishment and growth of rudist communities (Steuber 1997; Steuber *et al.* 1998). Rather different rates of production in species that shared the same environment (*Vaccinites ultimus*, *Yvaniella alpani*, Table 1) indicate that controlling factors cannot be easily inferred from the presently available data, and that species-specific effects may have been important. Further constraints on such factors could be provided if growth and production rates are determined from various specimens of a single species, derived from the full range of palaeogeographic and environmental settings of that species' occurrence.

While this has not yet been achieved, data obtained from several shells of *Vaccinites cornuvaccinum* and *Gorjanovicia* cf. *costata* from the Santonian–Lower Campanian of central Greece provide further clues concerning the controlling factors of carbonate production within a single species. All analysed specimens have been collected from various types of rudist formations (Fig. 6) found in a single depositional sequence of transgressive calcareous deposits (Steuber 1999*b*). They follow unconformably over a folded basement of Jurassic–Cretaceous limestones, early Cretaceous flysch-type deposits or ophiolites along the western margin of the Pelagonian continental fragment. At the basal levels, redeposited laterites and mixed calcareous–siliciclastic mud–wackestones are devoid of rudists. Thin-shelled miliolid foraminifera, ostracods and microgastropods are abundant in these deposits, and probably lived as epizoans in sea grass meadows. Associations of *Vaccinites cornuvaccinum* are found upsection, and thrived in a calm lagoonal environment with a remarkably poor calcareous microfauna. Geniculate or curved morphotypes of *Vaccinites* dominate, and the orientation of most specimens with the commissure parallel to the former sediment surface indicates that they are preserved in life position. Also, bioerosion is frequently restricted to the commissural parts of the shells that projected above the sediment surface after the death of individuals. Geniculate and curved morphotypes are generally common among hippuritid rudists and, in the central Greek rudist formations discussed here, resulted from occasional toppling of vertically growing shells and subsequent reorientation of the direction of growth to resume vertical growth. The mode of curvature and the random orientation of rudist shells has been interpreted to reflect toppling due to insufficient sediment support rather than dislocation by storms or currents (Steuber 1999*b*). Consequently, toppling resulted from vertical growth at much higher rates than ambient sedimentation. Broad-conical morphotypes of associated radiolitid rudists also indicate low sedimentation rates and calm, protected environments. These lagoonal deposits were sheltered from the surf of the open ocean by offshore shoals on which compact-shelled radiolitids (*Gorjanovicia*) formed mounds that can be traced laterally for several tens of metres (Fig. 6). Dense growth fabrics of vast numbers of vertically growing shells are embedded in bioclastic packstones that consist almost exclusively of mechanically and bioerosively reworked debris of shells. High-energy deposits are found at correlative levels in all studied sections, and locally consist of bioclastic rud- or grainstones on which rudists did not settle. Occasionally, *Vaccinites cornuvaccinum* formed clusters of mutually encrusting, vertically growing shells on the crests of such bioclastic shoals. Bioerosion, largely by clionid sponges, was intense within the *Gorjanovicia* mounds and produced large amounts of silt-sized sediments, which were exported to the deeper shelf. This is indicated by several tens of metres of deep-subtidal limestones composed essentially of silt-sized bioclasts in the topmost parts of all studied sections.

The studied shells of *Vaccinites cornuvaccinum* have been collected both from lagoonal deposits and from those of the landward slopes of the sheltering offshore shoals (Steuber 1997). Unfavourable preservation rendered geochemical analyses of shells that formed clusters at the top of high-energy shoals impossible, but a single specimen of the compact-shelled radiolitid *Gorjanovicia* from this environment was sufficiently well preserved. It was collected from life position and grew on large fragments of shells on the seaward slope of a rudist mound. The curvature of the shell indicates several reorientations during rapid growth of up to 54 mm a^{-1} (Steuber 1996).

In this depositional system, carbonate production expressed as $CaCO_3$ precipitated annually per square centimetre of shell surface is significantly lower in three specimens of *Vaccinites cornuvaccinum* (3.9 to 4.7 g) when compared to *Gorjanovicia* (11.4 g). This difference is also seen in potential community production (8.0 to 11.7 versus 28.5 kg m^{-2} a^{-1}). The maximum value of potential community production was probably almost reached in the dense growth fabrics of *Gorjanovicia*, but certainly not in associations of geniculate morphotypes of *V.*

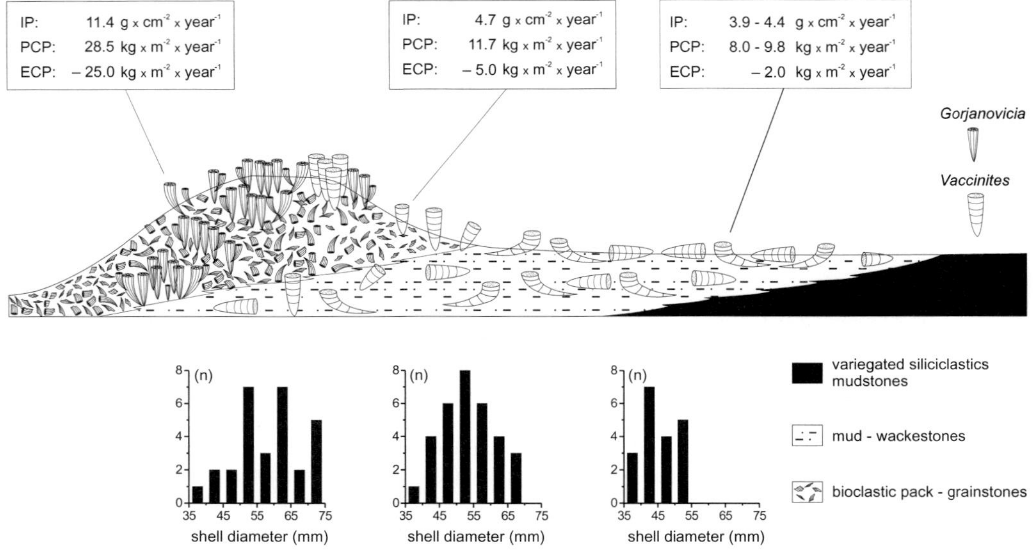

Fig. 6. Sketch of distribution of morphotypes and growth fabrics of *Vaccinites cornuvaccinum* and *Gorjanovicia* cf. *costata* in depositional environments at the western margin of the Pelagonian micro-continent in central Greece. Not to scale: horizontal distance is several kilometres so that vertical scale is exaggerated. Figures are given for carbonate production by individuals (IP), potential community production (PCP) and estimated community production (ECP). Analysed shells to which production values refer are indicated. Estimated community production accounts for the observed density of shells in different environments, in contrast to potential community production which assumes densely packed growth fabrics. Histograms at bottom of figure show size (diameter) distribution of shells of *V. cornuvaccinum* in environments shown above.

cornuvaccinum. Annual estimated community production by the latter species is 2.0 kg m^{-2} a^{-1}. Vertically growing shells of *V. cornuvaccinum* from the landward slopes of high-energy rudist mounds also lived in open growth fabrics, and mutual contact of shells was not observed. Consequently, only a fraction of potential community production (Table 1) is estimated to have occurred in these communities. This results in a steep gradient in estimated $CaCO_3$ production between restricted lagoonal settings and more turbulent environments of bioclastic shoals. This is plausible, as food supply was probably more favourable in turbulent offshore environments which potentially provided a large flux of plankton. Also, high turbulence swept away biodeposits (faeces and pseudofaeces) while their accumulation in restricted lagoons may have been inimical for benthic communities. The bioclastic shoals with *Gorjanovicia*, therefore, most probably owe their existence to a high *in situ* carbonate production, even though much of the produced carbonate was exported to the deeper shelf as is shown by the thick succession of limestones essentially composed of silt-sized bioclasts which overlie the *Gorjanovicia* shoals.

Differences in growth rates and $CaCO_3$ production of *Vaccinites cornuvaccinum* are small and probably not significant when cylindrical versus geniculate morphotypes (Table 1, Fig. 5a), or different depositional environments (Fig. 6) are compared. Owing to insufficient preservation, data from shells of this species from clusters in the high-energy environments are not available, but the size distribution of this species reveals a distinct gradient (Fig. 6). Predominantly curved or geniculate shells from lagoonal environments do not exceed 55 mm in diameter. Cylindrical morphotypes which grew upright on the landward slopes of bioclastic shoals have a mean diameter of 50–55 mm, and maximum values are up to 70 mm. The largest shells are found in specimens which grew on the crests of bioclastic shoals (maximum diameter 75 mm, mean diameter 55–60 mm). This size distribution may reflect a gradient in the availability of food and oxygen. Other physical conditions (temperature, salinity) were apparently less important, as the isotopic compositions of the shells from the described range of environments are similar (Steuber 1999*a*). The small size of curved or geniculate shells may also be due to frequent destabilization, toppling and subsequent reorientation of the growth axis. This shows that life position was unfavourable during prolonged periods of growth.

Discussion

Comparison of rudist $CaCO_3$ production with modern bivalve and coral reef associations

Numerous studies have focused on the biomass and carbonate production of modern coral reefs (e.g. Barnes & Chalker 1990; Muscatine 1990). The biomass production of modern bivalve communities is also reasonably well known (e.g. Bahr 1976; Loo & Rosenberg 1983; Asmus 1987), but few data are available on the carbonate production of recent bivalves. Annual accumulation of 0.17 g $CaCO_3$ cm^{-2} of shell surface has been reported from modern oysters (Wilbur & Jodrey 1952). Two individuals of the giant clam *Tridacna gigas* precipitated 2.7 and 4.1 g $CaCO_3$ cm^{-2}, respectively (Bonham 1965). It is remarkable that production rates of rudists match those of the extant photosymbiotic *Tridacna*, whereas carbonate production in extant, fast-growing non-photosymbiotic oysters is lower by an order of magnitude. However, available data for comparison are limited. Among extant photosymbiotic bivalves, only *Tridacna* constructs suspiciously large shells, while other genera such as *Corculum* and *Fragum* are characterized by small and rather thin shells (Ohno *et al.* 1995). It is, therefore, certainly an oversimplification to relate shell size to the absence or presence of photosymbiosis. Differences in the composition of Cretaceous and modern sea-water may be a more important controlling factor of calcification (Stanley & Hardie 1998; Kleypas *et al.* 1999).

Recalculation of the data on biomass production and shell growth of extant communities of *Mytilus edulis* in the Wadden Sea of Germany (Asmus 1987), and in suspended cultures in western Sweden (Loo & Rosenberg 1983), results in production rates of 2 to 6.5 kg $CaCO_3$ m^{-2} a^{-1}. The living shell biomass of an intertidal reef of *Crassostrea virginica* was determined as 23.4 kg $CaCO_3$ m^{-2}, individuals reaching an age of five to seven years (Bahr 1976; Kirby *et al.* 1998). Such oyster reefs can consist of more than 4000 living individuals per square metre (Dame 1976). The spatfall of *Crassostrea virginica* precipitated 0.25 kg $CaCO_3$ m^{-2} within the first six months after settling (Dame 1976).

Gross carbonate production in modern coral reefs has been reported to be in the range of 1 to 35 kg m^{-2} a^{-1} (Chave *et al.* 1972; Stearn *et al.* 1977; Barnes & Chalker 1990), but less than 2.0 kg m^{-2} a^{-1} is characteristic for most coral reef environments (Heiss 1995). This comparison shows that $CaCO_3$ production in modern mussel and oyster beds may outpace that of coral reefs. However, the patchy occurrence of densely populated bivalve associations causes their contribution to the sedimentary budget of depositional systems to be low. This was demonstrably different with late Cretaceous rudist associations, which were widespread along the shelves of Tethys (e.g. Ross & Skelton 1993).

Although the biomass production of rudist communities cannot be estimated, it is obvious that dense associations required a considerable supply of energy resources. Such congregations may have been maintained in environments with a large influx of plankton and other suspended organic particles, or they may have relied on the recycling of energy and nutrients by endosymbiotic algae. Photosymbiosis in rudists has been addressed repeatedly (Philip 1972; Kauffman & Johnson 1988; Lewy 1995). Focusing this discussion on the Hippuritidae, i.e. the group for which high rates of carbonate production have been demonstrated here and which dominated many late Cretaceous rudist associations (Fig. 1), two important lines of evidence argue against photosymbiosis. (1) A peculiar system of pores and canals in the opercular left (upper) valve has been convincingly shown to have functioned as a filtering device for inhaled water currents (Skelton 1976), in contrast to the interpretation of admitting light to the mantle (Seilacher 1998). (2) The presence of species-rich hippuritid associations in environments with a significant siliciclastic background sedimentation, and thus inferred high nutrient supply, contrasts with less abundant and species-poor communities on intra-oceanic carbonate platforms, i.e. in inferred oligotrophic settings (Steuber & Löser 1999, in press). Dismissing the hypothesis of photosymbiosis in hippuritid rudists implies a high availability of appropriate food to sustain dense congregations. This suggests that they flourished in areas of high primary production.

Organism–environment feedbacks

Although frequently referred to as characteristic Cretaceous reef-builders, the growth fabrics formed by elevator rudist bivalves have been shown to be predominantly sediment-supported (Gili *et al.* 1995; Sanders & Pons 1999). Typical features of rudist formations such as rapid vertical growth associated with sediment baffling, the production of significant amounts of calcareous sediment by either mechanical or bioerosive breakdown of shells, the lack of bound, wave-resistant growth fabrics, as well as the production of large amounts of biodeposits (faeces and pseudofaeces) suggest several feedbacks between these organisms and their environment.

In calm inner shelf or lagoonal settings, low flux of plankton and suspended organic material limited the number of individuals which could successfully establish. Low turbulence precluded the transport of bioclastic material, and sedimentation was restricted to fine-grained sediment which accumulated only slowly. Limited support of vertically growing shells by low rates of ambient sedimentation induced frequent toppling so that the potential for rapid vertical growth was not advantageous in such environments. In contrast to radiolitid rudists, the range of morphotypes observed in hippuritids is rather narrow, and there is no evidence that growth rates were adjusted to ambient sedimentation rates. The size of such populations was also controlled by the amount of biodeposits (faeces, pseudofaeces) produced in relation to turbulence. A population of *Mytilus edulis* was estimated to eject 10 kg m^{-2} a^{-1} (dry weight) of faeces and pseudofaeces (Tsuchiya 1980), and similarly large amounts must be expected to have been produced by dense rudist associations. It has been demonstrated that presumed ejection sites of faeces and pseudofaeces moved from a commarginal position in early rudists to a position on top of the opercular left valve in more advanced groups of elevator ecological morphotypes (Skelton 1979; Steuber 1999*b*). Overhead ejection helped to remove biodeposits as far as possible in the presence of the merest current (Gili & LaBarbera 1998). In environments with low turbulence where these biodeposits were not removed, their accumulation probably induced anoxic conditions at the sediment surface. This precluded the settlement of rudist larvae, and explains why other epibenthic organisms are frequently rare or absent in many rudist communities (Sanders & Pons 1999; Steuber 1999*b*). The exclusion of other organisms is common in modern mussel beds, and benthic communities are impoverished below suspended cultures of *Mytilus* on hanging long lines (Asmus 1987). Biodeposition may have even been inimical for rudists themselves, but elevator morphotypes were able to avoid these adverse conditions by vertical growth which raised the commissure over the sediment surface and into better oxygenated levels. This, however, was impossible, when stabilization by sediment was not provided. Consequently, there are several potential negative feedbacks which limited the number of individuals in relation to turbulence and supply with energy resources.

Food supply and oxygenation of sea water were less limiting in more open and turbulent outer shelf settings. Additionally, biodeposits were more effectively removed by currents, so that dense growth fabrics formed which may also include other benthic organisms. Large amounts of sediment either produced within the rudist community or transported by currents, provided stabilization of vertically growing shells. Rudist shells which projected above the sediment surface certainly induced baffling of sediment by reduction of current velocity. Also, as a variable fraction of the sea floor was covered by projecting rudist shells, the area on which sediment could settle was reduced, which additionally enhanced sedimentation rates in the interstices between individuals. While this positive feedback provided stabilization of vertically growing shells, it did not result in significant, long-lived changes of the sea-floor topography, as these unbound fabrics were unprotected against destruction by storms, and were vulnerable to currents which removed the stabilizing sediment (Gili *et al.* 1995; Skelton *et al.* 1995). In such environments, the adaptation of rapid vertical growth was most advantageous, and density of colonization was probably limited only by space and availability of energy resources. Such associations were, however, always endangered by mechanical destruction by storms, or by migrating bioclastic bars, resulting in short-term sedimentation rates which could not be matched by vertical growth (Gili & Skelton 1999). In fact, associations of vertically growing shells preserved *in situ* are commonly overlain by sheets of bioclastic deposits (Skelton *et al.* 1995; Sanders & Pons 1999).

These examples address only two end-member examples of elevator rudist formations, which show a remarkable diversity in growth fabric and faunal composition (Ross & Skelton 1993; Gili *et al.* 1995). Variations in each of the controlling factors discussed had significant impact on the density of colonization, the formation of growth fabrics, the inclusion of other benthic organisms, and thus on the carbonate production of these benthic associations. In view of the short life spans of individuals which are evident from the reconstructed growth histories, a few years of stable conditions in environments in which larvae could successfully settle were sufficient for the establishment of a high-performance carbonate factory. The unbound growth fabrics were rather short-lived so that most of these factories have been destroyed by mechanical breakdown and bioerosion, but those which entered the fossil record testify to the prominent contribution of rudist bivalves to carbonate production during the late Cretaceous.

Funding by Deutsche Forschungsgemeinschaft grants Ste 670/2 and Ste 670/3 is gratefully acknowledged. I am indebted to P. W. Skelton and D. Sanders, who provided valuable comments on the manuscript.

References

AMICO, S. 1978. Recherches sur la structure du test des Radiolitidae. *Travaux du Laboratoire de Géologie Historique et de Paléontologie, Marseille*, **8**, 1–131.

ASMUS, H. 1987. Secondary production of an intertidal mussel bed community related to its storage and turnover compartments. *Marine Ecology – Progress Series*, **39**, 251–266.

BAHR, L. M. 1976. Energetic aspects of the intertidal oyster reef community at Sapelo Island, Georgia (USA). *Ecology*, **57**, 121–131.

BARNES, D. J. & CHALKER, B. E. 1990. Calcification and photosynthesis in reef-building corals and algae. *In*: DUBINSKY, Z. (ed.) *Coral Reefs*. Elsevier, Amsterdam, 109–131.

BONHAM, K. 1965. Growth rate of giant clam *Tridacna gigas* at Bikini Atoll as revealed by radioautography. *Science*, **149**, 300–302.

CHAVE, K. E., SMITH, S. V. & ROY, K. J. 1972. Carbonate production by coral reefs. *Marine Geology*, **12**, 123–140.

DAME, R. F. 1976. Energy flow in an intertidal oyster population. *Estuarine and Coastal Marine Science*, **4**, 243–253.

GILI, E. 1992. Palaeoecological significance of rudist constructions: A case study from Les Collades de Basturs (Upper Cretaceous, south-central Pyrenees). *Geologica romana*, **28**, 319–325.

—— & LABARBERA, M. 1998. Hydrodynamic behaviour of hippuritid rudist shells: ecological consequences. *Geobios, Mémoire spécial*, **22**, 137–146.

—— & SKELTON, P. W. 1999. Factors regulating the development of elevator rudist congregations. *The 1999 Lyell meeting – Organism–environment feedbacks in carbonate platforms and reefs*, Abstracts, p. 11.

——, MASSE, J.-P. & SKELTON, P. W. 1995. Rudists as gregarious sediment-dwellers, not reef-builders, on Cretaceous carbonate platforms. *Palaeogeography, Palaeoclimatology, Palaeoecology*, **118**, 245–267.

HEISS, G. A. 1995. Carbonate production by scleractinian corals at Aqaba, Gulf of Aqaba, Red Sea. *Facies*, **33**, 19–34.

KAUFFMAN, E. G. & JOHNSON, C. C. 1988. The morphological and ecological evolution of middle and Upper Cretaceous reef-building rudists. *Palaios*, **3**, 194–216.

KIRBY, M. X., SONIAT, T. M., & SPERO, H. J. 1998. Stable isotope sclerochronology of Pleistocene and recent oyster shells (*Crassostrea virginica*). *Palaios*, **13**, 560–569.

KLEYPAS, J. A., BUDDEMEIER, R. W., ARCHER, D., GATTUSO, J.-P., LANGDON, C. & OPDYKE, B. N. 1999. Geochemical consequences of increased atmospheric carbon dioxide on coral reefs. *Science*, **284**, 118–120.

LEWY, Z. 1995. Hypothetical endosymbiontic zooxanthellae in rudists are not needed to explain their ecological niches and thick shells in comparison with hermatypic corals. *Cretaceous Research*, **16**, 25–37.

LOO, L.-O. & ROSENBERG, R. 1983. *Mytilus edulis* culture: growth and production in western Sweden. *Aquaculture*, **35**, 137–150.

MUSCATINE, L. 1990. The role of symbiotic algae in carbon and energy flux in reef corals. *In*: DUBINSKY, Z. (ed.) *Coral Reefs*. Elsevier, Amsterdam, 75–88.

OHNO, T., KATOH, T. & YAMASU, T. 1995. The origin of algal-bivalve photo-symbiosis. *Palaeontology*, **38**, 1–21.

PHILIP, J. 1972. Paléoécologie des formations à rudistes du Crétacé supérieur – l'exemple du Sud-Est de la France. *Palaeogeography, Palaeoclimatology, Palaeoecology*, **12**, 205–222.

——, MASSE, J.-P. & CAMOIN, G. 1995. Tethyan carbonate platforms. *In*: NAIRN, A. E. M., RICOU, L.-E., VRIELYNCK, B. & DERCOURT, J. (eds) *The Ocean Basins and Margins, Vol. 8*. Plenum, New York, 239–265.

ROSS, D. J. & SKELTON, P. W. 1993. Rudist formations of the Cretaceous: a palaeoecological, sedimentological and stratigraphic review. *In*: Wright, P. (ed.) *Sedimentology Review/1*. Blackwell, London, 73–91.

SANDERS, D. & PONS, J. M. 1999. Rudist formations in mixed siliciclastic-carbonate depositional environments, Upper Cretaceous, Austria: stratigraphy, sedimentology, and models of development. *Palaeogeography, Palaeoclimatology, Palaeoecology*, **148**, 249–284.

SEILACHER, A. 1998. Rudists as bivalvian dinosaurs. *In*: JOHNSTON, P. A. & HAGGART, J. W. (eds) *Bivalves, an Eon of Evolution – Paleobiological studies honoring Norman D. Newell*. University of Calgary, Calgary, 423–436.

SKELTON, P. W. 1976. Functional morphology of the Hippuritidae. *Lethaia*, **9**, 83–100.

—— 1979. Gregariousness and proto-cooperation in rudists (Bivalvia). *In*: LARWOOD, G. & ROSEN, B. R. (eds) *Biology and Systematics of Colonial Organisms*. The Systematics Association, Special Volume, **11**, 257–279.

—— 1991. Morphogenetic versus environmental cues for adaptive radiations. *In*: SCHMIDT-KITTLER, N. & VOGEL, K. (eds) *Constructional Morphology and Evolution*. Springer, Berlin, 375–388.

——, GILI, E., VICENS, E. & OBRADOR, A. 1995. The growth fabric of gregarious rudist elevators (hippuritids) in a Santonian carbonate platform in the southern central Pyrenees. *In*: PHILIP, J. & SKELTON, P. W. (eds) *Palaeoenvironmental models for the benthic associations of Cretaceous carbonate platforms in the Tethyan realm*.

Palaeogeography, Palaeoclimatology, Palaeoecology, **119**, 107–126.

STANLEY, S. M. & HARDIE, L. A. 1998. Secular oscillations in the carbonate mineralogy of reef-building and sediment-producing organisms driven by tectonically forced shifts in seawater chemistry. *Palaeogeography, Palaeoclimatology, Palaeoecology*, **144**, 3–19.

STEARN, C. W., SCOFFIN, T. P. & MARTINDALE, W. 1977. Calcium carbonate budget of a fringing reef on the west coast of Barbados. Part I – Zonation and productivity. *Bulletin of Marine Science*, **27**, 479–510.

STEUBER, T. 1996. Stable isotope sclerochronology of rudist bivalves: Growth rates and Late Cretaceous seasonality. *Geology*, **24**, 315–318.

—— 1997. Hippuritid rudist bivalves in siliciclastic settings – functional adaptations, growth rates and strategies. *Proceedings of the 8th International Coral Reef Symposium, Vol. II*. 1761–1766.

—— 1999*a*. Isotopic and chemical intra-shell variations in low-Mg calcite of rudist bivalves (Mollusca: Hippuritacea) – disequilibrium fractionations and late Cretaceous seasonality. *Geologische Rundschau*, **88**, 551–570.

—— 1999*b*. Cretaceous rudists of Boeotia, central Greece. *Special Papers in Palaeontology*, **61,** 1–226.

—— & LÖSER, H. 1999. First results from a palaeontological data base of the Hippuritacea: Species richness and abundance patterns of Tethyan Cretaceous rudists in the central-eastern Mediterranean and Middle East. Fifth International Conference of Rudists, Abstracts and Field Trip Guides, *Erlanger Geologische Abhandlungen*, **SB 3**, 70–71.

—— & —— (in press) Diversity and abundance patterns of Tethyan Cretaceous rudist bivalves (Mollusca: Hippuritacea) in the central-eastern Mediterranean and Middle East, analysed from a palaeontological data base. *Palaeogeography, Palaeoclimatology, Palaeoecology*.

——, YILMAZ, C. & LÖSER, H. 1998. Growth rates of early Campanian rudists in a siliciclastic-calcareous setting (Pontid Mts., North-central Turkey). *Geobios, Mémoire spécial*, **22**, 385–401.

TAYLOR, J. D. & LAYMAN, M. 1972. The mechanical properties of bivalve (Mollusca) shell structures. *Palaeontology*, **15**, 73–87.

TSUCHIYA, M. 1980. Biodeposit production by the mussel *Mytilus edulis* L. on rocky shores. *Journal of Experimental Marine Biology and Ecology*, **47**, 203–222.

VAN DER VOO, R. 1993. *Paleomagnetism of the Atlantic, Tethys and Iapetus Oceans.* Cambridge University, Cambridge.

WILBUR, K. M. & JODREY, L. H. 1952. Studies on shell formation. I. Measurement of the rate of shell formation using Ca^{45}. *Biological Bulletin*, **103**, 269–276.

Spatial and temporal patterns of macroboring within Mesozoic and Cenozoic coral reef systems

CHRISTOPHER T. PERRY[1] & MARKUS BERTLING[2]

[1] *Department of Environmental and Geographical Sciences, Manchester Metropolitan University, John Dalton Building, Chester Street, Manchester, M1 5GD, UK (e-mail: c.t.perry@mmu.ac.uk)*

[2] *Geologisch-Paläontologisches Institut und Museum, Pferdegasse 3, D-48143 Münster, Germany*

Abstract: Macroboring of coral reefs has varied significantly through time, with the modern intensity and producer composition (usually dominated by sponges) as a rather recent phenomenon. Given the outstanding role and influence of bioerosion on framework morphology, community composition and sediment production, Modern conditions are therefore poor analogues for the structure and function of pre-Neogene reef systems. Modern and Neogene reef borer associations are mostly dominated by sponges, although marked spatial variations in the abundance of borer groups are evident within individual reef systems. Highest diversity typically characterizes low energy, shallow water back-reef or lagoon sites. This condition evolved gradually from the Late Triassic onwards, when scleractinians first built reefs.

Sponges appear to have played a subordinate role in Mesozoic coral-dominated buildups. Worms and barnacles dominate in the early Mesozoic (Triassic and Lower Jurassic), with a progressive increase in bivalve borers through the Jurassic. The paucity of data collected to date makes determination of the causes of temporal change in macroboring community composition difficult to constrain. Macroboring groups seem to have withstood biotic crises much better than their coral substrate and thus reef ecological evolutionary units are not applicable. There is some indication that macroborers may have radiated to colonize new ecological niches during the early stages of coral reef diversification. The development was nonetheless influenced by biotic changes in the marine realm, the strongest effects potentially resulting from switches in nutrient status and the origin or diversification of reef grazers.

The term 'bioerosion' was introduced by Neumann (1966) to describe the process by which biological activity destroys and/or denudes hard substrates. Within coral reef environments this process is dominated by the activities of grazers (primarily fish and echinoids) and borers. The organisms responsible for boring are classed as either microborers (including Cyanobacteria, Chlorophyta, Rhodophyta and Fungi) or macroborers, the most abundant of which are the sponges, bivalves and worms (primarily polychaetes and sipunculans). Macroborers play a key role in the modification of hard coral substrates and are responsible for extensive substrate destruction (Goreau & Hartman 1963; Acker & Risk 1985), abundant associated sediment production (Fütterer 1974; Moore & Shedd 1977) and the generation of secondary porosity within the reef structure (Ginsburg & Schroeder 1974; Schroeder & Zankl 1974). The process thus exerts an important influence over styles of carbonate preservation and patterns of reef framework modification.

Current knowledge

The fossil record of boring organisms can be determined through trace fossils, and reasonably good delineation of at least the responsible 'group' of borer is possible in most cases. Consequently, the importance of boring organisms within ancient reef environments can be inferred, and their contribution to framework diminution, sediment production, reef demise (e.g. Hallock 1988; Wood 1993) and local patterns of diagenesis (e.g. Pemberton *et al.* 1988; Rehman *et al.* 1994) considered. All too often, however, studies of ancient reef communities fail to acknowledge or describe anything more than the principal reef-building organisms. Associated organisms (including the borers) are, at best, often only mentioned in passing with little description of ichnotaxa or even the groups of borers present. Whilst this can, in part, be attributed to difficulties in identifying borings within fossil coral material due to different styles of preservation, there appear to have been only limited attempts at quantifying or even

From: INSALACO, E., SKELTON, P. W. & PALMER, T. J. (eds) 2000. *Carbonate Platform Systems: components and interactions*. Geological Society, London, Special Publications, **178,** 33–50. 0305–8719/00/$15.00

describing temporal variability of this important guild.

By contrast, excellent advances have recently been made in the recognition, identification and assessment of temporal and spatial distributions of microborers (see Vogel (1997) and references therein). Only a few studies describe macroboring of ancient reef-related environments in any detail (e.g. Kauffman & Sohl 1974; Fürsich *et al.* 1994; Edinger & Risk 1994; Weidlich 1996). For this reason, data relating to macroboring in coral-dominated buildups are scarce and much of the current understanding of fossil boring communities derives from observations of ancient hardgrounds (e.g. Fürsich 1979; Palmer 1982). These environments, however, are not characterized by the complex species interactions and the diverse substrate types and morphologies associated with coral reefs. They thus provide poor analogues for understanding patterns of macroboring within ancient coral reef environments.

The fossil record of macroborers

The organisms responsible for modern reef macroboring are well known, with activity dominated by sponges (primarily Clionidae), bivalves (Lithophaginae and Gastrochaenidae), sipunculans (e.g. Phascolosomatidae, Aspidosiphonidae), polychaete worms (mostly Cirratulidae, Eunicidae, Fabriciinae and Spionidae) and barnacles (e.g. *Lithotrya, Trypetesa*). The holes produced by these groups within hard, calcareous substrates typically have good potential for preservation, although later macroboring and/or abrasion can destroy the outermost surfaces of corals. Such erosional impacts on the shallower tiers of bioerosion, however mostly affect grazing traces (Bromley & Asgaard 1993). Additional preservation problems can occur due to subsequent dia-genetic alteration of corals, although this sometimes results in improved preservation where diagenetic dissolution exposes boring casts.

Commonly borings can be attributed to a specific group of producers (Table 1), enabling a system of trace fossil names (ichnotaxa) to be applied. Sometimes, such borings match the producers' body outline so well (mainly in the case of sponges, bryozoans and, to a lesser extent, bivalves) that numerous ichnospecies can be recognized (see Bromley & d'Alessandro (1984) for *Entobia*; Kelly & Bromley (1984) for *Gastrochaenolites*). It is tempting to relate ichnospecies to a particular producer species, but ichnology would not need ichnotaxa if this approach were without problems. Bromley (1970, 1978), for example, discusses the potential for different sponge species to produce comparable traces (*Entobia* isp.) and for identical species to produce different boring morphologies in different substrates. The biological implications of 'worm' and cirriped boring taxa are largely unknown. Consequently, the use of ichnotaxa at any level (ichnospecies or ichnogenus) as proxy for biological evolution will remain problematic and thus borer diversity will not fully be mirrored by the resultant trace fossil association (ichnocoenosis). Identification of boring traces should, among others, be based upon the reviews by Bromley (1970, 1978), Warme (1975), Bromley & d'Alessandro (1983, 1984, 1987), Kelly & Bromley (1984), Pleydell & Jones (1988) and Fürsich *et al.* (1994).

Scope of this study and data collection

The aim of this paper is to describe the initial results of an on-going study to quantify the history, distribution and development of coral reef macroboring organisms. They are responsible for significant substrate destruction and sediment generation and thus the fossil record of their evolution/distribution in fossil coral reefs has important implications for understanding changes in patterns and styles of carbonate production and preservation. Particular attention is given to (1) assessment of temporal and spatial trends in the composition of boring communities, (2) the occurrence and distribution of trace fossils from the Triassic and later, and (3) evidence of any trends in terms of overall rates of boring over time. A series of case studies is used to support current ideas and observations.

This study draws together data collected by different workers from a wide range of materials and sites. Consequently methodologies vary between studies depending upon (1) the nature of coral preservation, (2) the suitability or potential for particular sampling/analytical strategies, and (3) facies exposure (a constraint on spatial studies). In some cases, dissolution during diagenesis results in leaching of the original coral aragonite leaving the boring traces exquisitely preserved (often in three-dimensional form) within the moulds of the former corals (see Pleydell & Jones 1988; Perry 1996; Bertling & Insalaco 1998). This permits very accurate identification of the boring trace, but only enables semi-quantitative assessment of ichnospecies abundance. By contrast, in examples where whole coral heads/blocks can be recovered, serial sections analysed using either X-ray techniques or point counting of traced surface sections can provide very accurate estimates of

Table 1. *Macroborers and their borings in coral reef environments, along with simple description of trace fossil morphologies*

Animal group	Trace fossil	Boring morphology
Porifera	*Entobia*	Complex, branched networks of chambers, multi-apertured, numerous ichnospecies.
	Uniglobites	One large chamber with small appendages, single aperture.
	Dendrina	Stellate with radii hardly separated, single aperture.
	Dictyoporus	Irregular meshwork without chambers, multi-apertured.
Polychaeta	*Trypanites*	Narrow cylindrical, round cross-section, blunt end, single aperture.
	Caulostrepsis	Cylindrical with dumbbell-shaped cross-section, single aperture.
	Lapispecus	Cylindrical with a vein running along one side, multi-apertured.
	Meandropolydora	U-shaped cylindrical in complex loops, multi-apertured.
	Spirichnus	Spiral cylindrical, single aperture.
	Cunctichnus	Cylindrical with several short branches, single aperture.
Bivalvia	*Gastrochaenolites*	Flask-shaped, single aperture, numerous ichnospecies.
	Phrixichnus	Flask-shaped with feather-like sculpture.
Cirripedia – Acrothoracica	*Rogerella*	Sock-shaped, narrow, oval or slit-like aperture.
– Thoracica	*Trypanites*	Simple cylindrical, oval cross-section, tapering, single aperture.
Phoronida	*Talpina*	Small, branched with cylindrical branches, multi-apertured.
	Conchotrema	Cylindrical, U-shaped and multi-apertured.
Bryozoa	*Ropalonaria*	Networks connecting elongate chambers, closely resembling producer morphology, multi-apertured.
	Terebripora	
	Spathipora	
	Iramena	
Sipuncula	*Trypanites*	Simple cylindrical, round cross-section, single aperture.

borer abundance (e.g. Klein *et al.* 1991; Risk *et al.* 1995; Perry 1998). However, such destructive sampling is not always possible (e.g. museum specimens or smooth outcrop sections) and even then accurate identification of ichnotaxa can be difficult if borings have become filled by sediment.

Data recovered by these different methodologies are thus not fully comparable (especially in relation to variations in boring intensity). Listing relative importance of inferred producer groups as percentages of the complete borer assemblage, however, should give an unbiased impression of the composition of the macroborer fauna. Table 2 lists the ages, locations and importance of borer groups drawn from all case studies available. The lack of data from fossil reefs is apparent and reflects the problems inherent in terms of quantifying and recognizing macroboring in the ancient (see above). Detailed examples from Cenozoic and Mesozoic coral reefs are hence restricted to localities where reasonable quantitative data are available, although reference is made to other available supporting material for each time period.

Results

Modern coral macroboring

There is an abundant literature documenting boring community composition, rates of substrate destruction and patterns of substrate colonization by borers within modern reef environments, although many of these studies have been restricted to analysis of samples collected from individual reef environments or from a limited range of water depths. Exceptions include studies undertaken in Bermuda (Bromley 1978), across the Great Barrier Reef (Sammarco & Risk 1990; Risk *et al.* 1995) and in north Jamaica (Perry 1998). Despite marked

Table 2. *Relative abundance of macroborers (by group) in reefs, compiled from studies with quantitative data*

			Environment		Facies			Average borer abundance (%)			
	Age	Locality	Clear	Silics.	'Lagoon'	Shallow	Deep	Sponge	Bivalve	Worm	Other
1.	Holocene	Discovery Bay, Jamaica	*		*	*	*	81.5	2.5	16.0	0.0
2.	Holocene	Great Barrier Reef, Australia	*		*	*		68.8	9.1	18.8	0.0
3.	Holocene	Eilat/Sinai, Red Sea	*			*	*	22.1	19.0	42.7	16.2
4.	Pleistocene	Falmouth Fm, Jamaica	*		*	*	*	64.7	8.2	25.8	1.3
5.	Pleistocene	Sinai, Red Sea	*			*	*	10.8	19.4	42.0	27.8
6.	Messinian	Cap Blanc, Mallorca	*		*	*	*	75.0	23.0	2.0	0.0
7.	Aquitanian	Bluff Fm, Grand Cayman	*		*			83.0	5.0	9.0	3.0
8.	Coniacian	Gosau Fm, Austria	*			*		34.0	42.9	23.0	0.1
9.	Turonian	Provence, France	*		*			6.2	34.1	53.6	5.4
10.	Aptian	Schrattenkalk, Allgäu, Germany	*			*		10.9	57.1	32.0	0.0
11.	Hauterivian	El Way, Chile	*		*			0.0	43.9	49.5	6.5
12.	Kimmeridgian	Kalkrieser Berg, N Germany		*		*		0.0	82.3	17.7	0.0
13.	Kimmeridgian	Celtiberic Chains		*		*		19.1	51.8	29.6	0.0
14.	Oxfordian	Ardennes, France	*				*	48.7	24.4	22.7	4.3
15.	Oxfordian	Osterwald, N Germany	*			*		0.0	76.5	23.5	0.0
16.	Oxfordian	Wesergebirge, N Germany		*		*		0.0	89.7	10.3	0.0
17.	Oxfordian	Hannover, N Germany	*		*			0.0	28.8	71.1	0.1
18.	Oxfordian	Liesberg, Switzerland		*			*	1.5	44.5	53.9	0.0
19.	Callovian	Qualeh Doktar Fm, Iran		*	*			2.3	45.0	49.1	3.6
20.	Bathonian	Kachchh, India		*	*			0.0	53.7	44.8	1.5
21.	Bajocian	Cerro Jaspe, Chile		*	*			2.3	59.7	38.0	0.0
22.	Bajocian	Badamu Fm, Iran		*	*			0.5	31.8	59.4	8.3
23.	Pliensbachian	High Atlas, Morocco		*			*	2.8	31.0	39.5	26.7
24.	Pliensbachian	High Atlas, Morocco		*			*	3.3	11.0	50.6	35.2
25.	Rhaetian, Tr.	Adnet, Austria		*	*			11.1	25.2	60.3	3.3
26.	Norian, Tr.	Nayband Fm, Iran		*	*	*		0.0	27.3	57.0	15.6
27.	Norian, Tr.	Dachsteinkalk, Austria	*		*	*		0.0	0.0	0.0	0.0
28.	Ladinian, Tr.	Wettersteinkalk, Austria	*		*	*	*	0.0	0.0	0.0	0.0

Locality numbers are as given in Fig. 4. 'Environment' generally clear or siliciclastically influenced. 'Facies' simplified as nearshore/lagoon, shallow fore-reef, deep fore-reef. References: 1. Perry (1998); 2. Risk *et al.* (1995); 3. Klein *et al.* (1991); 4. Perry (in prep.); 5. Klein *et al.* (1991); 6. Perry (1996); 7. Pleydell and Jones (1988); all others Bertling (in prep.).

differences in reef geomorphology and biogeographic region, these and other studies confirm that modern coral reef macroboring is typically dominated by the activities of sponges (in particular Clionidae). However, it is also clear from these studies that there are marked spatial variations in boring community composition within individual modern reef environments. Macintyre (1984) and Hutchings (1986) provide useful reviews of mechanisms, processes and products of bioerosion (including macroboring) within modern coral reef environments.

Studies from the Caribbean, e.g. the fringing/narrow barrier reef system at Discovery Bay, north Jamaica (Perry 1998; see Fig. 1), clearly illustrate spatial variability. Average boring community composition for the reef (all sites) is dominated by sponges, with bivalves, sipunculans and polychaetes of secondary importance (Table 2). Fore-reef environments are similarly dominated by the activity of clionid sponges, although the degree of substrate infestation increases markedly with increasing water depth (Fig. 1). Comparable high rates of substrate infestation were noted at deep fore-reef sites in Jamaica by Goreau & Hartman (1963). By contrast, more diverse boring assemblages characterize the back-reef and lagoon. Coral samples collected from back-reef sites are again dominated by sponges, although both bivalves (such as the live-coral borer *Lithophaga bisulcata*), sipunculans (e.g. *Phascolosoma perlucens*) and polychaetes (e.g. *Eunice mutilata*) are also common (Fig. 1). Similar diverse assemblages of sponges, bivalves and worms characterize lagoon patch reef frameworks.

Studies undertaken across the very different reef system of the Great Barrier Reef (Risk *et al.* 1995) using samples of *Acropora formosa* also document a macroboring community dominated by sponges (Table 2, Fig. 2). However, spatial patterns of macroboring are somewhat different to those recognized in Jamaica. Mean bioerosion (in terms of percentage framework removed) decreases across the shelf, and there are clear spatial variations in the relative abundance of groups of borers. The importance of sponges and bivalves decreases slightly moving across the shelf, whilst worms become progressively more important. Differences may reflect contrasts in (1) the overall structure of the shelf reef system, (2) the lack of a marked bathymetric gradient such as highlighted in the Jamaican example, or (3) the fact that sampling involved collection of live coral branches. Broadly comparable trends were observed by Sammarco & Risk (1990) across the same shelf sites using samples of *Porites lobata*.

Bromley (1978) reported a predominance of sponge and bivalve borers within coral substrates collected from the more northerly reefs around Bermuda. Although estimates of relative percentage abundance for individual groups/species were not given, clear spatial variations in the relative abundance of borers were evident, with only barnacles present in samples from the very high-energy site. More diverse assemblages of sponges, bivalves and worms

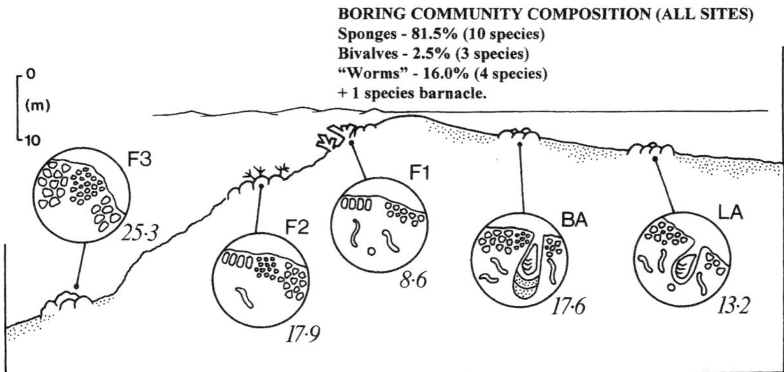

Fig. 1. Holocene: Discovery Bay, north Jamaica. Schematic diagram showing spatial (facies-related) patterns of macroboring across the reef profile. LA, lagoon: sponges 55.1%, bivalves 6.3%, 'worms' 38.5%; BR, back-reef: sponges 77.6%, bivalves 10.8%, 'worms' 11.6%; F1, shallow fore-reef: sponges 90.1%, bivalves 0.3%, 'worms' 8.5%; F2, fore-reef terrace: sponges 98.2%, bivalves 0.5%, 'worms' 1.3%; F3, fore-reef slope: sponges 97.9%, bivalves 1.6%, 'worms' 0.5%. Numbers in italics denote average percentage framework removed by macroboring within each facies setting. Source: Perry (1998).

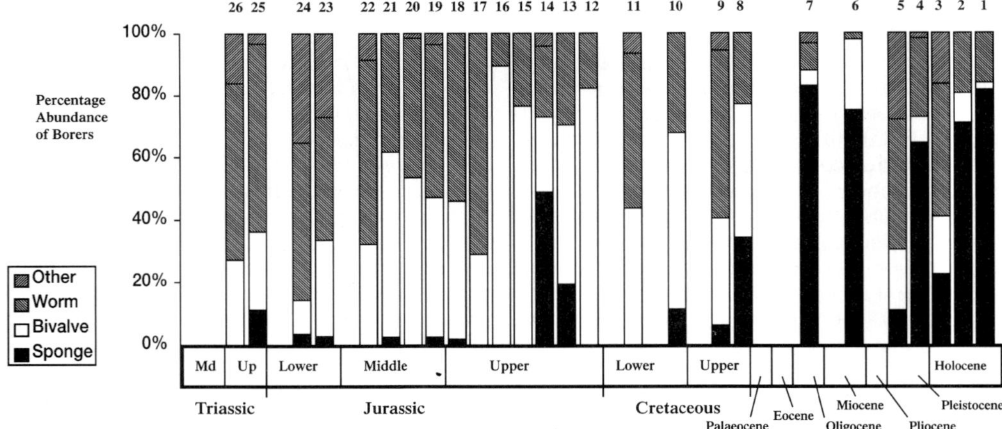

Fig. 2. Bar chart showing average macroboring community composition from sites where quantitative data are available. Numbers along the top refer to sites/data listed in Table 2. Note that various methods of data collection have been necessitated by variations in styles of preservation and sample accessibility.

characterize other settings (lagoon and open terrace). A contrasting study from the Sinai/Red Sea (Klein *et al.* 1991) found polychaetes and sipunculans as the dominant borers in both recent and Pleistocene samples (Table 2, Fig. 2). Sponges account for only between 15.6% and 32.1% of the boring community (based on analysis of cut coral sections). The authors also report that overall intensity of bioerosion is low (around 3% of the substrate).

In the Indo-Pacific, the region studied best is the atoll of Moorea, French Polynesia (Peyrot-Clausade *et al.* 1992, 1995; Chazottes *et al.* 1995). Depending on the duration of exposure of experimental substrates, polychaetes and sipunculans are the dominant macroborers, with bivalves much less important and sponges remarkably insignificant. Most bioerosion is performed by fish and a single species of sea urchin whose abundance is inversely related to macroborers. In live corals, however, the bivalve *Lithophaga laevigata* is the only macroborer. Dead substrate is bored at an increasing intensity with time.

Numerous other studies of macroboring from Holocene reefs have been published. Many detail either work done on (1) the distribution of individual species of borers (e.g. sipunculans (Rice & Macintyre 1982), sponges (Pang 1973; Rützler 1974)), (2) boring within individual species of coral (e.g. MacGeachy & Stearn 1976; Peyrot-Clausade *et al.* 1992; Hutchings *et al.* 1992), or (3) boring within a restricted range of reef environments (e.g. Hein & Risk 1975; Peyrot-Clausade & Brunel 1990). Numerous experimental studies have also been undertaken which examine rates and styles of substrate colonization within different reef environments (e.g. Davies & Hutchings 1983; Kiene 1985, 1989; Sammarco & Risk 1990; Peyrot-Clausade *et al.* 1992, 1995; Kiene & Hutchings 1993; Chazottes *et al.* 1995; Reaka-Kudla *et al.* 1996). Results indicate mutual interactions of grazers and borers during bioerosion, grazers normally removing more substrate than borers. A succession taking several years from microborers via small polychaetes to bivalves, sipunculans and sponges has frequently been recorded. Long-term studies suggest increased bioerosion with time but no significant changes in the boring fauna once the locally common suite has developed. Controlling factors of macroboring, apart from the outstanding (mostly inhibiting but sometimes also promoting) effects of grazers, have not been identified with certainty in most cases (see Discussion). Although an enormous variability in macroboring regarding intensity and producers has been noted for individual reefs, a nearshore dominance of bivalves versus an offshore dominance of sponges has repeatedly been reported.

Pre-Holocene coral macroboring

Triassic. The history of reef macroboring in the Mesozoic cannot be understood without regard to the evolution of the substrate, i.e. the history of corals. After a global pause in reef-building following the end-Permian biotic crisis, the first scleractinians appeared simultaneously in the Late Anisian on the northern margin of the

Tethys (Italian Alps, Hungary and south China: Fois & Gaetani 1984; Stanley 1988). Their diversity rose throughout the Middle and Late Triassic, paralleled by an increasing percentage of (mostly dendroid) colonial forms (Riedel 1991). During the Ladinian and Karnian, corals mainly inhabited deep-water mounds, while the structure of the reef ecosystem was reorganized. Palaeozoic holdover groups disappeared, and from the Norian onwards, scleractinians are found in reefs as builders (Flügel & Stanley 1984; Stanley 1988). This change might have been triggered by two factors: (1) an initial radiation of grazing gastropods and sea urchins, or (2) the consumption of nutrients by cyanobacterial mats. Both would have reduced growth of fleshy macroalgae and perhaps aided the establishment of an oligotrophic regime (Riedel 1991; Wood 1995). Following the origin of dinophytes, scleractinians probably acquired zooxanthellae as symbionts in the Norian (Stanley 1988; Stanley & Swart 1995; Wood 1995), facilitating rapid growth and regeneration. During the Late Rhaetian, a major extinction phase terminated the Triassic evolutionary unit, and corals suffered heavy losses as well (Sheehan 1985; Riedel 1991).

Published records of bioerosion mirror the stratigraphical distribution of reefs. A non-scleractinian deeper-water reef in the Anisian of the Dolomites, Italy, was sparsely bored by cirripedes (Senowbari-Daryan *et al.* 1993). The bivalve reefs of the German Muschelkalk (Ladinian) suffered very little bioerosion, by phoronids only (Hagdorn 1997). As might be expected, the same situation prevailed during the Carnian. Patch reefs of the Cassian Formation lagoon in northern Italy had no scleractinians and very little macroboring, which was again exclusively due to cirripedes (Fürsich & Wendt 1977). Sponges of the genus *Aka* were not boring at this time but were present encrusting calcareous sponges (Reitner & Keupp 1991). However, the presence of bivalve macroborers from the Norian of Upper Austria is well established (Frech 1890) where they were present within patch reefs in shallow fore-reef environments (Wurm 1982). The Alpine upper Rhaetian seems to exhibit an energy-related spatial variation in macroboring. No borers were found in the deep fore-reef (Kuss 1983), moderate bioerosion was recorded from lagoon environments with bivalves dominating (Stanton & Flügel 1989; Bernecker *et al.* 1999), and intensity was strongest in shelf-edge reefs (Roniewicz 1974; Michalik 1982).

Our own studies modify this picture. In the Ladinian Wettersteinkalk (northern Tyrol), various environments are represented (Henrich 1983) but only microborings were found. The Norian Steinplatte reef system in Salzburg, Austria (Zankl 1969), shows almost no bioerosion (abundance of borings less than 0.01/cm^2 of coral substrate) but bivalves and sponges were present. Reefs of the penecontemporaneous Nayband Formation (northern Iran) were moderately bored (0.14 borings/cm^2) with ‘worms’ dominating over bivalves and cirripedes (Table 2, Fig. 2). The upper Rhaetian reefs of Adnet, Austria, represent various environments with siliciclastic influx, from protected lagoon to open shelf (Bernecker *et al.* 1999), and were slightly bored (0.09 borings/cm^2) mainly by worms, with bivalves, sponges and cirripedes less important (Table 2). Reefs in pure carbonate facies at this time lacked sponges and cirripedes but bioerosion was more intensive (0.11 borings/cm^2).

Jurassic. The Jurassic arguably makes up a unit of its own in reef development, but is often united with the Cretaceous in this respect (Sheehan 1985). The first Jurassic reefs in the Sinemurian of southeast Spain, Morocco and British Columbia consisted of rigid structures with varying degrees of bioerosion. The Canadian occurrence, an offshore patch reef in agitated water, shows moderate macroboring caused by bivalves and worms (Stanley & MacRoberts 1993; Stanley & Beauvais 1994). Too little is known about the Moroccan localities (Dresnay 1971), and the Spanish ‘reefs’ are probably hydrozoan meadows within a lagoon rather than a true reef (Turnšek *et al.* 1975). Our study of these Spanish samples shows bivalve, worm and sponge borings (Fig. 3). Contemporaneous bivalve reefs were intensively bored mostly by worms; sponges seem not to have been present. Moroccan reefs of the middle and late Lias have been better sampled and investigated in terms of bioerosion. At the Pliensbachian localities, various worms dominated over bivalves as producers of weak bioerosion (0.08 and 0.12 borings/cm^2, respectively). Cirripedes were very important again (similar to the early reefs in the Triassic), and very few sponges were present (Table 2, Fig. 2).

Judging from the literature, Middle Jurassic reefs seem to have been intensively bored exclusively by bivalves, independent of age, locality or facies (Wullschleger 1966, 1971; Hallam 1975; Lathuilière 1982; Prinz 1986). Our own results, however, show a less homogeneous pattern (Table 2, Fig. 2). Bioerosion was moderate (0.10–0.24 borings/cm^2) in the Bajocian of northern Iran and Chile as well as in the Bathonian and Callovian of Kachchh, India. Bivalves and worms were approximately equally important, with

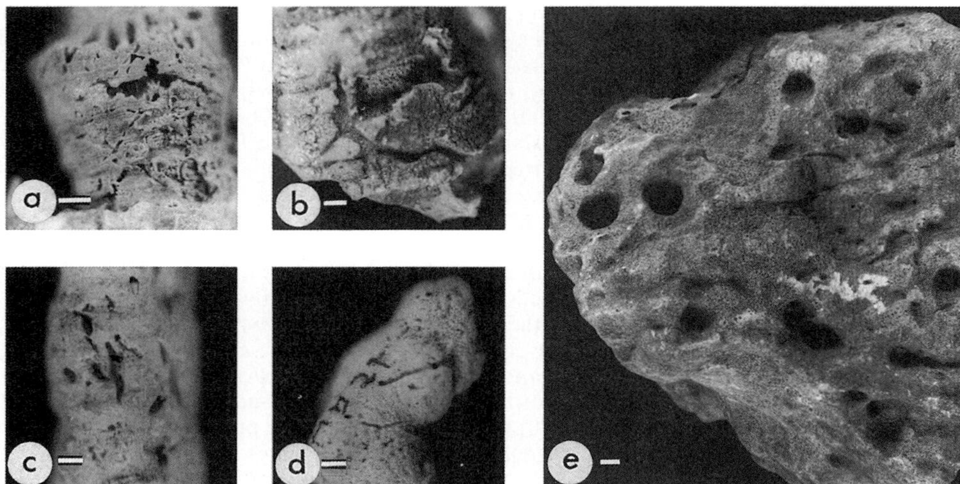

Fig. 3. Boring traces from the Lias of Morocco (specimens of collection L. Beauvais at Museum National d'Histoire Naturelle, Paris: (**a**) *Entobia* isp. in *Archaeosmilia*, middle Sinemurian, El Zibrina (No. R 11599); (**b**) *Entobia* isp. in Procyclolites, lower Toarcian, Col de Taililout (no. 5374); (**c**) *Rogerella pattei* Zapfe in *Rodinosmilia*, upper Pliensbachian, Beni Tadjit (No. Me 001); (**d**) *Meandropolydora sulcans* Voigt in *Actinastrea*, upper Pliensbachian, Beni Tadjit (No. MD 21); (**e**) *Gastrochaenolites* isp. and *Trypanites weisei* Mägdefrau in *Goldfussastrea*, lower Toarcian, Col de Moussaoua (No. R 11634). All scale bars 2 mm.

cirripedes, phoronids and sometimes sponges contributing far less to substrate removal. All localities studied, however, represent coastal high-energy settings with strong siliciclastic influx and this might explain the limited sponge bioerosion. The occurrence reported from the Bathonian of Morocco (Warme 1977) forms an exception inasmuch as macroboring was extensive and sponges were more important than worms. The locality represents a coral/microbial reef implying raised levels of nutrients and bacteria, i.e. ideal conditions for boring sponges.

Late Jurassic reefs are well represented in the rock and literature record. Unfortunately, upon reinvestigation the statements of the relevant authors regarding bioerosion are not always reliable because borings appear to have been misidentified. Bivalves seemingly dominated in all environments in Europe and the Gulf Coast of North America (Baria *et al.* 1982; Jansa *et al* 1982; Pisera 1987; Insalaco *et al.* 1997), whilst sponges were restricted to low-energy and turbid environments (presumably lagoons or deeper water). No clear pattern in terms of the control factors of overall boring intensity (e.g. sedimentation rate, light, energy) emerges from the descriptions of Insalaco *et al.* (1997). Our own studies, however, allow a more consistent interpretation (Bertling 1997, 1999; Bertling & Insalaco 1998). Judging from the Oxfordian in central and western Europe, bioerosion is strongest in microbial–coral reefs (0.33 and 0.59 borings/cm^2, respectively for the Oxfordian in northern France and the Kimmeridgian of the Celtiberic Chains) and in nearshore high-energy environments (0.38 borings/cm^2 at Hannover; Table 2; Fig. 2). Lowest overall values occur in turbid deep water facies of the Wesergebirge, Germany (0.03–0.06 borings/cm^2), with oolithic and open shelf environments ranging between (0.25±0.01 borings/cm^2). Bivalves and 'worms' were always important macroborers in varying proportions (Table 2). Sponges were significant only in reefs with microbial contribution (Oxfordian in northern France and Kimmeridgian of Celtiberic Chains; Table 2, Fig. 2) for the same likely reasons as given for the Middle Jurassic (Fig. 4). The only good data available for the Volgian are from bivalve (not coral) patch reefs in Dorset, England (Fürsich *et al.* 1994), where bioerosion was intense and dominated by sponges and bivalves. Macroboring in the sponge–coral reefs on the northern margin of the Tethys has unfortunately not been described.

Cretaceous. The Cretaceous is known as the system with rudist bivalves sustaining most calcification on the shelf. Contrary to widespread opinion, these heterotrophs never formed reefs (Kauffman & Sohl 1974; Gili *et al.* 1995; Skelton *et al.* 1995) although they occupied the 'typical' reef habitat and sometimes occurred in reefs built

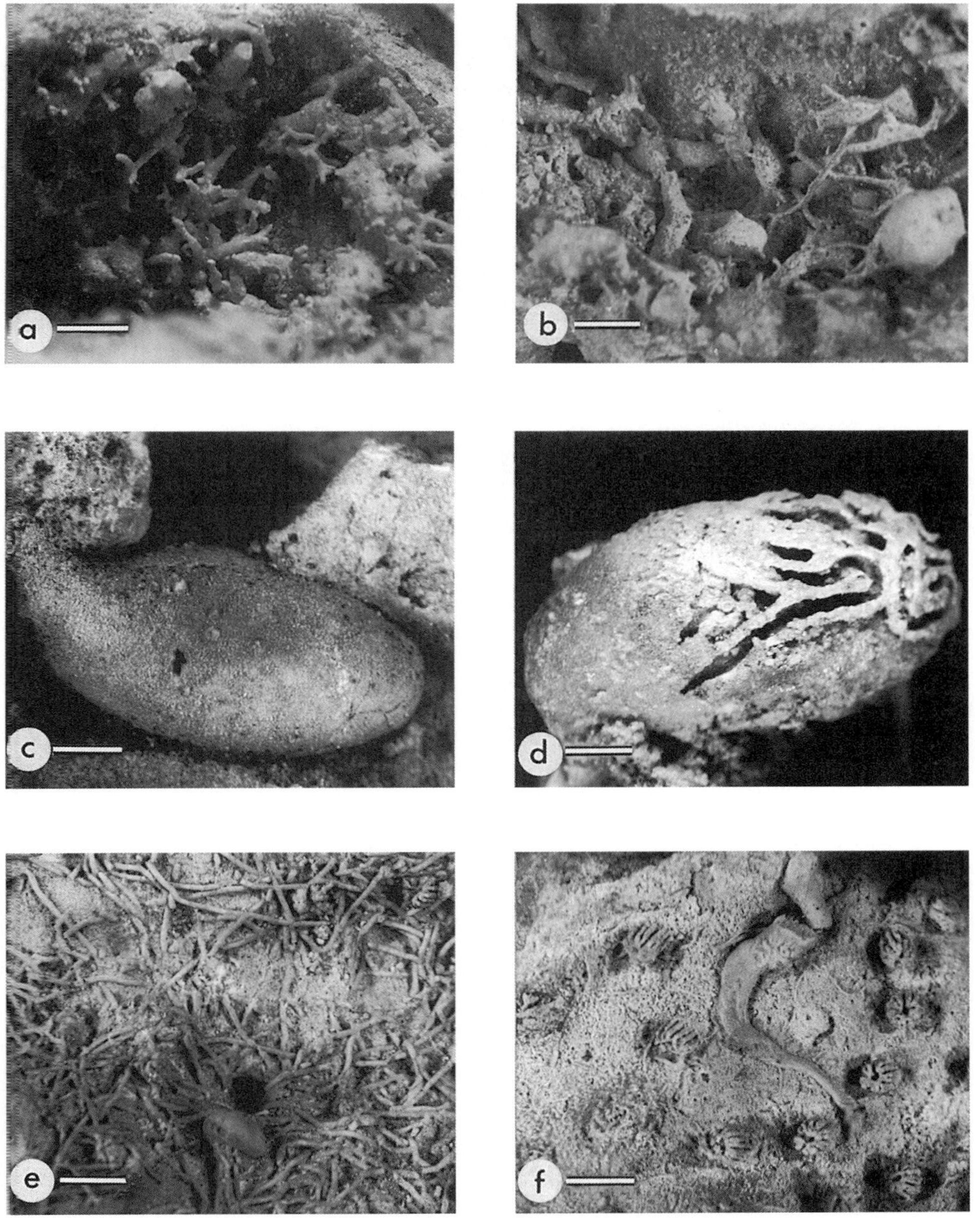

Fig. 4. Boring traces (mouldic preservation) from the Oxfordian of Novion-Porcien, Dept. Ardennes, France (specimens at the Geological Museum, Münster): (**a**) *Entobia retiformis* Fürsich et al. (No. B4B.8–5); (**b**) *Entobia* cf. *laquea* Bromley & d'Alessandro (No. B4B.8–21b); (**c**) *Gastrochaeolites dijugus* Kelly & Bromley (No. B4B.8–35); (**d**) *Meandropolydora sulcans* Voigt (No. B4B.8–51b); (**e**) *Talpina bromleyi* Fürsich et al. (No. B4B.8–12); (**f**) *Caulostrepsis cretacea* Voigt (No. B4B.8–31). All scale bars 2 mm.

by other organisms. With global temperatures and sea-level high, ocean circulation was slack and shelf waters were further enriched with nutrients by the leaf litter of angiosperms which revolutionized the land flora at this time (Wood 1993). Scleractinians were thus driven (with some exceptions) into special habitats on the shelf edge in the mid-Cretaceous (Scott 1988).

For the early Cretaceous, information on coral reef macroborers is patchy. The only quantitative data come from the Hauterivian of Chile, in a high-energy coastal setting, where bivalves and worms were codominant over cirripedes. Overall borer abundance was low (0.13 borings/cm^2) and sponges were remarkably absent (Table 2, Fig. 2). Comparable environments were studied in the Barremian of Provence, France, and eastern Serbia (Masse 1977; Turnšek and Mihajlović 1981) with similar results. With scleractinian reefs moving to the shelf edge in the Aptian/Albian, the pattern changed. Published data, however, are available only for the Lower Albian of Arizona (Hartshorne 1989) and the Lower Aptian of the Bavarian Alps (Scholz 1984; Baron-Szabo 1997). In both cases, patch reefs in high-energy settings were sampled, thus warranting comparability with earlier localities. Sponges were significant macroborers, perhaps rivalling bivalves. No other producers of 'intense' bioerosion are listed. Our own study of the Bavarian occurrence, however, indicates the important contributions of worms (Table 2).

Macroboring in the Late Cretaceous is hitherto mostly known from rudist associations (Philip 1972; Bein 1976; Bass 1984; Skelton *et al.* 1995; (and especially) Kauffman & Sohl 1974). In all instances, bioerosion is reportedly intense with sponges dominating over worms and bivalves. Some variability between samples has been noted as well as a pronounced habitat preference of some borers (Kauffman & Sohl 1974). Bioerosion in coral reefs was mentioned from the Maastrichtian of Bosnia and the United Arab Emirates (Turnšek & Polšak 1978; Metwally 1996), the latter occurrence being remarkable in the lack of sponges and importance of cirripedes, although overall macroboring was limited. The single other published profile with bored corals is in the Coniacian of the Bavarian Alps where 'intense' bioerosion was caused by bivalves, sponges and worms (Baron-Szabo 1997; Sanders & Baron-Szabo 1997) in a turbid lagoon. Our own studies of the samples show a rather varied suite of ichnotaxa causing moderate boring (0.18 borings/cm^2), with bivalves and sponges dominant over worms (Table 2, Fig. 2). Finally a good set of data exists from the Turonian of Uchaux, Provence, France, in a coastal high-energy environment with siliciclastic disturbance. Bioerosion was intense (0.44 borings/cm^2) with 'worms' strongly dominating over bivalves (sponges and cirripedes were relatively unimportant; Table 2, Fig. 2), perhaps reflecting the unfavourable conditions.

Palaeogene. Rather little is known about Palaeogene reefs but corals dominated in most environments. Paleocene and Eocene occurrences are rare, contrasting to a global climax in the Oligocene. This coincided with a sea-level highstand and associated oligotrophic oceanic conditions (Wood 1995). This may have been further aided by the radiation of nutrient-consuming diatoms in the Paleocene (Wood 1993). At the end of the system, a continuous decline of the biota affected reefs as well.

Authors working on Paleocene reefs provide some information on macroboring. In Bosnia, a shallow nearshore reef-tract without siliciclastic influx was intensively attacked by bivalves as well as (subordinately) 'worms' and sponges (Babić & Zupaniĉ 1981). The Danian aphotic reef at Fakse, Denmark, suffered little bioerosion, mostly by sponges and bryozoans (Bernecker & Weidlich 1996). Contemporaneous high-energy reefs in Egypt were bored predominantly by bivalves but some sponges and 'worms' do occur. Bioerosion increased with proximity to the coast (Schuster 1996). The sponge *Aka* is reported as a coral borer for the first time in its history (Schroeder 1986; Reitner & Keupp 1991) from the Thanetian. The only information on Eocene reef borers comes from shallow patch reefs in calm settings of the Spanish Pyrenees. Bioerosion is reported as frequent and as mainly caused by bivalves (Gaemers 1978).

During the Oligocene, macroboring obviously varied in composition and intensity but the distribution pattern is not easily explained. The crest of a shallow fringing reef in Liguria, Italy, was hardly bored, and more by microborers than by sponges and bivalves (Pfister 1985). A strong siliciclastic influx onto a nearshore biostrome in Piemonte, Italy, is unlikely to be responsible for the almost complete lack of bioerosion (Pfister 1980). In contrast to these European localities, Upper Oligocene patch and barrier reefs on Puerto Rico yield numerous ichnotaxa. Sponges dominated over bivalves and worms, and boring was more extensive in protected (lagoonal) sites, and reduced in higher-energy (shelf-edge) sites (Frost *et al.* 1983; Edinger & Risk 1994). Upper Oligocene or Lower Miocene reefs on Grand Cayman exhibit bioerosion structures particularly well, owing to leaching of coral skeletons after micrite infill of borings (Pleydell & Jones 1988). A wide range of ichnotaxa shows a marked preponderance of sponge borings and significant similarity to modern distributions. Bivalves and 'worms' were unimportant in the sites examined (Table 2, Fig. 2). Two facies types (lagoon and patch reef) are delineated, but estimates of the degree of macroboring were not made.

Neogene. The Neogene was a time of important scleractinian reef development, although no data for the Pliocene have been published. Macroboring in middle and late Miocene patch reefs on Puerto Rico was dominated by sponges with both bivalves and 'worms' subordinate (Edinger & Risk 1994). The only detailed study of bioerosion to date is from Upper Miocene reefs of Mallorca (Perry 1996). Superb exposures, combined with leaching of the aragonite coral heads, permitted detailed assessment of ichnospecies distribution, relative abundance and variations in the degree of substrate infestation. The average ichnocoenosis is comparable to that reported for many Holocene reefs, being dominated by *Entobia* (Table 2, Fig. 5). Bivalve (*Gastrochaenolites*) and worm borings (*Trypanites* and *Maeandropolydora*) are only locally abundant. One point of particular interest is the high relative abundance of bivalve borings mostly made by the live-coral borer *Lithophaga bisulcata*. Dense bands of *Gastrochaenolites torpedo* occur around the upper surfaces of coral heads within the upper fore-reef. Both diversity of ichnospecies and the degree of substrate infestation increase within back-reef and lagoon (Fig. 5). We have also observed a dominance of *Entobia* borings within Miocene reefs of Santa Pola, southern Spain, although data on borer abundance have not been collected.

Pleistocene. Despite good exposures of well constrained Pleistocene reef facies in many areas of the world (e.g. Barbados, Red Sea), there are limited published data on Pleistocene coral macroboring. In a recent study, Perry (in press) examines coral samples collected from a range of Pleistocene reef facies (back-reef, lagoon, shallow fore-reef) exposed on the north Jamaican coast (Table 2, Fig. 6). Boring communities are broadly comparable to those observed within nearby Holocene reefs (Perry 1998) and are dominated by *Entobia* isp. The trace types *Gastrochaenolites* isp. and *Trypanites* isp./*Maeandropolydora* isp. are only locally abundant within back-reef and lagoonal facies (Fig. 6). Individual ichnospecies can be identified within much of the material with reasonable confidence because many of the borings remain unfilled by sediment. Some degree of facies preference amongst ichnospecies is evident.

Reports of Pleistocene macroboring from other localities are more limited in terms of either description or facies constraint. Jones & Pemberton (1988) report abundant *Lithophaga* sp. preserved within their borings (*Gastrochaenolites torpedo* Kelly & Bromley) in coral heads collected from lagoon facies from Grand Cayman. The authors do not refer to other borings preserved within this material, but they do document

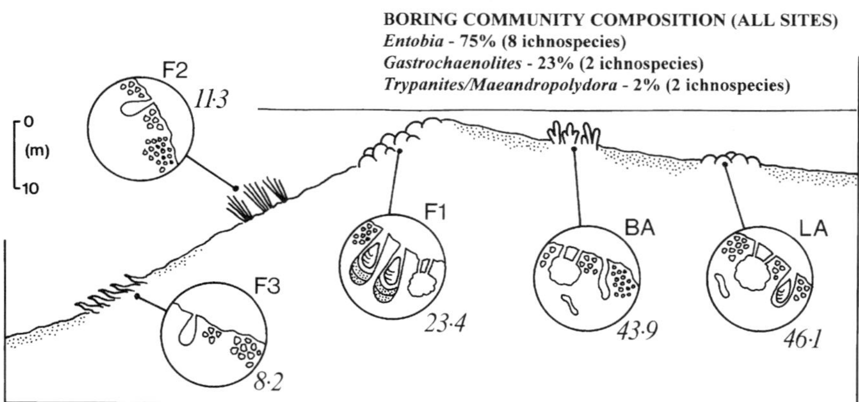

Fig. 5. Upper Miocene: Cap Blanc-Cala Pi section, Mallorca. Schematic diagram showing spatial (facies-related) patterns of macroboring across the reef profile. LA, lagoon facies: *Entobia* 67.3%, *Gastrochaenolites* 27.1%, *Trypanites/Maeandropolydora* 5.6%; BR, back-reef facies: *Entobia* 85.3%, *Gastrochaenolites* 9.8%, *Trypanites/Maeandropolydora* 4.9%; F1, reef crest facies: *Entobia* 56.3%, *Gastrochaenolites* 42.9%, *Trypanites/Maeandropolydora* 0.8%; F2, mid fore-reef facies: *Entobia* 94.2%, *Gastrochaenolites* 5.8%, *Trypanites/Maeandropolydora* 11.3%; F3, deep fore-reef facies: *Entobia* 91.9%, *Gastrochaenolites* 8.1%, *Trypanites/Maeandropolydora* 0.0%. Numbers in italics denote average percentage surface area of framework bored within each facies setting. Source: Perry (1996).

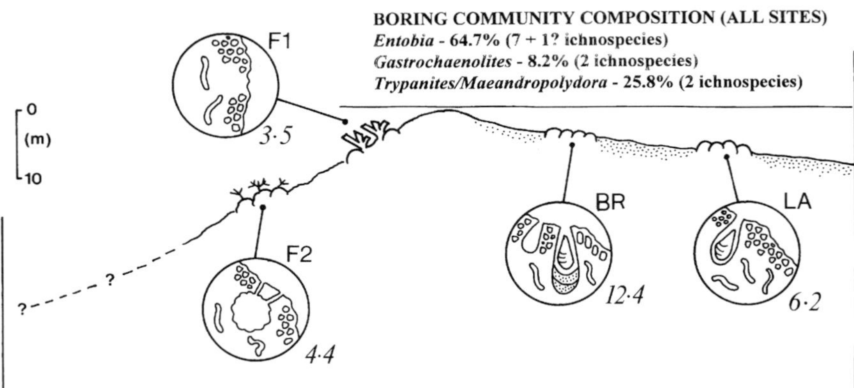

Fig. 6. Pleistocene: Falmouth Formation, north Jamaica. Schematic diagram showing spatial (facies-related) patterns of macroboring across the reef profile. LA, lagoon facies: *Entobia* 38.8%, *Gastrochaenolites* 3.5%, *Trypanites/Maeandropolydora* 52.7%; BR, back-reef facies: *Entobia* 58.3%, *Gastrochaenolites* 23.6%, *Trypanites/ Maeandropolydora* 18.1%; F1, shallow fore-reef facies: *Entobia* 73.0%, *Gastrochaenolites* 5.5%, *Trypanites/ Maeandropolydora* 21.5%; F2, fore-reef terrace facies: *Entobia* 88.9%, *Gastrochaenolites* 0%, *Trypanites/Maeandropolydora* 11.1%. Numbers in italics denote average percentage framework removed by macroboring within each facies setting. Source: Perry (in press).

intense substrate alteration in coral heads exhibiting high rates of bivalve infestation. Many of the borings they document are up to 15 cm in length and in places these show clear evidence of having been emplaced in living coral.

Klein *et al.* (1991) report macroboring in *Porites* corals collected from Pleistocene reef deposits in the Gulf of Eilat, Red Sea, but samples were not ascribed to a specific facies. Ichnotaxa were not used in the study but the borings pictured in their figure 1 can be assigned to *Trypanites,* suggesting dominance of sipunculans and/or polychaetes (Table 2, Fig. 2). Borings attributed to the activity of sponges constitute a surprisingly limited percentage of overall macroboring compared with that generally observed within Holocene reefs (although figures compare favourably with their Holocene data, also from the Red Sea; see above).

Discussion

Temporal changes in the macroboring subcommunity

The mass extinction at the end of the Palaeozoic had doubtful effects on reef macroborers. Late Permian coral reefs were very rare globally and exhibit no consistent pattern regarding bioerosion. Whereas in Turkey only microborers were found (J. Balog, pers. comm. 1996), rugose corals within sponge-dominated localities in eastern Oman were dominantly bored by sponges with other macroborers absent (Weidlich 1996).

The Triassic was a system with parallel evolution of reefs, scleractinians and their borers. In an initial lag phase, cirripedes and worms were the only macroborers. As corals became reef-builders in the Norian, boring bivalves appeared immediately in surprising size and abundance. Just before the end of this evolutionary unit, reefs became bored by sponges after they had been absent for more than 30 Ma. This is much more time than solitary animals required to redevelop the macroboring habit. The Rhaetian extinction of most corals rapidly deprived them of their habitat and precluded their establishment in the coral borer suite.

Jurassic macroboring can probably be summarized as a rather homogenous pattern, despite the lack of information for the last stage. It was initially dominated by cirripedes and ‘worms’ but already during the Lias, bivalves became the most important phylum. They competed with ‘worms’ but currently we do not understand which factors governed the success of one or the other group. Together they removed substantial amounts of reef framework in the Middle Jurassic for the first time in Earth history. Clionid sponges became important in coral–microbial reefs from the Middle Jurassic onwards but did not attain a remarkable position in ‘normal’ settings. An important event in this respect probably was the end-Jurassic sea-level fall in Europe, where most reefs were located. Increased input of decaying terrestrial organic matter during the lowstand provided the substratum for bacteria as nutrients necessary for the boring sponges (Bertling 1997). Overall macroboring intensity constantly rose during the Jurassic, probably for two reasons: (1) reef organisms built more strongly calcified skeletons, and (2) the most vigorous grazing gastropods (Patelloidea and Neritoidea) diversified together with phymosomatoid sea urchins (Steneck 1983), thus increasing the pressure of 'accidental predation' on the smaller reef fauna. The suggestion of Fürsich *et al.* (1994) that the Late Jurassic was a time of borer radiation cannot be substantiated based on our data.

Data for the Cretaceous are very scarce currently, partially due to the rarity of coral reefs, although a distinct pattern emerges, even from this reduced set. Overall bioerosion intensity was continuously high, even in previously unfavourable environments. The Cretaceous was the time when boring sponges invaded coral reefs at a large scale, more or less independent of the facies. Comparable to the situation today, their organization precluded intense colonization of areas with higher sedimentation rates. Their importance certainly increased but dominance over other macroborers (bivalves, worms) was never attained. Minor boring groups do not show any significant changes. The 'Mesozoic marine revolution' (Vermeij 1977) caused by a rise of predatory fish and crustaceans thus is not expressed in macrobioerosion. Arguably, sponges as colonial organisms suffer less from an attack by a biting predator than do solitary borers but this has hitherto not been substantiated. Rather, they might have favoured the nutrient-enriched Cretaceous shelf waters, as they depend largely on bacteria as their diet. The radiation of grazing sea urchins and the resulting dramatic increase in herbivory (Steneck 1983) seem not to have been expressed in bioerosion either.

The style of macroboring in the early Palaeogene was identical to the Cretaceous situation. From the patchy data available, it becomes obvious that the boundary drawn by Sheehan (1985) at the end of the Mesozoic as terminating a unit in reef evolution did not exist for borers. Rather, the major change occurred during the Oligocene, perhaps as a response to the origin of grazing reef fish (Steneck 1983; Bellwood 1996). Sponges seem to have become the most important macroborers regionally in the Caribbean and continued to do so unaffected by the end-Palaeogene extinction phase (Edinger & Risk 1994). From the late Oligocene onwards, hardly any changes occurred despite the origin of parrot fish (Scaridae) in the Middle Miocene (Bellwood 1996) and the almost simultaneous rise of corallinacean rhodophytes as reef binders (Steneck 1983). Today, grazers control boring to an unprecedented extent but it is currently unknown when exactly this situation originated.

Spatial patterns of macroboring within individual reefs

Environmental controls on the distribution and intensity of macroboring are hard to determine. A vast array of experiments has been performed to date, but have largely been unable to identify the effects of single ecological factors. In addition, far-reaching conclusions have been drawn from short-term experiments of less than three years' duration which have only limited applications for palaeontologic or taphonomic approaches. Of the numerous ecofactors proposed as controls, only five seem to match the data sets unequivocally: (1) increased plankton productivity triggers bioerosion because most borers are suspension-feeders (e.g. Pang 1973; Rützler 1975; Highsmith 1980; Edinger & Risk 1994); (2) less vital corals are more susceptible to boring because their ability to reject settling borer spat decreases (e.g. Otter 1937; McCloskey 1970; Bromley 1978; Highsmith 1980; Peyrot-Clausade & Brunel 1990); (3) reduced sedimentation rates are favourable because the borehole apertures are not smothered as easily (Otter 1937; Wilkinson 1983); (4) the less grazing pressure, the more borers can develop because their spat is not accidentally scraped off (e.g. Sammarco *et al.* 1987; Kiene & Hutchings 1994; Risk *et al.* 1995); (5) the longer a substrate is exposed, the more intense is macroboring because larvae have more time to settle (e.g. McCloskey 1970; Peyrot-Clausade & Brunel 1990; Kiene & Hutchings 1994).

Other factors with less convincing evidence include degree of encrustation, water energy and water depth. In addition, various substrate characteristics, including density of the skeleton (Highsmith 1981) as well as size and shape, have previously been suspected to influence boring. A number of modern studies have demonstrated substrate variations (e.g. Perry 1998); however, the relevance of specific factors (e.g. density, shape) is by no means clear. Some factors exert different controls for different groups of macroborers, especially substrate shape and hydrodynamics. However, even on the level of borer groups, opposing findings have been published recently (e.g. Rützler 1975; Bromley 1978; Kiene & Hutchings 1994). Explanations for the spatial distributions of borers within reef systems consequently remain complex and thus interpretations of temporal patterns of bioerosion remain tentative.

Although several trends in the spatial and temporal distribution of macroborers become apparent from this study, much more work could and should be undertaken. This not only refers to studies of ancient reef environments but also to controlled experiments in modern settings aiming at the identification of individual factors governing bioerosion. Both will have major implications for understanding the evolution of this reef guild whose importance is second only to the reef-builders themselves, and which exert significant control on patterns and rates of reef accretion and biodegradation.

Conclusions

1. Modern and Neogene boring communities are typically dominated by sponges, regardless of water depth or biogeographic area, with polychaetes, sipunculans and bivalves of secondary importance. Marked spatial variations in the abundance of these groups of borers are evident within individual reef systems, with the most diverse boring communities typically occurring in low-energy, back-reef or lagoon environments.
2. Although data on ancient (pre-Quaternary) reef macroboring remain patchy, marked changes in macroboring community composition have occurred through the Mesozoic/Cenozoic period of coral reef construction. Preliminary data suggest major innovations in coral macroboring early during peaks of reef development.
3. Boring bivalves appear in the Norian (although are reported earlier from hardgrounds), bivalves and worms were codominant from Bajocian at least to the Late Cretaceous, and sponges became the dominant reef borers in the early Miocene. Seen independently of other groups, cirripedes were important in the Late Triassic and Early Jurassic.
4. At present there is limited evidence to enable accurate determination of the nature of community change through time (e.g. step-like in response to extinction events, or progressive as individual groups of borers have evolved). However, the basic pattern does not seem to link in with major marine extinction events, as boring communities apparently crossed system boundaries unaffected.
5. Overall boring intensity appears to have increased through time independent of environment. Precipitous increases seem to have occurred during the Late Triassic in bioclastic facies, and during the Late Jurassic in siliciclastic facies. We are currently unable to assess whether this is an artifact of insufficient sampling or a real phenomenon.

Financial support has come from a number of areas for various aspects of this study. C. T. P. acknowledges previous support from the NERC (Grant GT4/93/245/G), Manchester Geographical Society and Manchester Metropolitan University. M. B. gratefully received an ARC travel grant from the DAAD; he thanks museum curators and individuals at institutes for permission to study their samples as well as for logistical support: M. Amler (Marburg), R. Baron-Szabo (then Berlin), M. Bernecker (Erlangen), F. T. Fürsich (Würzburg), R. Henrich (Bremen), A. von Hillebrandt (Berlin), R. Leinfelder (then Stuttgart), M. Nose (Stuttgart), C. Perrin (Paris), B. Senowbari (Erlangen) and H. Zankl (Marburg). T. Palmer and M. Wilson are thanked for reviews of the manuscript.

References

ACKER, K. L. & RISK, M. J. 1985. Substrate destruction and sediment production by the boring sponge *Cliona caribbaea* on Grand Cayman Island. *Journal of Sedimentary Petrology*, **56**, 705–711.

BABIĆ, L. & ZUPANIĈ, J. 1981. Various pore types in a Paleocene reef, Banija, Yugoslavia. *In:* TOOMEY, D. F. (eds) *European Fossil Reef Models*. Society of Economic Palaeontologists and Mineralogists, Special Publications, **30**, 473–482.

BARIA, L. R., STOUDT, D. L., HARRIS, P. M. & CREVELLO, P. D. 1982. Upper Jurassic reefs of the Smackover Formation, United States Gulf Coast. *AAPG Bulletin*, **66**, 1449–1482.

BARON-SZABO, R. C. 1997. Die Korallenfazies der ostalpinen Kreide (Helvetikum: Allgäuer Schrattenkalk; Nördliche Kalkalpen: Brandenberger Gosau) Taxonomie, Palökologie. *Zitteliana*, **21**, 3–97.

BASS, M. 1984. Macroborings and epizoans on Late Cretaceous rudists from West Coast active-margin environments. *AAPG Bulletin*, **68**, 452.

BEIN, A. 1976. Rudistid fringing reefs of Cretaceous shallow carbonate platform of Israel. *AAPG Bulletin*, **60**, 258–272.

BELLWOOD, D. R. 1996. The Eocene fishes of Monte Bolca: the earliest coral reef fish assemblage. *Coral Reefs*, **15**, 11–19.

BERNECKER, M. & WEIDLICH, O. 1996. The Danian (Paleocene) coral limestone of Fakse, Denmark: a model for ancient aphotic, azooxanthellate coral mounds. *Facies*, **22**, 103- 138.

——, —— & Flügel, E. 1999. Response of Triassic reef coral communities to sea-level fluctuations, storms and sedimentation: evidence from a spectacular outcrop (Adnet, Austria). *Facies*, **40**, 229–280.

BERTLING, M. 1997. Bioerosion Late Jurassic reef corals – implications for reef evolution. *Proceedings of the Eighth International Coral Reef Symposium*, **2**, 1663–1668.

—— 1999. Late Jurassic reef bioerosion – the dawning of a new era. *Bulletin of the Geological Society of Denmark*, **45**, 173–176.

—— & Insalaco E. 1998. Late Jurassic coral/microbial reefs from the northern Paris Basin – facies, palaeoecology and palaeobiogeography. *Palaeogeography, Palaeoclimatology, Palaeoecology*, **139**, 139–175.

BROMLEY, R. G. 1970. Borings as trace fossils and *Entobia cretacea* Portlock, as an example. *In*: CRIMES, T. P. & HARPER, J. C. (eds) *Trace Fossils. Geological Journal* Special Issue, **3**, 49–90.

—— 1978 Bioerosion of Bermuda reefs. *Palaeogeography, Palaeoclimatology, Palaeoecology*, **23**, 169–197.

—— & ASGAARD, U. 1993. Endolithic community replacement on a Pliocene rocky coast. *Ichnos*, **2**, 93–116.

—— & D'ALESSANDRO, A. 1983. Bioerosion of the Pleistocene of southern Italy: ichnogenera *Caulostrepsis* and *Maeandropolydora*. *Rivista Italiana di Paleontologia e Stratigrafia*, **89**, 283–309.

—— & —— 1984. The ichnogenus *Entobia* from the Miocene, Pliocene and Pleistocene of southern Italy. *Rivista Italiana di Paleontologia e Stratigrafia*, **90**, 227–296.

—— & —— 1987. Bioerosion of the Plio-Pleistocene transgression of southern Italy. *Rivista Italiana di Paleontologia e Stratigrafia*, **93**, 227- 296.

CHAZOTTES, V., LE CAMPION-ALSUMARD, T. & PEYROT-CLAUSADE, M. 1995. Bioerosion rates on coral reefs: interactions between macroborers, microborers and grazers (Moorea, French Polynesia). *Palaeogeography, Palaeoclimatology, Palaeoecology*, **113**, 189–198.

DAVIES, P. J. & HUTCHINGS, P. A. 1983. Initial colonization, erosion and accretion on coral substrate. *Coral Reefs*, **2**, 27–35.

DRESNAY, R. DU. 1971. Extension du dévellopement des phénomènes récifaux jurassiques dans le domaine atlasique marocain, particulièrement au Lias moyen. *Bulletin de la Societé Géologique de France, 7me. Série,* **8**, 46–56.

EDINGER. E. N. & RISK, M. J. 1994. Oligocene-Miocene extinction and geographic restriction of Caribbean corals: roles of turbidity, temperature, and nutrients. *Palaios*, **9**, 576–598.

FLÜGEL, E. & STANLEY, G. D. 1984. Reorganization, development and evolution of post-Permian reefs and reef organisms. *Palaeontographica Americana*, **54**, 177–186.

FOIS, E. & GAETANI, M. 1984. The recovery of reef-building communities and the role of cnidarians in carbonate sequences of the middle Triassic (Anisian) in the Italian Dolomites. *Palaeontographica Americana*, **54**, 191–200.

FRECH, F. 1890. Die Korallen der juvavischen Triasprovinz (Zlambachschichten, Hallstätter Kalke, Rhaet). *Palaeontographica*, **37**, 1–116.

FROST, S. H., HARRIS, P. M. & HARBOUR, J. L. 1983. Late Oligocene reef-tract carbonates, southwestern Puerto Rico. *In*: HARRIS, P. M. (ed.) *Carbonate Buildups – a core workshop*. Society of Economic Palaeontologists and Mineralogists, Core Workshop, **4**, 491–518.

FÜRSICH, F. T. 1979. Genesis, environments and ecology of Jurassic hardgrounds. *Neues Jahrbuch für Geologie und Paläontologie, Abhandlungen*, **158**, 1–63.

—— & WENDT, J. 1977. Biostratinomy and palaeoecology of the Cassian Formation (Triassic) of the Southern Alps. *Palaeogeography, Palaeoclimatology, Palaeoecology*, **22**, 257–323.

——, PALMER, T. J. & GOODYEAR, K. L. 1994. Growth and disintegration of bivalve- dominated patch reefs in the Upper Jurassic of England. *Palaeontology*, **37**, 131–171.

FÜTTERER, D. K. 1974. Significance of the boring sponge *Cliona* for the origin of fine-grained material of carbonate sediments. *Journal of Sedimentary Petrology*, **44**, 79–84.

GAEMERS, P. A. M. 1978. Biostratigraphy, palaeoecology and palaeogeography of the mainly marine Ager Formation (Upper Paleocene–Lower Eocene) in the Tremp Basin, central – south Pyrenees, Spain. *Leidse Geologische Mededelingen*, **51**, 151–231.

GILI, E., MASSE, J.-P. & SKELTON, P. W. 1995. Rudists as gregarious sediment-dwellers, not reef-builders on Cretaceous carbonate platforms. *Palaeogeography, Palaeoclimatology, Palaeoecology*, **118**, 245–267.

GINSBURG, R. N. & SCHROEDER, J. H. 1974. Growth and submarine fossilization of algal cup reefs, Bermuda. *Sedimentology*, **20**, 575–614.

GOREAU, T. F. & HARTMAN, W. D. 1963. Boring sponges as controlling factors in the formation and maintenance of coral reefs. *American Association for the Advancement of Science*, **75**, 25–54.

HAGDORN, H. 1997. Riffe aus dem Muschelkalk. *Fossilien*, **1997** (3), 180–190.

HALLAM, A. 1975. Coral patch reefs in the Bajocian (Middle Jurassic) of Lorraine. *Geological Magazine*, **112**, 383–392.

HALLOCK, P. 1988. The rate of nutrient availability in bioerosion: consequences to carbonate buildups. *Palaeogeography, Palaeoclimatology, Palaeoecology*, **63**, 275–291.

HARTSHORNE, P. M. 1989. Facies architecture of a Lower Cretaceous coral-rudist patch reef, Arizona. *Cretaceous Research*, **10**, 311–336.

Hein, F. J. & Risk, M. J. 1975. Bioerosion of coral heads: inner patch reefs, Florida reef tract. *Bulletin of Marine Science*, **25**, 133–138.

HENRICH, R. 1983. Der Wettersteinkalk am Nordwestrand des tirolischen Bogens in den Nördlichen Kalkalpen: der jüngste Vorstoß einer Flachwasserplattform am Beginn der Obertrias. *Geologica et Palaeontologica*, **17**, 137–177.

HIGHSMITH, R. C. 1980. Geographic patterns of coral bioerosion: a productivity hypothesis. *Journal of Experimental Marine Biology and Ecology,* **46**, 177–196.

—— 1981. Coral bioerosion: damage relative to skeletal density. *The American Naturalist*, **117**, 193–198.

HUTCHINGS, P. A. 1986. Biological destruction of coral reefs: a review. *Coral Reefs*, **4**, 239–252.

——, KIENE, W. E., CUNNINGHAM, R. B. & DONNELLY, C. 1992. Spatial and temporal patterns of non-colonial boring organisms (polychaetes, sipunculans and bivalve molluscs) in *Porites* at Lizard Island, Great Barrier Reef. *Coral Reefs*, **11**, 23–31.

INSALACO, E., HALLAM, A. & ROSEN, B. 1997. Oxfordian (Upper Jurassic) coral reefs in Western Europe: reef types and conceptual depositional model. *Sedimentology*, **44**, 707–734.

JANSA, L. F., TERMIER, G. & TERMIER, H. 1982. Les biohermes a algues, spongiaires et coraux des séries carbonatées de la flexure bordière du <<paleoshelf>> au large du Canada oriental. R*evue de Micropaléontogie*, **25**, 181–219.

JONES, B. & PEMBERTON, S. G. 1988. *Lithophaga* borings and their influence on the diagenesis of corals in the Pleistocene Ironshore Formation of Grand Cayman Islands, British West Indies. *Palaios*, **3**, 3–21.

KAUFFMAN, E. G. & SOHL, N. F.. 1974. Structure and evolution of Antillean Cretaceous rudist frameworks. *Verhandlungen der Naturforschenden Gesellschaft in Basel*, **84**, 399–467.

KELLY, S. R. A. & BROMLEY, R. G. 1984. Ichnological nomenclature of clavate borings. *Palaeontology*, **27**, 793–807.

KIENE, W. E. 1985. Biological destruction of experimental coral substrates at Lizard Island (Great Barrier Reef, Australia). *Proceedings of the Fifth International Coral Reef Symposium,* **5**, 339–344.

—— 1989. A model of bioerosion on the Great Barrier Reef. *Proceedings of the Sixth International Coral Reef Symposium*, **3**, 449–454.

—— & HUTCHINGS, P. A. 1993. Long-term bioerosion of experimental coral substrates from *Proceedings of the Seventh International Coral Reef Symposium*, **1**, 397–403.

—— & —— 1994. Bioerosion experiments at Lizard Island, Great Barrier Reef. *Coral Reefs,* **13**, 91–98.

KLEIN, R., MOKADY, O. & LOYA Y. 1991. Bioerosion in ancient and contemporary corals of the genus *Porites*: patterns and palaeoenvionmental implications. *Marine Ecology Progress Series*, **77**, 245–251.

KUSS, J. 1983. Faziesentwicklung in proximalen Intraplattform-Becken: Sedimentation, Palökologie und Geochemie der Kössener Schichten (Ober-Trias, Nördliche Kalkalpen). *Facies*, **9**, 61–172.

LATHUILIÈRE, B. 1982. Bioconstructions bajociennes a madréporaires et faciès associés dans l'Ile Crémieu (Jura du Sud; France). *Geobios*, **15**, 491–504.

MCCLOSKEY, L. R. 1970. The dynamics of a community associated with a marine scleractinian coral. *International Revue der gesamten Hydrobiologie*, **55**, 13–81.

MACGEACHY, J. K. & STEARN, C. W. 1976. Boring by macro-organisms in the coral *Montastrea annularis* on Barbados reefs. *Internationale Revue der gesamten Hydrobiologie*, **61**, 715- 745.

MACINTYRE, I. G. 1984. Preburial and shallow-subsurface alterations of modern scleractinian corals. *Paleontographica Americana*, **54**, 229–244.

MASSE, J.-P. 1977. Les constructions à Madrépores des calacaires urgoniens (Barrémien- Bédoulien) de Provence (SE de la France). *Memoires de la Bureau de la Recherche Géologique et Minèralogique*, **89**, 322–335.

METWALLY, M. H. M. 1996. Maastrichtian scleractinian corals from the Western flank of the Oman Mountains, U. A.E. and their paleoecological significance. *Neues Jahrbuch für Geologie und Paläontologie, Monatshefte,* **1996** (6), 375–388.

MICHALIK, J. 1982. Uppermost Triassic short-lived bioherms complexes in the Fatric, western Carpathians. *Facies*, **6**, 129–146.

MOORE, C. H. & SHEDD, W. W. 1977. Effective rates of sponge bioerosion as a function of carbonate production. *Proceedings of the Third International Coral Reef Symposium,* **2**, 499–505.

NEUMANN, A. C. 1966. Observations on coastal erosion in Bermuda and measurements of the boring rate of the sponge *Cliona lampa*. *Limnology and Oceanography*, **11**, 92–107.

OTTER, G. W. 1937. Rock-destroying organisms in relation to coral reefs. *British Museum's Great Barrier Reef Expedition 1928–29, Scientific Reports*, **1** (12), 323–352.

PALMER, T. J. 1982. Cambrian to Cretaceous changes in hardground communities. *Lethaia*, **15,** 309–323.

PANG, R. K. 1973. The ecology of some Jamaican excavating sponges. *Bulletin of Marine Science*, **23**, 227–243.

PEMBERTON, S. G., JONES, B. & EDGECOMBE, G. 1988. The influence of *Trypanites* in the diagenesis of Devonian stromatoporoids. *Journal of Paleontology*, **62**, 22–31.

PERRY, C. T. 1996. Distribution and abundance of macroborers in an Upper Miocene reef system, Mallorca, Spain: implications for reef development and framework destruction. *Palaios*, **11**, 40–56.

—— 1998. Macroborers within coral framework at Discovery Bay, north Jamaica: species distribution and abundance, and effects on coral preservation. *Coral Reefs*, **17**, 277–287.

—— 2000. Macroboring of Pleistocene coral communities, Falmouth Formation, Jamaica. *Palaios*, in press.

PEYROT-CLAUSADE, M. & BRUNEL, J.-F. 1990. Distribution patterns of macroboring organisms on Tuléar reef flats (SW Madagascar). *Marine Ecology Progress Series*, **61**, 133–144.

——, HUTCHINGS, P. & RICHARD, G. 1992. Temporal variation of macroborers in massive *Porites lobata* on Moorea, French Polynesia. *Coral Reefs*, **11**, 161–166.

——, LE CAMPION-ALSUMARD, T., HUTCHINGS, P., LE CAMPION, J., PAYRI, C. & FONTAINE, M.-F. 1995. Initial bioerosion and bioaccretion on experimental substrates in high islands and atoll lagoons. *Oceanologica Acta*, **18**, 531–541.

PFISTER, T. 1980. Paläoökologie des oligozänen Korallenvorkommens von Cascine südlich Acqui (Piemont, Norditalien). *Jahrbuch des Naturhistorischen Museums Bern*, **7**, 247–262.

—— 1985. Coral fauna and facies of the Oligocene fringing reef near Cairo Montenotte (Liguria, northern Italy). *Facies*, **13**, 175–226.

PHILIP, J. 1972. Paléoécologie des formations à rudistes du Cretacé supérieur – l'exemple du sud-est de la France. *Palaeogeography, Palaeoclimatology, Palaeoecology*, **12**, 205–222.

PISERA, A. 1987. Boring and nestling organisms from Upper Jurassic coral colonies from northern Poland. *Acta Palaeontologica Polonica*, **32**, 83–104.

PLEYDELL, S. M. & JONES B. 1988. Borings of various faunal elements in the Oligocene-Miocene Bluff Formation of Grand Cayman, British West Indies. *Journal of Paleontology*, **62**, 348–367.

PRINZ, P. 1986. Mitteljurassische Korallen aus Nordchile. *Neues Jahrbuch für Geologie und Paläontologie, Monatshefte,* **1986**, 736–750.

REAKA-KUDLA, M. L., FEINGOLD, J. S. & GLYNN, P. W. 1996. Experimental studies of rapid bioerosion of coral reefs in the Galapagos islands. *Coral Reefs,* **15**, 101–107.

REHMAN, J., JONES, B., HAGAN, T. H. & CONIGLIO, M. 1994. The influence of sponge borings on aragonite-to-calcite inversion in late Pleistocene *Strombus gigas* from Grand Cayman, British West Indies. *Journal of Sedimentary Research*, **A64**, 174–179.

REITNER, J. & KEUPP, H. 1991. The fossil record of the haplosclerid excavating sponge *Aka* de Laubenfels. *In:* Reitner, J. & Keupp, H. (eds) *Fossil and Recent Sponges*. Springer, Berlin, 102–120.

RICE, M. E. & MACINTYRE, I. G. 1982. Distribution of Sipuncula in the coral reef community, Carrie Bow Cay, Belize. *In*: Rützler, K. & Macintyre, I. G. (eds) *The Atlantic Barrier Reef Ecosystem at Carrie Bow Cay, Belize I: structure and communities*. Smithsonian Contributions to Marine Sciences, **12**.

RIEDEL, P. 1991. Korallen in der Trias der Tethys: Stratigraphische Reichweiten, Diversitätsmuster, Entwicklungstrends und Bedeutung als Rifforganismen. *Mitteilungen der Gesellschaft der Bergbaustudenten Österreichs*, **37**, 97–118.

RISK, M .J., SAMMARCO, P. W. & EDINGER, E. N. 1995. Bioerosion in *Acropora* across the continental shelf of the Great Barrier Reef. *Coral Reefs*, **14**, 79–86.

RONIEWICZ, E. 1974. Rhaetian corals of the Tatra Mts. *Acta Palaeontologica Polonica*, **24**, 97–116.

RÜTZLER, K. 1974. The burrowing sponges of Bermuda. *Smithsonian Contributions to Zoology*, **165**, 1–32.

—— 1975. The role of burrowing sponges in bioerosion. *Oecologia,* **19**, 203–216.

SAMMARCO, P. W. & RISK, M. J. 1990. Large-scale patterns in internal bioerosion of *Porites*: cross continental shelf trends on the great Barrier reef. *Marine Ecology Progress Series*, **59**, 145–156.

——, —— & ROSE, C. 1987. Effects of grazing and damselfish territoriality on internal bioerosion of dead corals: indirect effects. *Journal of Experimental Marine Biology and Ecology*, **112**, 185–199.

SANDERS, D. & BARON-SZABO, R. C. 1997. Coral-rudist bioconstructions in the Upper Cretaceous Haidach section (Gosau Group; Northern Calcareous Alps, Austria). *Facies*, **36**, 69–90.

SCHOLZ, H. 1984. Bioherme und Biostrome im Allgäuer Schrattenkalk (Helvetikum, Unterkreide). *Jahrbuch der Geologischen Bundesanstalt Wien*, **127**, 471–499.

SCHROEDER, J. H. 1986. Diagenetic diversity in Paleocene coral knobs from the Bir Abu El-Husein Area, S-Egypt. *In*: SCHROEDER, J. H. & PURSER, B. H. (eds) *Reef Diagenesis.* Springer, Berlin, 132–158.

—— & ZANKL, H. 1974. Dynamic reef formation: a sedimentological concept based on studies of Recent Bermuda and Bahama reefs. *Proceedings of the Second International Coral Reef Symposium*, **2,** 413–428.

SCHUSTER, F. 1996. Palaeoecology of Paleocene and Eocene corals from the Kharga and Farfara oases (Western Desert, Egypt) and the depositional history of the Paleocene Abu Tartur carbonate platform, Kharga Oasis. *Tübinger Geowissenschaftliche Arbeiten, Reihe A*, **31**, 1–96.

SCOTT, R. W. 1988. Evolution of Late Jurassic and Early Cretaceous reef biotas. *Palaios*, **3**, 184–193.

SENOWBARI-DARYAN, B., ZÜHLKE, R., BECHSTÄDT, T. & FLÜGEL, E. 1993. Anisian (Middle Triassic) buildups of the northern Dolomites (Italy): the recovery of reef communities after the Permian/Triassic crisis. *Facies*, **28**, 181–256.

SHEEHAN, P. M. 1985. Reefs are not so different – they follow the evolutionary pattern of level-bottom communities. *Geology*, **13**, 46–49.

SKELTON, P. W., GILI, E., VICENS, E. & OBRADOR, A. 1995. The growth fabric of gregarious rudist elevators (hippuritids) in a Santonian carbonate platform in the southern Central Pyrenees. *Palaeogeography, Palaeoclimatology, Palaeoecology*, **119**, 107–126.

STANLEY, G. D. 1988. The history of early Mesozoic reef communities: a three-step process. *Palaios*, **3**, 170–183.

—— & BEAUVAIS, L. 1994. Corals from an Early Jurassic coral reef in British Columbia: refuge on an oceanic island reef. *Lethaia*, **27**, 35–47.

—— & MCROBERTS, C. A. 1993. A coral reef in the Telkwa Range, British Columbia: the earliest Jurassic example. *Canadian Journal of Earth Sciences*, **30**, 819–831.

—— & SWART, P. K. 1995. Evolution of the coral-zooxanthellae symbiosis during the Triassic: a geochemical approach. *Paleobiology,* **21**, 179–199.

STANTON, R. W. & FLÜGEL, E. 1989. Problems with reef models: the Late Triassic Steinplatte 'reef' (northern Alps, Salzburg/Tyrol, Austria). *Facies,* **20**, 1–138.

STENECK, R. S. 1983. Escalating herbivory and resulting adaptive trends in calcareous algal crusts. *Paleobiology*, **9**, 44–61.

TURNŠEK, D. & MIHAJLOVIĆ, M. 1981. Lower Cretaceous cnidarians from eastern Serbia. *Slovenska Akademija Znanosti in Umetnosti, Razprave IV razreda*, **23**, 1–54.

—— & POLŠAK, A. 1978. Senonijske kolonijske korale iz biolititnega kompleksa v ore´´sju na Medvednici. *Slovenska Akademija Znanosti in Umetnosti, Razprave IV razreda*, **21**, 129–181.

——, SEYFRIED, H. & GEYER, O. F. 1975. Geologische und paläontologische Untersuchungen an einem

Korallenvorkommen im subbetischen Unterjura von Murcia (Süd-Spanien). *Slovenska Akademija Znanosti in Umetnosti, Razprave IV razreda*, **18**, 120–156.

VERMEIJ, G. J. 1977. The Mesozoic marine revolution: evidence form snails, predators and grazers. *Paleobiology*, **3**, 245–258.

VOGEL, K. 1997. Bioerosion in rezenten Riffbereichen: Experimente vor Inseln Bahamas und des Großen Barriereriffs. *Natur und Museum*, **127**, 198–208.

WARME, J. E. 1975. Borings as trace fossils and the processes of marine bioerosion. *In*: FREY, R. W. (ed.) *The Study of Trace Fossils*. Springer, New York, 181–227.

—— 1977. Carbonate borers – their role in reef ecology and preservation. *AAPG Studies in Geology*, **4**, 261–279.

WEIDLICH, O. 1996. Bioerosion in Late Permian Rugosa from reefal blocks (Hawasina Complex, Oman Mountains): implications for reef degradation. *Facies*, **35**, 133–142.

WILKINSON, C. R. 1983. Role of sponges in coral reef structural processes. *In*: Barnes, D. J. (ed.) *Perspectives on Coral Reefs*. Clouston, Manuka, 263–274.

WOOD, R. 1993. Nutrients, predation and the history of reef-building. *Palaios*, **8**, 526–534.

—— 1995. The changing biology of reef-building. *Palaios*, **10**, 517–529.

WULLSCHLEGER, E. 1966. Bemerkungen zum fossilen Korallenriff Gisliflue-Homberg. *Mitteilungen der Aargauischen Naturforschenden Gesellschaft*, **27**, 101–152.

—— 1971. Bemerkungen zum fossilen Korallenvorkommen Tiersteinberg-Limperg-Kei. *Mitteilungen der Aargauischen Naturforschenden Gesellschaft*, **28**, 251–292.

WURM, D. 1982. Mikrofazies, Paläontologie und Palökologie der Dachsteinriffkalke (Nor) des Gosaukammes, Österreich. *Facies*, **6**, 203–296.

ZANKL, H. 1969. Der Hohe Göll. Aufbau und Lebensbild eines Dachsteinkalk-Riffes in der Obertrias der nördlichen Kalkalpen. *Abhandlungen der Senckenbergischen Naturforschenden Gesellschaft*, **519,** 1–123.

Microfacies development in Late Archaean stromatolites and oolites of the Ghaap Group of South Africa

DAVID T. WRIGHT[1] & WLADYSLAW ALTERMANN[2]

[1]*Department Of Geology, University of Leicester, University Road, Leicester, LE1 7RH, UK (e-mail: dtw1@le.ac.uk)*

[2]*Institut für Allgemeine und Angewandte Geologie, Luisenstrasse 37, Ludwig Maximilians Universität, München, Germany (e-mail: Wlady.Altermann@iaag.geo.uni-muenchen.de)*

Abstract: Organism–environment feedbacks in Precambrian platformal carbonates and reefs were strongly influenced by the activities of diverse microbial ecosystems. Microfacies studies of representative platformal microbial carbonates, comprising cyanobacterial mat, stromatolites and giant ooids, from the Late Archaean Ghaap Group of South Africa have provided compelling evidence for an intimate relationship between taphonomic evolution, fabric development and mineralogy in rocks of the Gamohaan and Boomplaas formations. Cements, both in fold hinges and between the limbs of slump-folded and contorted, partially-degraded, pyritiferous stromatolitic laminae, were precipitated after deformation of organic fabrics, but before or during their compaction, indicating that cementation took place at the same time as anoxic organic degradation involving bacterial sulphate reduction. Bundles and strands of the organic remains of filamentous cyanobacteria, in varying states of degradation in both stromatolites and ooids, have been preserved by mineralization. Structural detail is usually best preserved in calcite, where cyanobacterial sheaths, 10 μm to 25 μm in diameter and hundreds of micrometres in length, can be clearly seen. Petrographic analysis of the microfabrics using cathodoluminescence reveals dolomicrite nucleated along the outer margin of some sheaths. Dolomicrospar and dolospar fabrics developed progressively in association with increasing sheath degradation, as evidenced by the sequential loss of structural detail, culminating in a xenotopic fabric comprising brown, inclusion-rich, anhedral crystals with irregular boundaries. Biogeochemical modelling supports a genetic link between bacterial sulphate reduction and (1) calcite precipitation in the contorted laminae, and (2) replacive dolomitization of the calcitic matrix in the stromatolites and ooids. The evidence indicates that anoxic organic diagenesis was an essential and major process in controlling carbonate precipitation and mineralogy in widespread microbialitic sediments of the Ghaap Group, a depositional environment analogous to many other Archaean, Proterozoic and, during periods of biotic stress, some Phanerozoic carbonate platforms.

In the Phanerozoic, marine carbonate production, facies and fabrics have been largely determined by metazoans, but in the Precambrian, organic influence was restricted to microbes. These dominated the shallow, non-siliciclastic shelf environment and, in the form of stromatolites and other microbialites, acted as major repositories for carbonates. Interactions between micro-organisms and their environments have generated a variety of facies and microfacies in platformal carbonates, both by sediment trapping and binding, and by *in situ* precipitation, but the specific roles of each are difficult to determine because of the complexity of biotic influences and water–sediment interactions. Moreover, the distinction between biotic and abiotic carbonates is complicated by the proximity of the latter to microbial structures. Carbonate mineralogy may also be complex, and interlayering of dolomite and limestone on a subcentimetre scale is frequently observed in ancient carbonates, particularly in stromatolites and other microbialites in which cyanobacteria played a dominant role in lithogenesis.

Ooids, common sediments of both modern and ancient carbonate platforms, display a wide range of nuclear and cortical fabrics. There is still no consensus as to their mode of formation, in particular whether they are exclusively chemical precipitates or whether microbes are involved in their accretion in some way (e.g. Suess & Futterer 1972; Davies *et al.* 1978; Heydari & Moore 1991; Thompson & Huber 1997; Wright 1997), but whichever model is invoked, it is undeniable that the internal fabrics of modern ooids, and by analogy ancient ooids, comprise abundant microbial material (e.g. Green *et al.* 1988).

This paper investigates the relationship between constituent microbes, microfacies and mineralogy in variably preserved stromatolites

From: INSALACO, E., SKELTON, P. W. & PALMER, T. J. (eds) 2000. *Carbonate Platform Systems: components and interactions*. Geological Society, London, Special Publications, **178,** 51–70. 0305–8719/00/$15.00

and ooids from the Late Archaean Ghaap Group, and provides new data on growth fabric development and dolomite formation as a response to biogeochemical interactions driven by taphonomic diagenesis. This has significant implications for assessing the role of microbial ecosystems in carbonate generation and accumulation.

Geological setting and depositional environment

Occurrences of well preserved Archaean platformal carbonates are rare, but an important exception is the Ghaap Group in the northern Cape Province of South Africa (Fig. 1), comprising the Schmidtsdrif and overlying Campbellrand subgroups with a combined thickness of >2000 m. The 2642 ± 3 Ma Schmidtsdrif Subgroup (Walraven & Martini 1995) includes the Vryburg, Boomplaas and Lokammona formations, and conformably to paraconformably underlies the Campbellrand Subgroup (Fig. 2). The Campbellrand Subgroup platform persisted some 80 Ma from 2588 ± 6 Ma BP to at least 2516 ± 4 Ma BP (Altermann & Nelson 1998) and comprises tidal flat and shallow marine deposits followed by subtidal carbonates as transgression

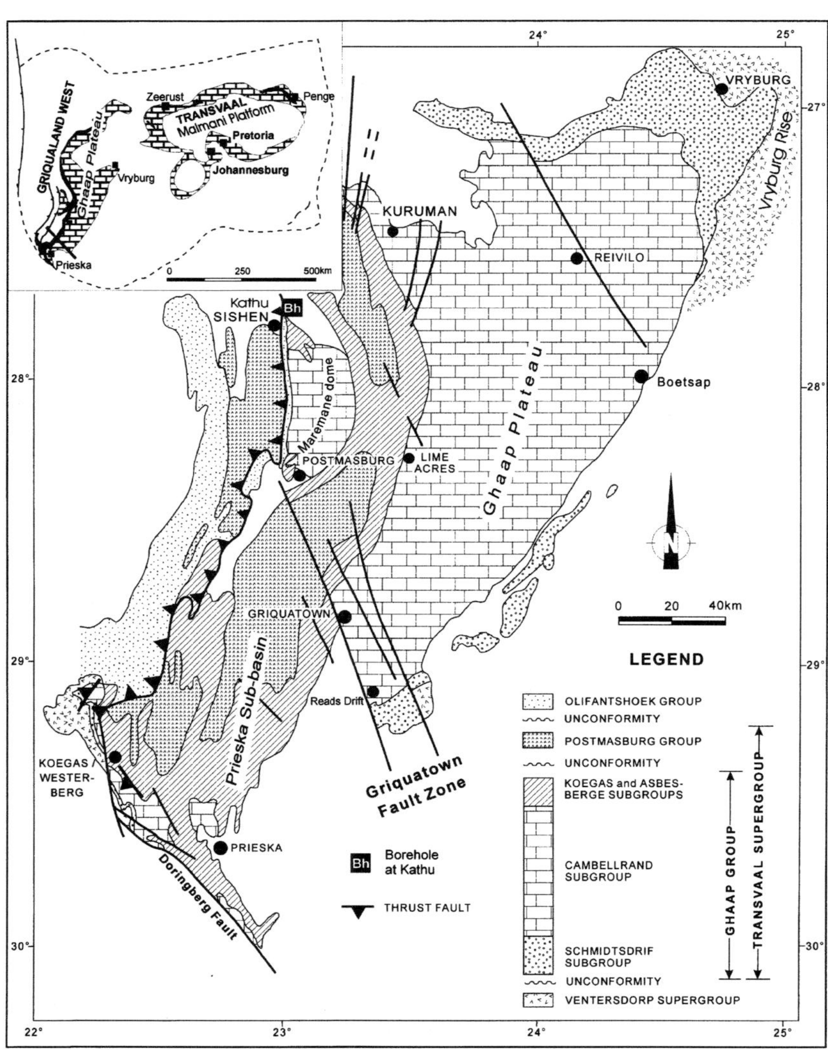

Fig. 1. Locality map of the Ghaap Group, Cape Province, Republic of South Africa.

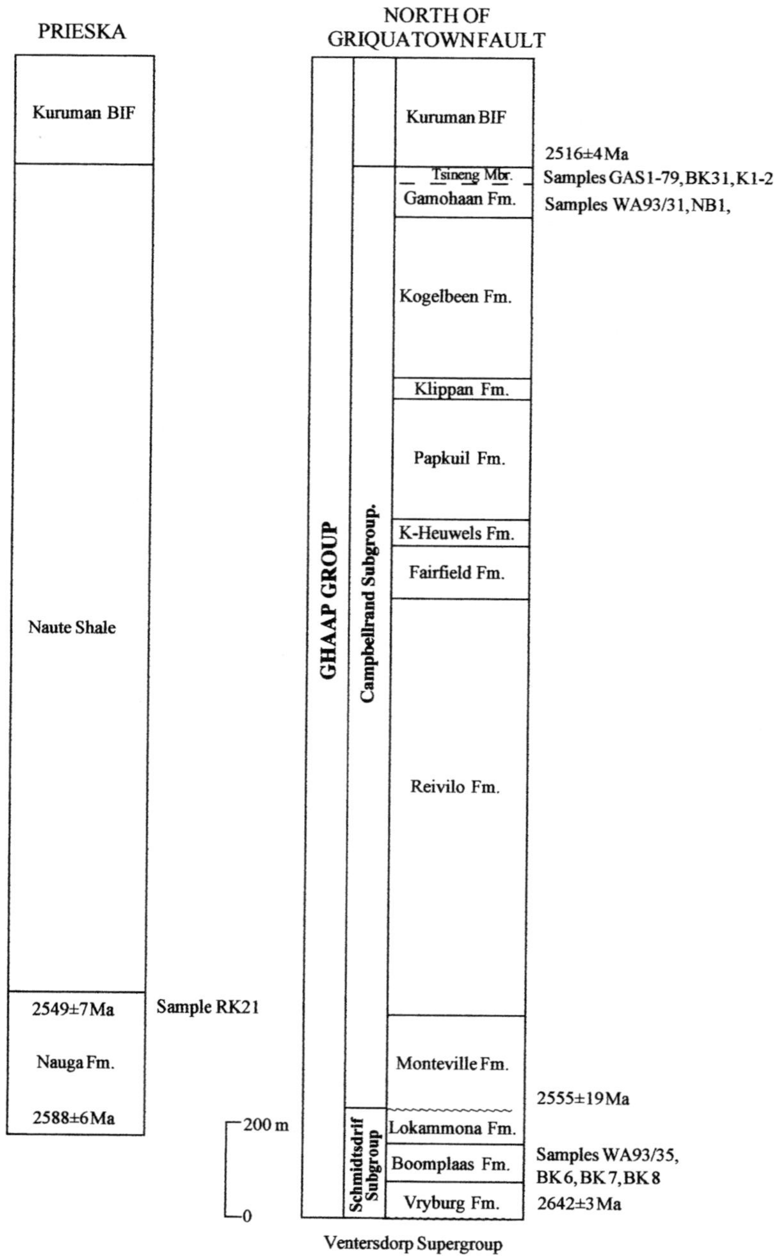

Fig. 2. Simplified stratigraphic setting of the Ghaap Group, Cape Province, Republic of South Africa.

extended from southwest to northeast for a distance of over 400 km, with abundant microbialitic carbonates, including shallow water and deep water stromatolites, and ooids.

The Nauga Formation crops out only in the southwestern part of the Campbellrand Subgroup, around the Prieska area (Fig. 1) and is the oldest part of the carbonate platform (2588 ± 6 to 2549 ± 7 Ma BP; Altermann & Nelson 1998). It is conformably covered by a shale sequence that correlates in age to the Monteville, Reivilo, Fairfield, Klipfontein Heuwels,

Papkuil, Klippan, Kogelbeen and Gamohaan formations in the Campbellrand Subgroup.

The Boomplaas Formation of the Schmidtsdrif Subgroup overlies the Vryburg Formation of siliciclastics, carbonates and lavas (2642 ± 3 Ma BP; Walraven *et al.* in press). It is covered by shales of the Lokammona Formation, in turn covered by the 2555 ± 19 Ma old Monteville Formation of the Campbellrand Subgroup. According to Beukes (1979, 1983) the Boomplaas Formation reaches about 100 m in thickness, and is typically developed as cyclic successions of stromatolites, followed by inversely graded oolites, carbonate sands and oolites. In the north (at Vryburg), graded, transported oolite beds are typically succeeded by stromatolites. The gross depositional environment of the formation can be interpreted as a carbonate platform developed over the transgressive, siliciclastic–volcanic Vryburg Formation. The platform was drowned and covered by the Lokammona shales during progressive transgression. Descriptions from other localities show that the Boomplaas Formation attains up to 185 m thickness and consists mainly of shales and deep microbial laminites with minor transported oolites, deposited in a shelf environment devoid of shallow platformal carbonates (Altermann & Siegfried 1997). The Gamohaan Formation (2516 ± 4 Ma BP; Altermann & Nelson 1998) of South Africa is *c.* 100 m thick, and forms the topmost part of Campbellrand Subgroup carbonates. Generally the Gamohaan Formation is interpreted as a transgressive (deepening-upward) cycle leading to the drowning of the Campbellrand carbonate platform, prior to the deposition of the banded iron formation (BIF) (Beukes 1987; Altermann & Herbig 1991; Altermann & Nelson 1998). The carbonates of the Nauga Formation are mainly peritidal, but are subtidal, passing into a shelf facies towards the top of the sequence (Altermann & Nelson 1998). A thick sequence of shales (Naute Shale Member) with some chert beds of great lateral continuity covers the Nauga Formation and is overlain by a thick BIF sequence.

Sample localities and stratigraphic position

The samples described herein were collected from different formations of the Ghaap Group carbonates, i.e. from the Boomplaas Formation of the Schmidtsdrif Subgroup, from the Nauga Formation and from the Gamohaan Formation of the Campbellrand Subgroup. Samples have been selected for study on the basis of preservational quality, the extent to which the fabrics represent common facies, and the information revealed by petrographic analysis.

Sample WA93/31 is from the Gamohaan Formation, and was collected at the White Bank locality, of the Kuruman Kop mountain, approximately 10 km NW of the Kuruman town. It is a sample of a coniform, thesauroid microbial mat (Fig. 3 a–c), deposited below fairweather wave base but above the storm wave base. These stromatolites were often described as *Conophyton*. A new detailed description and facies interpretation of these stromatolites is given by Sumner (1997). The detailed stratigraphic section at the Kuruman Kop locality was described by Halbich *et al.* (1992), and compared with drillcores and sections described by Beukes (1980) and Klein & Beukes (1989) from the same locality. The stratigraphic position of the sample corresponds to the zone 'e' of Halbich *et al.*, (1992, their fig. 10) and to zone '3' of Beukes (1980). Sample NB 1 was taken from the same horizon. The stratigraphic distance to the overlying BIF is around 60 m. A tuff layer around 45 m above this sample was dated at 2516 ± 4 Ma BP and the lower age limit of the sample is the age of the Monteville Formation of 2555 ± 19 BP Ma (Altermann & Nelson 1998).

Sample RK21 (Fig. 4a) is from the uppermost Nauga Formation carbonates at Prieska and of deep water stratiform microbial laminite, rolled up (coiled) by a storm wave or slump. Similar to samples GAS1–79 and WA93/31, the lamination consists of thin filamentous structures interpreted as remnants of sheaths of filamentous cyanobacteria. The sample GAS1–79 (Fig. 4b) is from the core, drilled by the SAMANCOR company and stored in Hotazel. It is from the Tsineng Member of the uppermost Gamohaan Formation, 48 m below the BIF and 558 m below the surface. Thus it is from the same stratigraphic horizon as sample WA93/31. The sample reveals two rolled up, but otherwise laterally flat, stratiform laminae that cover a slumped and disturbed coniform microbial mat deposited below the fairweather wave base. The facies interpretation mirrors that of sample WA93/31. Sample BK 31, illustrated in thin section (Fig. 7), is taken from the same section.

The sample Kl-2 (Fig. 10) was collected from borehole DI1 drilled on the Derby Farm, some 55 km south of Kuruman. The borehole penetrated the cherty, fossiliferous Tsineng Member of the uppermost Gamohaan Formation. The sample is from that member and its fossil content was described in detail by Klein *et al.* (1987). The facies was described as being transgressive over the carbonate platform and transitional to the overlying BIF, and the depth of

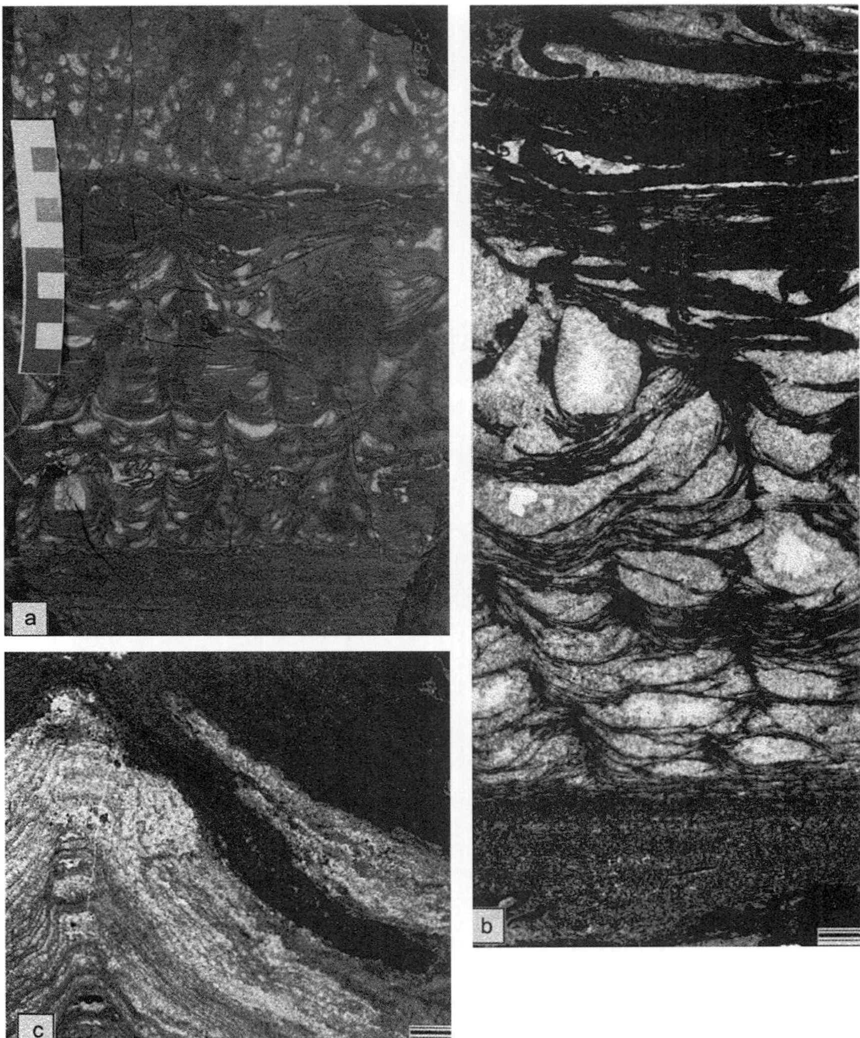

Fig. 3. Thesauroid mat with conical supports from the Gamohaan Formation. (**a**) Cross-view of the biostrome (scale in cm). The cements and the lack of sediment in the voids between the laminae can be readily observed. The shape of the voids suggests they were water-filled during cementation. (**b**) Same mat as in (a); in the upper part contorted and convoluted laminae, deformed in a soft state are visible. (**c**) Same mat as in (a); detail of the conical support showing the fine microbial lamination and the lack of micritic sediment in the laminae. Scale bars = 2 mm in (b) and (c).

deposition was estimated to be around 40 m. This sample shows unique diagenetic and preservational features and is therefore included in this investigation.

Sample WA93/35 is an oolitic sample from the Boomplaas Formation, collected at Kaffersfontein farm, at the Orange River, southwest of Griquatown, and approximately 300 km SW of Vryburg. At this locality the oolites are inversely graded, *in situ*, and directly bury domal and domical stromatolitic bioherms. Samples BK 6, 7 and 8 are also oolites from the Boomplaas Formation.

Stromatolitic fabrics

The Gamohaan Formation comprises a variety of microbialitic facies, including stromatolites,

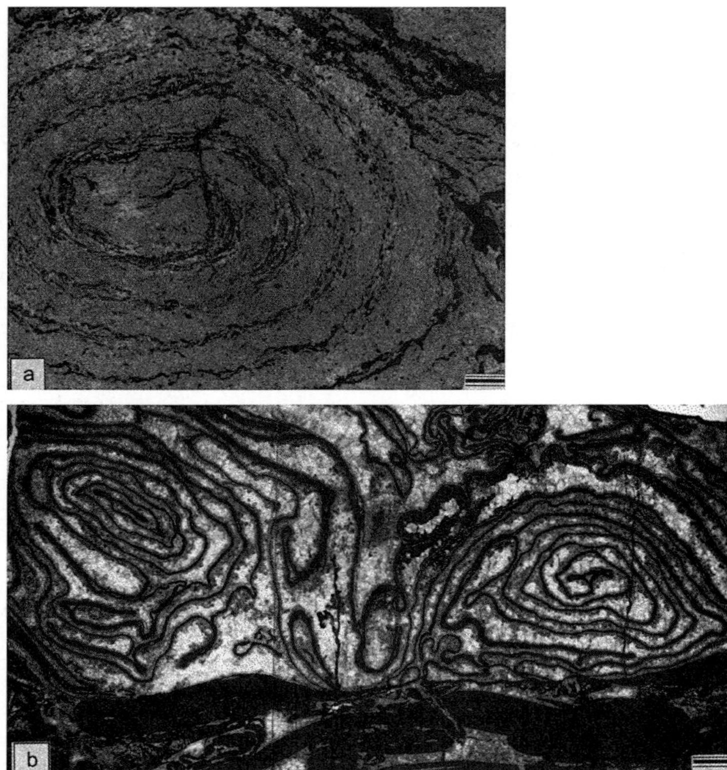

Fig. 4. Microbialites showing coiled microbial lamination rolled up in a carpet-like manner. (**a**) Nauga Formation at Prieska; (b) Gamohaan Formation. In (a), radial and fan-shaped cements are not recognizable, and anhedral, microsparitic, ferroan dolomite crystals penetrate the darker microbial laminae. In (b) such cements can be recognized due to less advanced dolomitization. Scale bar = 2 mm.

crinkly mat and interbedded laminated carbonates (Altermann & Wotherspoon 1995; Sumner 1997). The morphology and dimensions of the filamentous forms in the stromatolites, the association of the laminae with pyrite and the isotopic evidence for pyrite formation by bacterial reduction, together with ^{12}C-enriched carbon from organic matter in the same rocks (Strauss & Beukes 1996), the work of other researchers (referred to below) including the exceptional preservation of microfossils in rock samples from the Tsineng Member, and the fact that we know that stromatolites are built up by the actions of constituent cyanobacterial and other microbial communities, all provide compelling evidence for the organic nature of the laminae we describe, and that they are indigenous to and syngenetic with the Archaean sediments in which they are found. Fossil filamentous microbial sheaths, some exceptionally well preserved (*Siphonophycus transvaalensis* Beukes, Klein and Schopf, 1987) and often intermeshed, are reported from both cherts and carbonates in stromatolites of the upper part of the Formation (Klein *et al.* 1987). From the stromatolitic Kogelbeen Formation, underlying the Gamohaan Formation, further filamentous and coccoidal microfossils such as *Eomycetopsis* cf. *filiformis*, *Archaeotrichion* and *Eoentophysalis* were described by Altermann & Schopf (1995). This study focuses on conical stromatolites and contorted filamentous laminae from the Gamohaan Formation, on coiled microbial laminites from the Nauga and Gamohaan Formations, and on oolites from the Boomplaas Formation.

The coniform stromatolites of the Gamohaan Formation are 10–12 cm high, (Fig. 3a–c) comprising a calcitic column supporting bundles or strands of laminae ranging in thickness from <1–100 μm, and forming a complex framework with abundant, calcite-filled voids. The framework forms thesauroid mats (e.g. Hofmann & Masson 1994) with dish-shaped depressions between the upward protruding coni which are interconnected with each other laterally. The coni are often inclined, all in one direction, probably reflecting current activity. The upper part of such a lithostrome layer is often disturbed by slumps and convolute lamination and covered with shaly–marly intervals on which a new thesauroid layer starts to develop with a sharp base. The voids may have resulted from gas bubbles generated by organic degradation, which is consistent with deposition in a subtidal setting (Altermann & Wotherspoon 1995). The size of the voids increases conspicuously upward within the mat and they become more irregular in shape, which probably points towards some gas concentration within the upper level of the particular mat. However, most of the voids are not convex upward like gas bubbles but are rather concave, and therefore could have been filled with water during growth and diagenesis (Sumner 1997). The evidence suggests that the accretion mechanism of the stromatolites was *in situ* precipitation. Sumner (1997) described these lithostromes as cuspate microbialites and interpreted their setting, from their inferred stratigraphic position, as of deep open-marine environment.

The coiled laminites from the GAS1–79 core (Fig. 4a,b) show continuous, rolled up laminae on top of convoluted laminae. The rolls are somewhat flattened by compaction but the lamination is largely unbroken and space, now filled with cements, remained devoid of sediment fill after the rolling up of these laminae. One reason why the laminae of the mats did not completely collapse may be that the space between organic strands was water-filled. Alternatively, ongoing organic degradation may have produced gas that was trapped as bubbles and this space was later filled by cements in a manner similar to the coniform stromatolites of the Gamohaan Formation, and to fenestrae. These slump structures, roll structures and shale intercalations are interpreted as tempestites caused by storm waves, below the fairweather wave base. They provide evidence that the microbial mats were not thoroughly lithified but were soft within at least a few upper centimetres of the mat during microbial growth. The voids are trapped between the filmy laminae connecting the coni and forming the depressions.

Petrographic examination of thin sections of the upper part of coniform stromatolites shows that a calcitic, sparry matrix is traversed by elongate, thin, filamentous structures in various states of preservation 15μm to >25μm in diameter and many hundreds of micrometres in length, occurring in strands (Fig. 5a,b), bundles (Fig. 5c,d) or 'shrubs' (Fig. 6). The cement matrix comprises radial fibrous calcite nucleated on a fan-shaped growth surface and shows increasing crystal size inwards. The filaments are confidently interpreted as the moulds of sheaths of constituent, stromatolitic cyanobacteria. In the best preserved examples, the outline of the sheath can be clearly seen, and careful inspection may reveal the presence of tiny micritic crystals apparently attached to the outer surface. This is confirmed under cathodoluminescence, when these crystals luminesce a dark purple or red-brown colour (Figs. 5b, 6b), which strongly contrasts with the pale yellow of the calcitic matrix, facilitating identification and revealing a difference in mineralogy – the micritic crystals are of dolomite. These crystals are not found dispersed in the matrix where the sheaths are absent. In the interior of the stromatolitic column, the sheaths are less well defined, becoming increasingly encrusted by darkly luminescing crystals ranging in size from micrite to *c.* 25μm. Towards the base of the stromatolite, sheaths may grade laterally into micritic to microsparry laminae. Lower down, sheaths are barely discernible, but their former presence is indicated by replacive, turbid, brownish, inclusion-rich crystals up to 100 μm in size, producing a hypidiotopic, coarse crystalline texture. This culminates in a totally replacive, sparry fabric in which no remaining sheath material is visible (Fig. 5c,d). It thus appears that the replacive mineral conforms to the original distribution of sheaths of filamentous cyanobacteria involved in stromatolite construction, and that increased crystal size and abundance of this phase leads to decreased preservational detail.

The contorted laminae, interbedded with thesauroid beds, are encased in calcite. The original filamentous nature of the microflora is indicated by the organization of elongate strands and remnant sheaths parallel to the axis of the lamina bundle, which occasionally enclose thin, elongate, calcitic crystals (Fig. 7a, b). Such organic structures are discernible only in a calcite matrix which shows no associated mineralogical changes under cathodoluminescence; however, cathodoluminescence shows that less well preserved, contorted laminae are often associated with pyrite or dolomite, and here the filamentous nature of the sheaths is apparent only from the overall shape of the lamina bundles (Figs 8, 9). In many cases, pyrite or coarse dolomite crystals follow the outline of former lamina bundles, and constituent filaments or sheaths are rarely observed. Significantly, the presence of pyrite is restricted to the occurrence of microbial laminae, where it is often observed as disseminated grains or

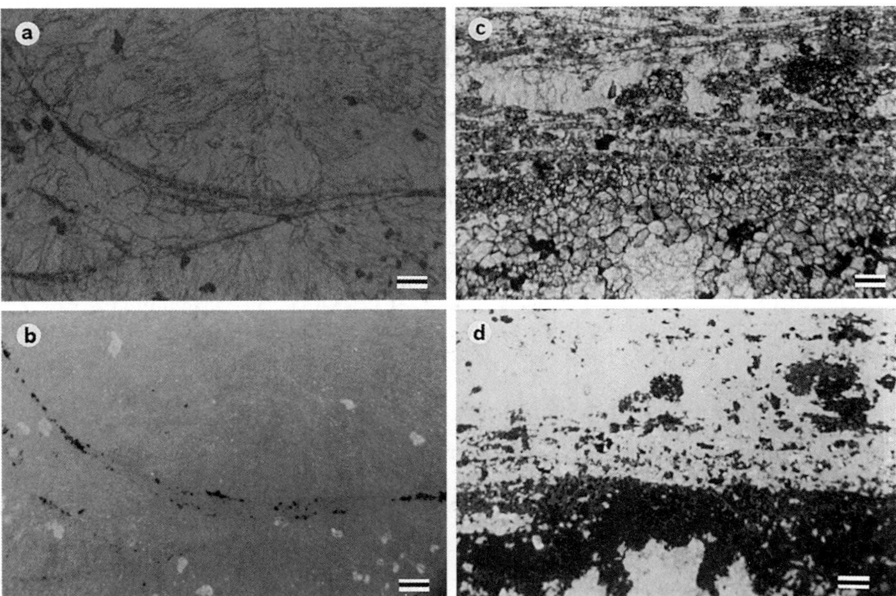

Fig. 5. Photomicrographs of matching couplets in plane polarized light (**a**, **c**) and under cathodoluminescence (**b**, **d**) of a section of the upper part of the stromatolites of the Gamohaan Formation (sample NB1), in which the sparry matrix is traversed by permineralized filamentous cyanobacterial sheaths. The outline of the sheaths can be clearly seen in (a), and careful inspection reveals the presence of tiny micritic crystals (dolomite) attached to the outer surface. Under cathodoluminescence (b), the dark purple colour of the dolomite, restricted to the surface of the sheaths, constrasts with the pale yellow of the calcitic matrix: no dolomite is found floating in the matrix. In the lower part of the stromatolite (c, d), relatively well preserved sheaths in the upper part of the photo are associated with dolomicrite, clearly seen under cathodoluminescence (d). In the centre of the photo, sheaths are recognizable but less well defined, and are associated with an increase in the size and abundance of dolomite crystals. In the lower part of the photo, sheaths are not preserved, but their former presence is indicated by abundant, coarse, turbid dolomite spar forming a hypidiotopic fabric. This sequence is interpreted to reflect progressive dolomitization in association with sheath degradation, and demonstrates how a massive, structureless, dolomitic fabric may develop, bearing no clue to its microbial origin. Scale bars = 200 μm.

nodules in the organic-rich matrix formed from the lamina bundle (Fig. 8a,b). The pyrite is genetically associated with anoxic organic degradation of the laminae, and grew as a reaction product from sedimentary Fe and H_2S which was promoted by bacterial sulphate reduction of organic material (Strauss & Beukes 1996). Where the filaments have been dolomitized (Fig. 9a,b), anhedral, microsparitic, ferroan dolomite crystals penetrate the darker microbial laminae and the space between them and the laminae may have been bound by micrite prior to recrystallization. The structural detail of microbial filaments is extremely poor and single sheaths are not recognizable: the presence of a former lamina bundle is revealed only by the replacive dolomite. The well preserved sheaths in the upper part of Fig. 5c, d are clearly recognizable but below, they become gradually less well defined, and are associated with an increase in the size and abundance of dolomite crystals. Structural detail is progressively lost until in the lower part of Fig. 5c, d, sheaths are not preserved. This sequence is interpreted to reflect progressive sheath degradation with taphonomic evolution.

The laminated microbial mat from the Tsineng Member (Fig. 10, sample Kl-2) exhibits probably the best preservation of Archean microbial sheaths ever encountered. The best preservation is surprisingly not within the early diagenetic cryptocrystalline quartz, but in the ferroan dolomite and dolomite (Fig. 10a–f), where sheaths of *Siphonophycus transvaalensis* are perfectly preserved (Klein *et al.* 1987). This unique mode of preservation is evidence that sheath preservation was dependent on fabric stabilization after early micritic dolomitization, sometimes with preserved aragonitic needles. However, the state of preservation is not

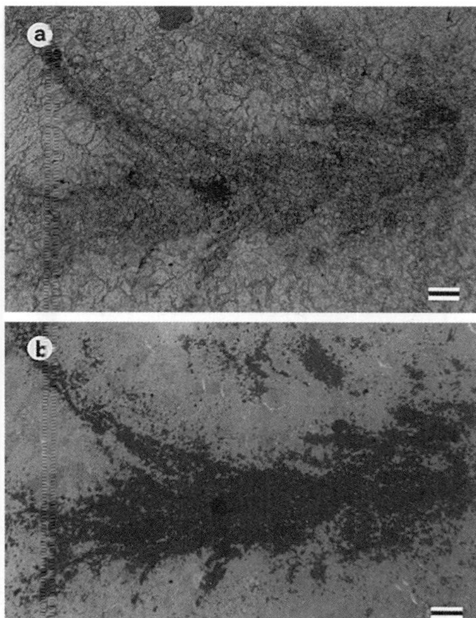

Fig. 6. Photomicrographs of a matching couplet in plane polarized light (**a**) and under cathodoluminescence (**b**) from sample WA 93/31d. Preserved cyanobacterial sheaths are organized here into shrub-like forms, and are preferentially dolomitized (dark red-brown luminescence) in a calcite cement matrix (yellow luminescence). Dolomitization is clearly confined to the sheaths, and is attributable to anoxic organic degradation involving sulphate-reducing bacteria. Scale bars = 200 μm.

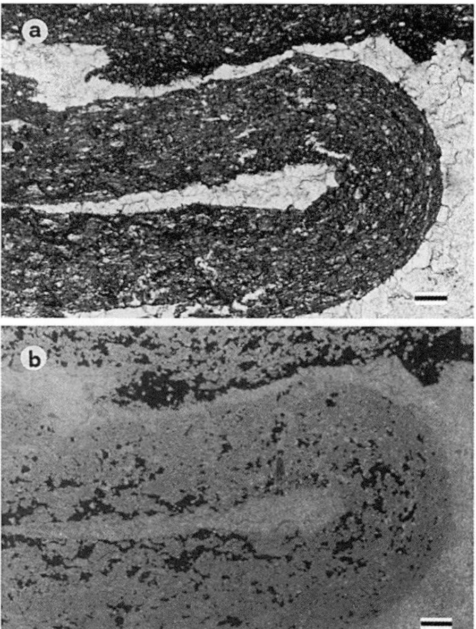

Fig. 7. Contorted laminae encased in calcite under plane polarized light (**a**) and cathodoluminescence (**b**), sample BK 31. The original filamentous nature of the microflora is indicated by the organization of elongate strands and remnant sheaths parallel to the axis of the lamina bundle, which occasionally enclose thin, elongate, calcitic crystals. The flexibility of the cyanobacterial strands shows that they were unmineralized prior to and during deformation of the mat, while the preservation of voids between uncompressed hinges and folds of the strands shows that cementation must have occurred immediately after folding but prior to compaction. Scale bars = 200 μm.

uniform throughout the sample, the largest part of it being fossil free. In the microsparitic anhedral dolomite within the lamination, the best preserved sheaths of *Siphonophycus transvaalensis* exhibit a detailed sheath structure of fine mineralic needles a few micrometres in length and 0.5 μm across. Klein *et al.* (1987) described these needles as of acicular morphology and probably originally of aragonitic composition (Fig. 10c). This detailed structure is less well preserved where the cyanobacterial mat has been silicified (Fig. 10d), and where the sheaths have disintegrated, the needles are evenly disseminated in the cryptocrystalline quartz matrix. A well defined transition between needles in well preserved sheaths and loose needles in the quartz matrix can be outlined. In parts of the thin section, dolomitic rhombohedra are dispersed within the quartz and show relatively well preserved sheath fragments with recognizable aragonite needles, although no sheaths are preserved in the surrounding quartz.

Later silicification in some parts of the sample (Schopf 1983) was only partly successful in preserving the microfossils. Recrystallization of the microsparitic anhedral dolomite to dolomite rhombohedra also affected parts of the fabric, following further disintegration of the sheaths. Subsequent silicification mineralized the sheaths in an even more advanced decompositional stage. In later stages, veinlets of calcite (Fig. 10a) penetrated the rock and recrystallization of the cryptocrystalline silica led to a total destruction of the cyanobacterial remains in most of the laminae of the rock. The sample exhibits in many places idiomorphic growth of cubic minerals, now replaced by quartz and preserved only as ghost structures (Fig. 10e).

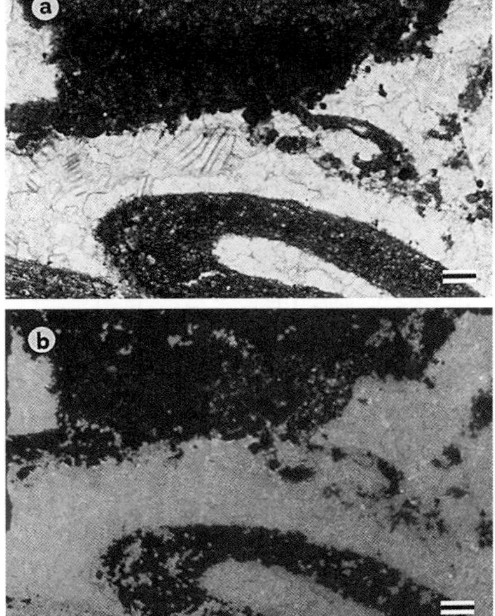

Fig. 8 Pyritized contorted laminae, under plane polarized light (**a**) and CL (**b**), sample BK 31. The presence of abundant authigenic pyrite is an indicator of active sulphate reduction associated with organic material in the early diagenetic environment. Calcite precipitation is known to be inhibited by the presence of sulphate but promoted by sulphate reduction. Scale bar = 200 μm.

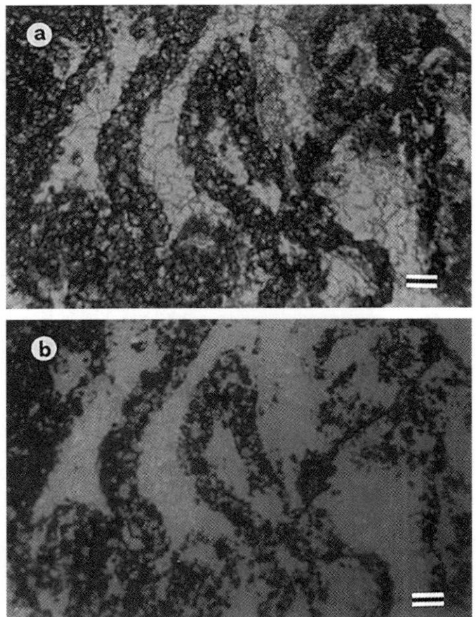

Fig. 9 Dolomitized contorted laminae, sample WA93/31. Partially dolomitized lamina bundles under plane polarized light (**a**) and CL (**b**). The close physical association between dolomite and poorly preserved cyanobacterial sheaths, contrasted with the absence of dolomite in the calcitic mineral matrix, strongly indicates that conditions favouring dolomite formation existed within these organic-rich laminae. The absence of pyrite suggests a location below the sulphate-reduction zone. Scale bars = 200 μm.

Mineralogy

Microprobe and X-ray diffraction analyses reveal that the main, yellow-luminescing fabric of the stromatolites and ooids is calcite, and that the darkly luminescing mineral is near-stoichiometric, but rather poorly ordered, dolomite. Chemical analyses of the darkly luminescing crystals attached to sheath material recorded a Mg: Ca ratio of *c.* 45: 55, whereas in coarser crystals, magnesium content increases to 48: 52. A trace of X-ray diffraction analysis (Fig. 11a,b,c) shows the major reflection peak for dolomite at 30.99° 2θ, corresponding to *d*(104) in all samples. The principal order reflection (015) at 35.3 2θ, together with subordinate dolomite peaks (110) and (113) appearing on the diffractometer trace in the 2θ region at 37.4° and 41.2°, show that in sample NB1 the dolomite is well ordered. However, in the finer-grained dolomites of sample 93/31d, the principal order reflection (015) at 35.3 2θ is absent, indicating that the dolomite is poorly ordered.

In the Kl-2 sample dolomite was analysed by a microprobe attached to a scanning electron microscope. CaO values range between 20 and 25% while MgO reaches up to 15% in the rhombohedra. FeO varies from undetectable to 12%. This is in accordance with the analyses by Klein *et al.* (1987), who found average values of CaO = 29.98%, FeO = 13.24%, MgO = 8.82% and MnO = 0.88% by weight.

Oolitic fabrics

The Boomplaas Formation occurs near the base of the Ghaap Group, above the siliciclastic basal Vryburg Formation, and comprises 100 m of cyclical carbonates, dated between 2641 Ma and 2555 Ma old, capped by either stromatolites or oolites on a carbonate ramp. Sample WA 95/35 is from an inversely graded oolite which directly overlies domal and domical stromatolites in a shallowing upward cycle, heralding an overall transgressive situation resulting in drowning of the Boomplaas carbonate platform and the deposition of the Lokammona Formation shales, with transported oolitic debris (normally

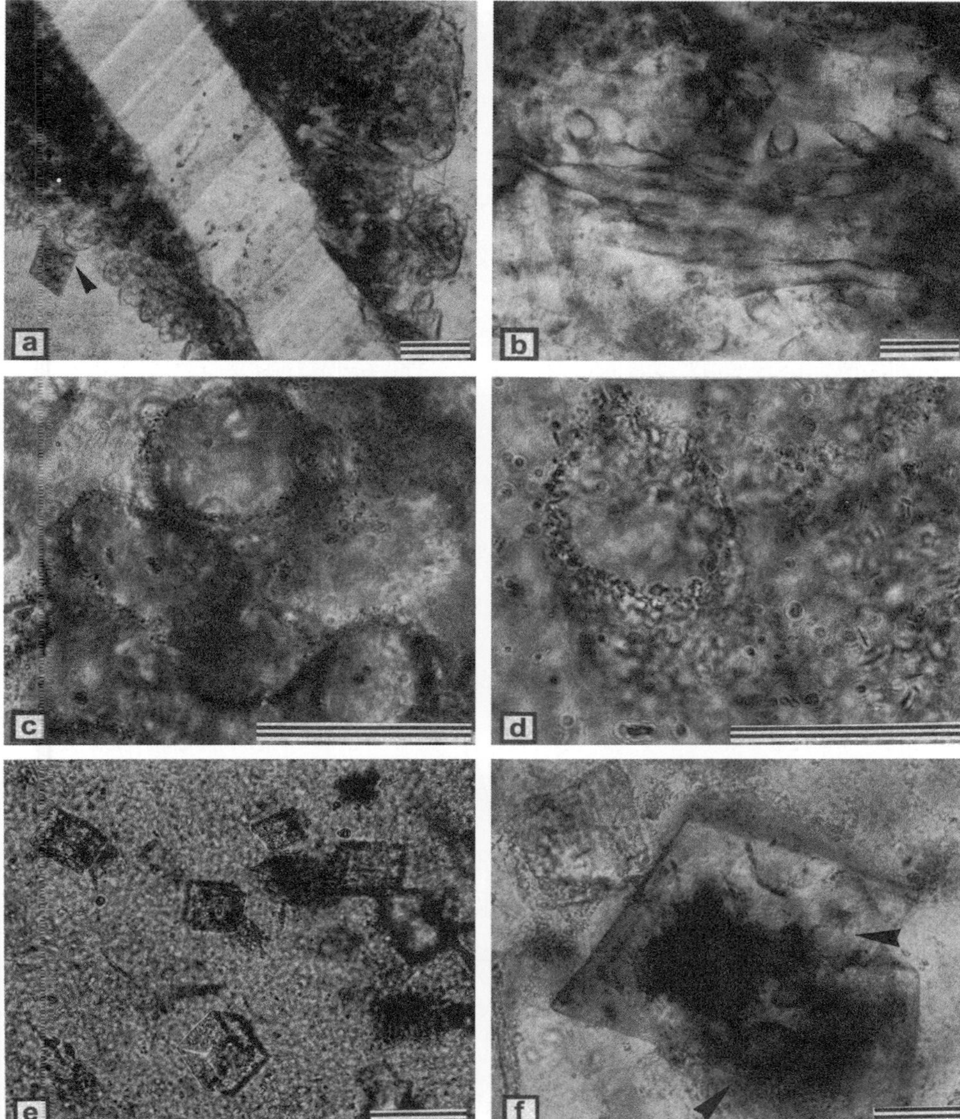

Fig. 10. Micrographs of the thin section KI-2 from the Tsineng Member. (**a**) A calcite vein cross-cutting dark dolomitic rhombohedra with preserved cyanobacterial sheaths, especially well visible in cross-section (circles) within the rhombohedron in the lower left corner. The silicified part of the section, surrounding the dolomite (light grey) has no preserved microfossils. Scale bar is 100 μm. (**b**) Excellent preservation of the *Siphonophycus transvaalensis* sheaths, in cross-sections (circular rings) and longitudinal sections, in microsparitic dolomite. Scale bar is 50 μm. (**c**), (**d**) Less well preserved sheaths, in cross-sections, embedded in cryptocrystalline quartz (scale bars are 25 μm). In (d) the needles of former aragonite within the sheath (on the left, circular cross-section and disseminated in the quartz matrix) are visible. The needles show high refractance and are thus difficult to focus on. (**e**) Cubic minerals replaced by quartz; the pseudomorphs are probably ghosts of evaporitic diagenetic minerals (halite?). Scale bar is 50 μm. (**f**) Preservation of cyanobacterial sheaths within later diagnetic dolomite rhombohedrons embedded in cryptocrystalline quartz barren of preserved microfossils. The sheath cross-sections are recognizable as dark to faint rings in the lower corner and in the right corner of the rhombohedron, or as a deformed ring between these two corners. Scale bar is 50 μm.

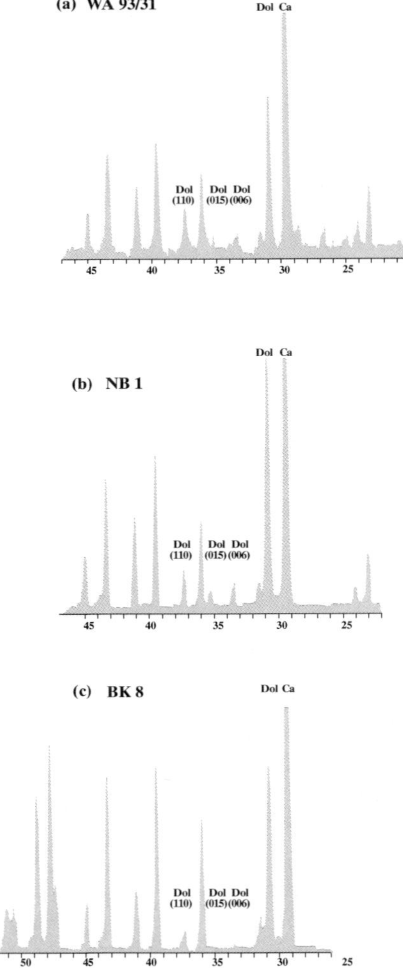

Fig. 11. X-ray diffractogram traces (<25° to >45° 2θ) of bulk powdered samples from (**a**) contorted laminae (sample WA 93/31), (**b**) a stromatolite from the Gamohaan Formation (sample NB1) and (**c**) giant ooids from the Boomplaas Formation (sample BK8). Three subordinate peaks are labelled (110), (015) and (006), and appear on the diffractometer trace in the 2θ region between 32° and 38°. Peak (015) is a principal order reflection, (006) is a lower order basal reflection which becomes attentuated with increasing Ca content, and (110) serves as a reference for noting changes in the intensity of order. The degree of ordering can be calculated by dividing the intensity of the (015) ordering peak by that of the diffraction peak of (110). The greater the ratio (i.e. the nearer to unity), the greater the degree of ordering (Goldsmith & Graf 1958). All samples comprise bimineralic carbonates. The traces demonstrate that all samples produced the principal order reflection (104), and are therefore true dolomites. Both (a) and (c) show subordinate ordering peaks. In the stromatolite sample, it is accompanied by subordinate (110) and (113) peaks, but the absence of the (015) peak at about 34.8 2θ indicates that the dolomite is poorly ordered.

graded and/or cross-bedded) and tuffites. Samples BK 6, 7 and 8 are from adjacent horizons in the Boomplaas Formation.

The giant ooids are up to 4 mm in diameter. Typically, a brown-coloured coat visible in hand specimen is seen in thin section to comprise a complex association of blocky, turbid dolospar and clearer calcite, with poor preservation of the nucleus. Mineralogy is clearly seen in cathodoluminescence – dolomite luminesces dark red-brown, contrasting with the bright yellow or orange of the calcite. Dolomicrite may be arranged in concentric laminae, but dolomite rhombohedra are often observed cross-cutting the laminae, or clustered in the area of the nucleus (Figs. 12a,b and 14a,b). The dolomite is intragranular, and is absent from the calcitic matrix cement (Fig. 13a,b). The ooids may show thin, calcitic, marine isopachous cement fringes (Fig. 14a,b), which typically enclose bimineralic cortical and nuclear fabrics of varying textures. Occasionally, fine, concentric cortical lamination is observed (Fig. 15a,b) and here, the primary structure of the cortex is clearly displayed by the preservation of closely packed, sheath-like concentric laminae *c.* 5 μm wide, resembling the sheaths of filamentous cyanobacteria. This unusually good preservation is confined to calcitic ooids, but where cross-cutting dolomite has destroyed the cortical fabric, relict areas of calcite may persist (Fig. 16a,b).

The original mineralogy of the ooids is difficult to ascertain. The concentric ooids generally preserve little original lamination, often showing cross-cutting and overprinting by dolospar but also calcite spar; however, there is a close relationship between fabric preservation and mineralogy. Generally, concentric lamination is not preserved, but is overprinted by blocky, crystalline dolospar. Larger crystals (≥150 μm) are generally inclusion-rich and brown coloured, indicating residual organic material. Dolomite is not found outside the shell of the ooids indicating that it must be related to internal diagenetic processes.

Taphonomic evolution of cyanobacterial sheaths and fabric development

In modern algal mats and stromatolites, which are usually dominated by cyanobacteria, the living mat community is typically just a few millimetres thick. In the absence of bioturbation, burial of dead material initiates anoxic conditions, and organic material is subjected to anaerobic microbial degradation, including sulphate reduction just below the sediment–water interface (Golubic 1976; Glenn & Arthur 1988). Biochemical degradation, together with burial and compaction, leads to the progressive

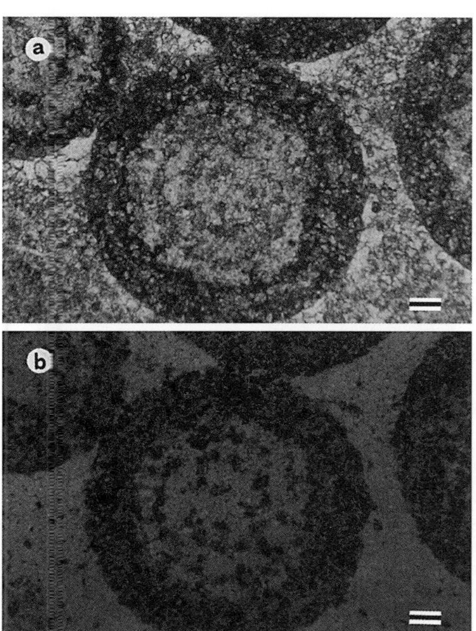

Fig. 12. Giant ooid, sample WA 95/35, in thin section under plane polarized light (**a**) and cathodoluminescence (**b**). The ooids typically comprise a complex association of blocky, turbid dolospar and clearer calcite, with degraded organic material. Mineralogy is clearly seen in cathodoluminescence: dolomite luminesces dark red-brown, contrasting with the orange of the calcite. Dolomicrite occurs in concentric laminae, with cross-cutting dolomite rhombohedra. Scale bars = 200 μm.

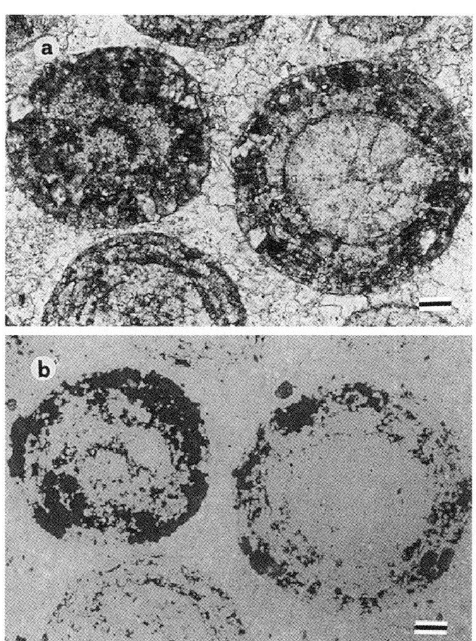

Fig. 13. Giant ooids, sample BK 7, in thin section under plane polarized light (**a**) and cathodoluminescence (**b**). The intragranular dolomite is absent from the calcitic matrix cement, strongly suggesting that conditions favouring dolomite formation existed within these organic-rich grains. Scale bars = 200 μm.

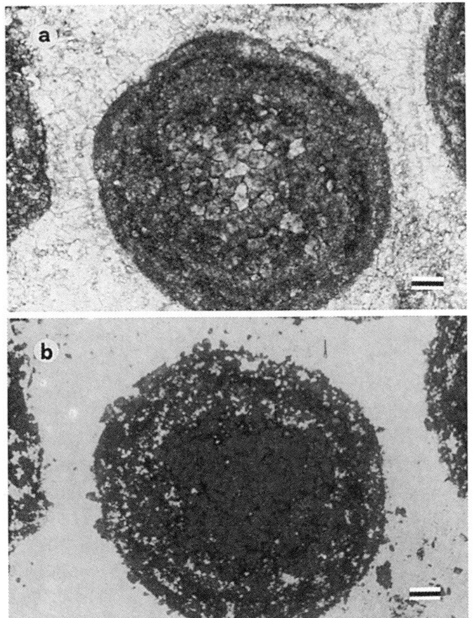

Fig. 14. Sample BK 8, in thin section under plane polarized light (**a**) and cathodoluminescence (**b**). The ooids possess thin, calcitic, marine isopachous cement fringes, which typically enclose bimineralic cortical and nuclear fabrics of varying textures. Relict concentric lamination is overprinted by dolomicrite, while dolospar rhombohedra are clustered in the nucleus. Intragranular dolomitization has progressed to a point where the original form of the ooid is barely recognizable. Scale bars = 200 μm.

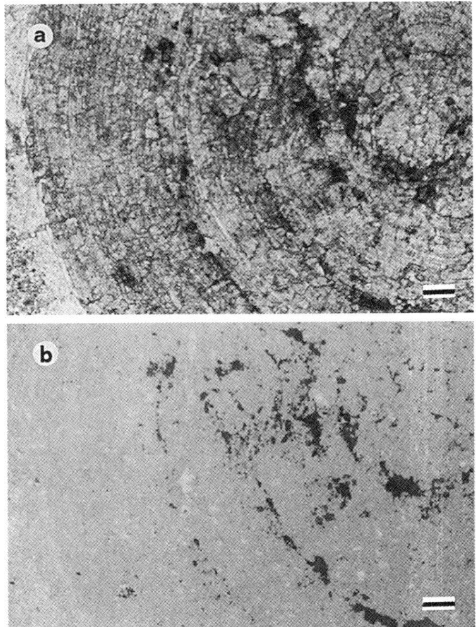

Fig. 15. Sample BK 6 in thin section under plane polarized light (**a**) and cathodoluminescence (**b**), showing fine, concentric, cortical lamination in the ooids; the primary structure of the calcitic cortex is clearly displayed by the preservation of closely packed, concentric laminae *c.* 5 μm wide. Significantly, dolomite is virtually absent, and where present either picks out the concentric laminae or forms small, dark clots. An interpretation by this model predicts that the dolomite indicates former coated surfaces and pockets of endolithic microbes, subjected to degradation by bacterial sulphate reduction. Scale bars = 100 μm.

destruction of organic fabrics: the cell contents are the first to be destroyed, but the polysaccharides of cell walls, mucilage and sheath material are much more resistant and, having a higher preservational potential, may persist longer at greater depths of burial and even survive lithification. Degradation usually results eventually in total loss of organic material. However, well documented Recent examples have shown that just below the growth surface of stromatolitic communities, calcium carbonate may permineralize cyanobacterial cells and sheaths during microbial degradation (e.g. calcification of *Entophysalis* colonies in hypersaline lakes in California, and *Scytonema* filaments in Andros Island in the Bahamas; Horodyski & Vonder Haar 1975; Monty 1976), thereby preserving their morphology. Crystals may be of magnesium calcite, suggesting that during precipitation, magnesium from sheath degradation may become incorporated in the crystal lattice (e.g. Gebelein & Hoffman 1973; Gerdes & Krumbein 1987). There is thus an intimate relationship between taphonomic evolution and fabric development.

In the grainstones, euendolithic boring would have created pore space in the ooids, and early, anoxic, organic degradation removed sulphate, with concomitant changes in pore-water chemistry.

The biochemistry of anoxic organic degradational processes, especially bacterial sulphate reduction, has profound effects on ambient waters (Wright 1997, 2000), and can generate conditions favourable to carbonate precipitation, including dolomite. Within the zone of sulphate reduction, organic matter is consumed, sulphate removed, and sulphide released into ambient waters with metabolic CO_2 (Berner 1980):

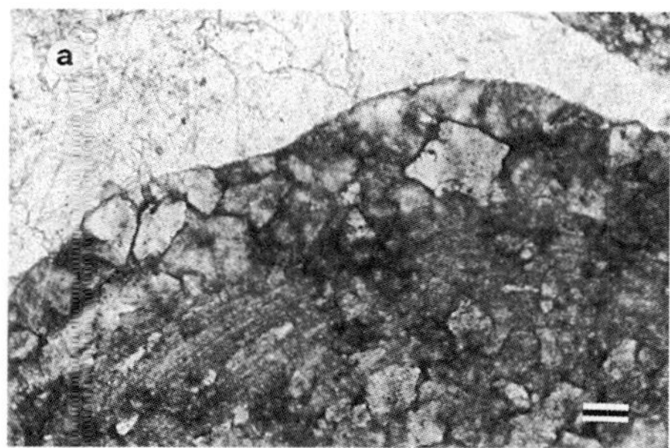

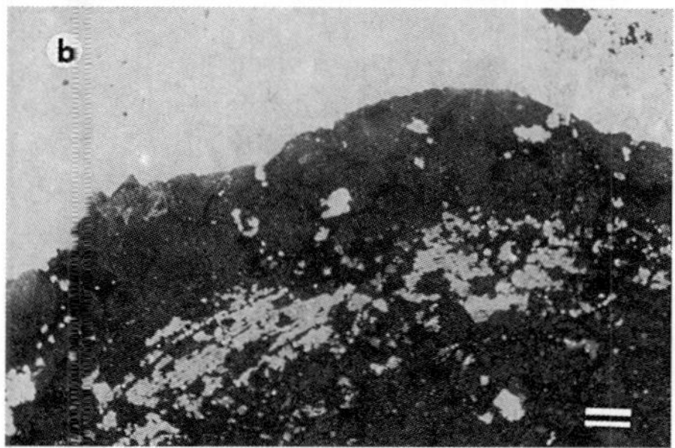

Fig. 16. Sample BK 7 in thin section under plane polarized light (**a**) and cathodoluminescence (**b**). Although cross-cutting intragranular dolomite has largely destroyed the cortical fabric, relict areas of calcite preserving early fabrics may persist, clearly seen in (b). Scale bars = 100 μm.

$$2CH_2O + SO_4^{2-} \rightarrow H_2S + 2HCO_3^-$$

The sulphide may combine with available iron to produce pyrite and/or pyrrhotite (e.g. Last & De Decker 1990), as illustrated by the pyrite-rich contorted and coiled laminae. Sulphate-reducing bacteria and other microbes oxidize cyanobacterial and other organic matter to support their metabolism, producing ammonia from enzymatic breakdown of proteins, which is rapidly absorbed by ambient waters, thereby increasing pH and carbonate alkalinity to levels necessary for dolomite formation (Berner 1980; Durand 1980; Slaughter & Hill 1991):

$$2NH_3 + CO_2 + H_2O \rightarrow 2NH_4^+ + CO_3^{2-}$$
$$NH_3 + H_2O \rightarrow NH_4^+ + OH^-$$
$$OH^- + HCO_3^- \rightarrow H_2O + CO_3^{2-}$$

Sulphate reduction may also lead to the release of free Mg^{2+} ions (Slaughter & Hill 1991):

$$2CH_2O + MgSO_4^0 \rightarrow 2HCO_3^- + H_2S + Mg^{2+}$$

Bacterial sulphate reduction in association with cyanobacterial degradation can thus create conditions favourable to dolomite precipitation, by increasing the activity of carbonate and magnesium ions and removing sulphate.

The question now is: do these taphonomic and biochemical responses to organic degradation dominated by anoxic bacterial sulphate reduction provide a realistic analogue for Archaean counterparts? The answer to this question depends largely on whether sulphate was present in appreciable quantities in Archaean sea water.

Sulphate concentrations in Archaean sea water and the antiquity of sulphate-reducing bacteria

A variety of models have been developed for Archaean sea-water composition. Walker (1983) and Holland (1984) proposed that sea-water chemistry has been more or less constant throughout most of geological time, though with lower oxygen and higher Fe/Mn concentrations in Archaean oceans. Sumner & Grotzinger (1996) have argued that the kinetics of calcite precipitation in Archaean sea water may have been inhibited by high concentrations of Fe^{2+} in an oxygen-poor ocean. In contrast, Kempe and Degens (1985) proposed a 'soda ocean' model, with high bicarbonate concentrations and a pH of 9–11. However, it is worth noting that at such a high pH, the basic form (CO_3^{2-}) will predominate; moreover, the more saline the solution, the higher the concentration of CO_3^{2-} relative to HCO_3^- at a fixed pH (Pytkowicz 1983).

In modern sediments, anoxic bacterial sulphate reduction, in addition to enhancing carbonate precipitation, leads to the fractionation of sulphur isotopes ^{32}S and ^{34}S, often to an extreme extent (e.g. Chambers & Trudinger 1979), with residual sulphate in ambient waters being enriched in ^{34}S because it is kinetically easier to break bonds between lighter atoms than between heavier ones. Bacterial sulphate reduction gives off hydrogen sulphide as a byproduct, which in the presence of iron (abundant in Archaean oceans) leads to the precipitation of iron sulphide as pyrite enriched in ^{32}S with respect to estimated values in contemporaneous marine sulphate. However, the rarity of sedimentary sulphate deposits during the Archaean hampers an accurate assessment of sulphate–sulphide isotopic compositions. Variable environmental factors such as temperature, organic substrate and low sulphate concentrations can produce similar isotopic signatures to sulphides derived from volcanic and hydrothermal sources, although abiological sulphate reduction under normal sedimentary conditions has not so far been experimentally achieved (Trudinger 1992). Evidence of isotopic discrimination, and therefore for the possible presence of sulphate-reducing bacteria, is found in a number of Archaean sediments (see below for examples), and there is biochemical evidence for suspecting that a sulphate reduction pathway existed from the earliest stages of microbial evolution (Trudinger 1992). A biogenic origin for pyrite must therefore be assessed using all relevant evidence and criteria.

A major shift in the sulphur isotope record around 2.5 Ga BP was interpreted by Lambert & Donelly (1992) to represent a major increase in sulphate levels at about this time, while Schidlowski (1987) argued that bacterial sulphate reduction became significant around 2.8 Ga BP. Evidence for significant sulphate concentrations in late Archaean sea water is provided by Kakegawa *et al.* (1998), who report pyrite with $\delta^{34}S$ –6.3 to +7.1 interpreted to be formed by bacterial sulphate reduction of sea-water sulphate in the 2.5 Ga Mount McRae Shale of the Hamersley Basin, Western Australia. They go on to propose that the data indicate that sulphate concentration in the 2.5 Ga BP sea water was around one-third of its present value in the modern ocean, and that the activity of sulphate-reducing bacteria was generally higher. Eastoe *et al.* (1990), in studying the isotopic signatures of Early Proterozoic volcanogenic sulphides, proposed that marine sulphate was confined to an upper water layer, a conclusion consistent with the comparatively high activities of sulphate-reducing bacteria reported by Kakegawa *et al.* (1998).

In the studied samples from the Gamohaan and Nauga Formations, the intimate association of disseminated and nodular pyrite with degraded organic material indicates a genetic association, and the absence of evidence for volcanic activity in adjacent strata further supports a biogenic origin. Direct support comes from Strauss & Beukes (1996) who recorded sulphur isotope values of $\delta^{34}S$ −1.3 to +23.6 and carbon isotope values of $\delta^{13}C$ −31 to −43 in these minimally altered, organic-rich samples from the Transvaal Supergroup (including the Campbellrand Subgroup). A summary of the available evidence thus suggests that marine sulphate concentrations, at least in shallow water, were adequate to sustain widespread bacterial sulphate reduction in microbial communities of the platform carbonates of the Late Archaean Transvaal Group, and therefore that the taphonomic evolution of modern benthic microbial biofabrics and associated biochemical mediation of ambient waters provide a process analogue for their Archaean counterparts.

Discussion

Early marine cements are volumetrically important, and constitute up to 80% of some microbialite structures. However, it is clear from the data presented above that organic diagenesis exerted a strong control on both fabric development and mineralogy in the microbial carbonates of the Campbellrand platform. Moreover, different stages in the taphonomic sequence may be preserved according to the timing of mineralization. Microbial degradation processes influence primary carbonate fabrics, and these may be preserved where dolomitization has not occurred: preservational quality is proportional to the degree of dolomitization. Significantly, where early dolomicrite has not been subject to neomorphic changes, preservation of microbial sheaths might be very good, and to our knowledge the Klein *et al.* (1987) example is the only reported case of exceptional preservation of identifiable Archean microfossils in carbonate.

In the contorted and coiled laminae, the flexibility of the cyanobacterial strands shows that they were unmineralized prior to and during deformation of the mat. However, the preservation of voids between uncompressed, slump-generated folds and hinges of the strands shows that cementation must have occurred immediately after slump-folding but prior to compaction, and presumably fast enough to prevent collapse and amalgamation of separate strands. During this time, the loss of structural detail indicates that organic degradation had taken place. The presence of abundant authigenic pyrite is an indicator of active sulphate reduction associated with organic material in the early, diagenetic environment. Calcite precipitation is known to be inhibited by the presence of sulphate but promoted by sulphate reduction (e.g. Walter 1986), and so sites near to intensive anoxic organic degradation would be favourable to calcite precipitation. Bubbles of gas, including carbon dioxide and methane, expelled during organic diagenesis may have been in part responsible for deforming the laminae, thus forming the sites of preferred precipitation.

The origin of dolomite in all the cases discussed appears to be linked in the stromatolites to the degradation of cyanobacterial sheaths. Organic diagenesis is also implicated in dolomite formation in the ooids, and in both cases, the luminescent carbonate reveals a reducing environment during early diagenesis. Anoxic microbial degradation by bacterial sulphate reduction has been shown to be actively involved in dolomite formation on a large scale by removing kinetic inhibitors that normally operate in the marine environment (Wright 1997, 1999). Recent experiments have also demonstrated the involvement of sulphate-reducing bacteria in dolomite formation (Vasconcelos & McKenzie 1997).

The Gamohaan and Nauga stromatolites and microbialites display textural and mineralogical features consistent with *in situ* dolomitization of calcium carbonate in association with microbial degradation of cyanobacterial matter. The variable structural detail of the sheaths indicates their differential degradation prior to stabilization of the fabric, but the general absence of pyrite here indicates that sulphate was no longer present in the interstital waters, having been consumed by sulphate-reducing bacteria, and that degradation was by other microbes, probably methanogens, which occur beneath sulphate-reducing bacteria in typical depth-stratified communities (e.g. Talbot & Kelts 1986). The style of encrustation of the sheaths by dolomicrite appears to be analogous to the calcification associated with partially degraded cells in modern cyanobacterial mats and stromatolites where mineralization forms moulds of felted groups of sheaths, similar to those seen in the Gamohaan stromatolites. The close physical association between early-formed dolomite and less well preserved cyanobacterial sheaths, contrasted with the absence of dolomite in the calcitic mineral matrix, strongly indicates that conditions favouring dolomite formation existed within these organic-rich laminae. Support for this hypothesis is provided by Gebelein & Hoffman (1973) who suggested from the known ability of cyanobacteria to concentrate magnesium in their sheaths, and from calculations based on laboratory experiments, that the slow release of magnesium from the long-term degradation of the sheaths could produce dolomitic laminae 1 mm thick from cyanobacterial layers 2 mm thick.

Although controversy persists over the origin of ooids, the close association of ooids and endolithic and epilithic microbes, including cyanobacteria, has led to the suggestion that degradation of concentric filamentous cyanobacteria produces organic membranes which form the substrate for the growth of crystals and mineral laminae constituting the oolitic envelope. Wright (1997) has proposed that degradation of intra-ooidal microbial material can result in the replacement of calcium carbonate by dolomite. The relationships between preserved microstructure and mineralogy allow the interpretation of early ooid microfabrics and mineralogy. Cross-cutting of concentric laminae by blocky dolospar indicates progressive diagenetic loss of early concentric lamination.

The uniform luminescence of calcite in both the cement matrix and the fabric of the ooids, whether this is preserved as fine concentric laminae or as coarse spar, suggests that these fabrics are largely unaltered, and were precipitated at the same time in the same conditions, preceding the development of dolomite. In the ooid grainstones, the restriction of dolomite to intragranular pore space (dolomite is not found in the matrix cement), and the uniform luminescence of the calcite in the ooids, indicate that the necessary conditions were generated within the allochems during early diagenesis, and were not related to later permeating or circulating dolomitizing fluids. The dolomite formed, within the ooids and was confined to them. The presence of inclusions and the brown coloration indicate a close association with intragranular organic material. Euendolithic boring and anoxic organic degradation would have created pore space in the ooids, removed sulphate from porewaters and released magnesium, with concomitant changes in pore-water chemistry favourable to dolomite formation.

The original mineralogy may have been aragonite, calcite or high-Mg calcite. Aragonite is unstable at normal temperatures and pressures, is susceptible to leaching by meteoric pore waters and to inversion by neomorphism to calcite spar (e.g. Bathurst 1975). There is some evidence for oomoulds, now filled with calcite spar (e.g. Fig. 13), while the presence of relict, concentric inclusions indicating the former structure also suggest that Mg-calcite/aragonite may have been the original mineralogy. However, while the evidence is not conclusive, organic degradation is progressive, and may have first promoted calcite replacement of Mg-calcite/aragonite prior to total degradation of intragranular cyanobacterial sheaths, because relict sheaths are occasionally preserved in calcite (Fig. 16). Dolomite was probably formed during sheath degradation, and thus post-dated the calcite.

Implications for controls on carbonate production through time

There is general consensus that microbial ecosystems dominated shelf environments from Archaean through Proterozoic times. Since the Phanerozoic, metazoans have successfully colonized the majority of available environments, and carbonate-secreting organisms have dominated marine carbonate production. This leads us inevitably to a discussion of the role and scale of microbial processes in diagenesis and carbonate production through time, and of carbonate precipitation itself: there is a stark dichotomy between overwhelmingly biogenic carbonate precipitation throughout the Phanerozoic, and unresolved processes of carbonate precipitation in the Precambrian. In other words, despite supersaturation of surface sea water with calcite, aragonite and dolomite, inorganic precipitation of these phases from the water column does not normally follow (e.g. Leeder 1982). Although precipitation is favoured thermodynamically, kinetic inhibitors operating at the molecular scale prevent carbonate precipitation in normal sea water. No microbes have been found to be 'obligate calcifiers' (Riding 1982). So we should ask: in the absence of metazoans, how can the kinetic inhibitors to marine carbonate production be overcome?

It has been shown that sulphate ions inhibit calcite as well as dolomite precipitation (Walter 1986; Slaughter and Hill 1991), that Mg^{2+} ions inhibit calcite precipitation (Berner 1975; Berner *et al.* 1978) and that some humic compounds and phosphate inhibit aragonite precipitation (Berner *et al.* 1978). Moreover, the concentration of CO^{3-} ions in sea water is very low, and their activity, due to complexing with cations, is even lower. Importantly, Berner *et al.* (1978) have shown that changing pCO_2 for fixed degrees of supersaturation has negligible effect upon calcite or aragonite precipitation. Yet aragonite (now pseudomorphed), calcite and dolomite are all found in Archaean and Proterozoic marine rocks. It is clear that in order to promote precipitation, kinetic inhibitors must be overcome.

One effective way to overcome the kinetic inhibitors is to raise carbonate alkalinity: in Precambrian microbially dominated shelf ecosystems, the potential existed for sulphate-reducing bacteria to do this on a large scale, and at the same time remove inhibitor SO^{4-} ions. Examples of modern environments in which benthic microbial communities dominate are rare, but can provide a natural laboratory for testing the validity of the model proposed here (e.g. Wright 1999).

Conclusions

The microfabrics of microbialitic sediments, including ooids, of the Campbellrand and Schmidstdrif Subgroups were largely controlled by anoxic organic diagenesis operating in a reducing geochemical environment. It is argued here that anoxic microbial degradational processes, by modifying ambient water chemistry, created the geochemical conditions necessary for carbonate precipitation, and that the degree and type of organic degradation was a major control on carbonate mineralogy. Where bacterial sulphate reduction was active, iron (as pyrite) was removed and calcite was precipitated, whereas below the sulphate reduction zone, dolomite formation was favoured.

Microbial ecosystems dominated the depositional environments of Archaean, Proterozoic and some Phanerozoic carbonate platforms, and taphonomic processes associated with sulphate reduction and organic diagenesis were widespread. If our analysis is correct, anoxic microbial diagenesis acted not only as the 'engine house' of carbonate production, but also provides a 'process analogue' for the generation of cements and dolomitization wherever such conditions prevailed.

We thank B. Schopf and CSEOL/PPRG for making the Kl-2 sample available for microscopy and SAMANCOR Ltd, South Africa, for granting access to their core magazine and for sampling permission. We also thank B. Kamber for providing samples.

References

ALTERMANN, W. & HERBIG, H. G. 1991. Tidal flats deposits of the Lower Proterozoic Campbell Group along the southwestern margin of the Kaapvaal Craton, Northern Cape Province, South Africa. *Journal of African Earth Science*, **13**, 415–435.

—— & NELSON, D. R. 1998. Sedimentation rates, basin analysis and regional correlations of three Neoarchean and Palaeoproterozoic sub-basins of the Kaapvaal craton as inferred from precise U-Pb zircon ages from volcaniclastic sediments. *Sedimentary Geology*, **120**, 225–256.

—— & SCHOPF, J. W. 1995. Microfossils from the Neoarchean Campbell Group, Griqualand West Sequence of the Transvaal Supergroup, and their paleoenvironmental and evolutionary implications. *Precambrian Research*, **75**, 65–90.

—— & SIEGFRIED, H. P. 1997. Sedimentology and facies development of an Archean shelf–carbonate platform transition in the Kaapvaal Craton, as deduced from a deep borehole at Kathu, South Africa. *Journal of African Earth Science*, **24**, 391–410.

—— & Wotherspoon, J. McD. 1995. The carbonates of the Transvaal and Griqualand West Sequences of the Kaapvaal craton, with special reference to the Lime Acres limestone deposit. *Mineralogica Deposita*, **30**, 124–134.

BATHURST, R. G. C., 1975. *Carbonate Sediments and their Diagenesis*. Amsterdam, Elsevier.

BERNER, R. A. 1975. The role of magnesium in the crystal growth of calcite and aragonite from seawater. *Geochimica et Cosmochimica Acta*, **39**, 489–504.

—— 1980. *Early Diagenesis: a Theoretical Approach*. Princeton University, Princeton.

—— WESTRICH, J. T., GRABER, J., SMITH, J. & MARTENS, C. S. 1978. Inhibition of aragonite precipitation from supersaturated seawater: a laboratory and field study. *American Journal of Science*, **278**, 816–837.

BEUKES, N. J. 1979. Litostratigrafiese onderverdeling van die Schmidtsdrif-Subgroep van die Ghaap-Groep in Noord-Kaapland. *Transactions of the Geological Society of South Africa*, **82**, 313–327.

—— 1980. Stratigrafie en litofasies van die Campbellrand-Subgroep van die Proterofitiese Ghaap-Groep, Noord-Kaapland. *Transactions of the Geological Society of South Africa*, **83**, 141–170.

—— 1983. Paleoenvironmental setting of iron formations in the depositional basin of the Transvaal Supergroup, South Africa. *In*: TRENDALL, A. F. & MORRIS, R. C. (eds) *Iron Formation: Facts and Problems*. Developments in Precambrian Geology, **6**, Elsevier, Amsterdam, 131–210.

—— 1987. Facies relations, depositional environments and diagenesis in a major Early Proterozoic stromatolitic carbonate platform to basinal sequence, Campbellrand Subgroup, Transvaal Supergroup, Southern Africa. *Sedimentary Geology*, **54**, 1–46.

CHAMBERS, L. A. & TRUDINGER, P. A. 1979. Microbial fractionation of stable sulfur isotopes: a review and critique. *Geomicrobiology Journal*, **1**, 249–293.

DAVIES, P. J., BUBELA, B. & FERGUSON, J. 1978. The formation of ooids. *Sedimentology*, **25**, 703–730.

DURAND, B. 1980. Sedimentary organic matter and kerogen. Definition and quantitative importance of kerogen. *In*: Durand B. (ed.) *Kerogen*. Technip, Paris, **27**, 13–34.

EASTOE, C. J., GUSTIN, M. S., HURLBUT, D. F. & ORR, R. L. 1990. Sulfur isotopes in Early Proterozoic volcanogenic massive sulfide deposits: new data from Arizona and implications for ocean chemistry. *Precambrian Research*, **46**, 353–364.

GEBELEIN, C. D. & HOFFMAN, P. 1973. Algal origin of dolomite laminations in stromatolitic limestone. *Journal of Sedimentary Petrology*, **43**, 603–613.

GERDES, G. & KRUMBEIN, W. E. 1987. *Biolaminated Deposits*. Lecture Notes in Earth Science, **9**, Springer, Berlin.

GLENN, C. R. & ARTHUR, M. A. 1988. Petrology and major element geochemistry of Peru margin phosphorites and associated minerals: authigenesis in modern organic-rich sediments. *Marine Geology*, **80**, 231–267.

GOLDSMITH, J. R. & GRAF, D. L. 1958. Structural and compositional variations in some natural dolomites. *Journal of Geology*, **66**, 678–693.

GOLUBIC, S. 1976. Organisms that build stromatolites. *In*: WALTER, M. R. (ed.) *Stromatolites*. Developments in Sedimentology, **20**, Elsevier, Amsterdam, 113–126.

GREEN, J. W., KNOLL, A. H. & SWETT, K. 1988. Microfossils from oolites and pisolites of the Upper Proterozoic Eleonore Bay Group, central East Greenland. *Journal of Paleontology*, **62**, 835–852.

HALBICH, I. W., LAMPRECHT, D. F., ALTERMANN, W. & HORSTMANN, U. E. 1992. A carbonate–banded iron formation transition in the early Proterozoikum of South Africa. *Journal of African Earth Science*, **15**, 217–236.

HEYDARI, E. & MOORE, C. H. 1991. *A paleoceanographic and paleoclimatic model for the formation of bimineralic ooids of the Smackover Formation in the Mississippi Salt Basin and implications for global change in calcite-aragonite ooid formation*. Geological Society of America, Abstracts with Programs, **23**.

HOFMANN, H. J. & MASSON, M., 1994. Archean stromatolites from Abitibi greenstone belt, Quebec, Canada. *Bulletin of the Geological Society of America*, **106,** 424–429.

HOLLAND, H. D. 1984. *The Chemical Evolution of the Atmosphere and Oceans*. Princeton, New Jersey, Princeton University Press.

HORODYSKI, R. J. & VONDER HAAR, S. P. 1975. Recent calcareous stromatolites from Laguna Mormona (Baja California), Mexico. *Journal of Sedimentary Petrology*, **45**, 680–696.

KAKEGAWA, T., KAWAI, H. & OHMOTO, H. 1998. Origins of pyrite in the ~2.5 Ga Mt. McRae Shale, the Hamersley District, Western Australia. *Geochimica et Cosmochimica Acta*, **62**, 3205–3220.

KEMPE, S. & DEGENS, E. T. 1985. An early soda ocean? *Chemical Geology*, **53**, 95–108.

KLEIN, C. & BEUKES, N. J. 1989. Geochemistry and sedimentology of a facies transition from limestone to iron-formation in the early Proterozoic Transvaal Supergroup, South Africa. *Economic Geology*, **84**, 1733–1774.

——, —— & SCHOPF, J. W. 1987. Filamentous microfossils in the early Proterozoic Transvaal Supergroup: their morphology, significance and palaeoenvironmental setting. *Precambrian Research*, **36**, 81–94.

LAMBERT, I. B. & DONNELLY, T. H. 1992. Global oxidation and a supercontinent in the Proterozoic: evidence from stable isotopic trends. *In*: SCHIDLOWSKI, M. *ET AL* (eds) *Early Organic Evolution: Implications for Mineral and Energy Resources*. Springer, Berlin, 408–414.

LAST, W. M. & DE DECKER, P. 1990. Modern and Holocene carbonate sedimentology of two saline volcanic maar lakes, southern Australia. *Sedimentology*, **37**, 967–981.

LEEDER, M. R., 1982, *Sedimentology: Process and Product*. Allen & Unwin.

MONTY, C. L. V. 1976. The origin and development of cryptalgal fabrics. *In*: WALTER, M. R. (ed.) *Stromatolites*. Developments in Sedimentology, **20**, Elsevier, Amsterdam, 193–249.

PYTKOWICZ, R. M. 1983. *Equilibria, Nonequilibria and Natural Waters*. Wiley, New York.

RIDING, R. 1982. Cyanophyte calcification and changes in ocean chemistry. *Nature*, **299**, 814–815.

SCHIDLOWSKI, M. 1987. Evolution of the early sulphur cycle. *In*: RODRIGUEZ-CLEMENTE, R. & TARDY, Y.

(eds) *Geochemistry and Mineral Formation in the Earth Surface*. Consejo Superior de Investigaciones Cientificas, Madrid, 29–49.

SCHOPF, J. W. (ed.) 1983. *Earth's Earliest Biosphere, its Origin and Evolution*. Princeton University, Princeton.

SLAUGHTER, M. & HILL, R. J. 1991. The influence of organic matter in organogenic dolomitization. *Journal of Sedimentary Petrology*, **61**, 296–303.

STRAUSS, H. & BEUKES, N. J. 1996. Carbon and sulfur isotopic compositions of organic carbon and pyrite in sediments from the Transvaal Supergroup, South Africa. *Precambrian Research*, **79**, 57–71.

SUESS, E. & FUTTERER, D. 1972. Aragonitic ooids: experimental precipitation from seawater in the presence of humic acid. *Sedimentology*, **19**, 129–139.

SUMNER, D. Y. 1997. Late Archean calcite–microbe interactions: two morphologically distinct microbial communities that affected calcite nucleation differently. *Palaios*, **12**, 302–318.

—— & GROTZINGER, J. P. 1996. Were kinetics of Archean calcium carbonate precipitation related to oxygen concentration? *Geology*, **24**, 119–122.

TALBOT, M. R. & KELTS, K. 1986. Primary and diagenetic carbonates in the anoxic sediments of Lake Bosumtwi, Ghana. *Geology*, **14**, 912–916.

THOMPSON, J. B. & HUBER, J. A. 1997. *Geological Society of America, 1997 Annual Meeting, Abstracts with Programs*. Geological Society of America, **29**, Boulder.

TRUDINGER, P. A. 1992, Bacterial sulfate reduction: current status and possible origin. *In*: SCHIDLOWSKI, M. *et al.* (eds) *Early Organic Evolution: Implications for Mineral and Energy Resources*. Springer, Berlin, 367–377.

VASCONCELOS, C. & MCKENZIE, J. A. 1997. Microbial mediation of modern dolomite precipitation and diagenesis under anoxic conditions (Lagoa Vermelha, Rio de Janeiro, Brazil). *Journal of Sedimentary Petrology*, **67**, 378–390.

WALKER, J. C. G. 1983. Possible limits on the composition of the Archean ocean. *Nature*, **302**, 518–520.

WALRAVEN, F. & MARTINI, J. 1995. Zircon Pb evaporation age determinations of the Oak tree Formation, Chuniesport Group, Transvaal Sequence: implications for Transvaal-Griqualand West basin correlations. *South African Journal of Geology*, **98**, 58–67.

——, BEUKES, N. J. & RETIEF, E. A. (in press) 2.64 Ga zircons from the Vryburg Formation, Griqualand West: implications for the age of the base of the Transvaal Supergroup and accumulation rate of the carbonate–iron formation succession. *Precambrian Research*.

WALTER, L. M. 1986. Relative efficiency of carbonate dissolution and precipitation during diagenesis: a progress report on the role of solution chemistry. *In*: GAUTIER, D. L. (ed.) *Roles of Organic Matter in Sediment Diagenesis*. SEPM Special Publication No. **38**, 1–11.

WRIGHT, D. T. 1997. An organogenic origin for widespread dolomite in the Cambrian Eilean Dubh Formation, north western Scotland. *Journal of Sedimentary Research*, **67**, 54–64.

—— 1999. A microbial role for dolomite formation in distal ephemeral lakes of the Coorong region, South Australia. *Microbial Mediation in Carbonate Diagenesis* special issue of *Sedimentary Geology*, **126**, 147–157.

—— 2000. Benthic microbial communities and dolomite formation in marine and lacustrine environments – a new dolomite model. *In*: GLENN, C. R., LUCAS J. & PREVOT-LUCAS, L. (eds) *Marine Authigenesis: From Global to Microbial*. SEPM, Special Publication No. **66**, 7–20.

Reefs and coral carpets in the northern Red Sea as models for organism–environment feedback in coral communities and its reflection in growth fabrics

BERNHARD RIEGL & WERNER E. PILLER

Institut für Geologie und Paläontologie, Karl-Franzens-Universität Graz, Heinrichstrasse 26, 8010 Graz, Austria (e-mail: Bernhard.Riegl@kfunigraz.ac.at; Werner.Piller@kfunigraz.ac.at)

Abstract: Coral framework construction and resultant growth fabrics in response to environmental factors were studied in the northern Red Sea, and the Gulfs of Suez and Aqaba. The dependence of growth fabric types on sea-floor topography, oceanography and the ecology of constituent coral species was investigated. Five types of coral frameworks and their growth fabrics were differentiated: *Acropora* reef framework (platestone to mixstone facies); *Porites* reef framework (domestone facies); *Porites* carpet (columnar pillarstone facies); faviid carpet (mixstone facies); *Stylophora* carpet (thin pillarstone facies). Two non-framework community types were found: *Stylophora–Acropora* community and soft coral communities. Reef frameworks and resultant growth fabrics show a clear ecological zonation along depth and hydrodynamic exposure gradients. Coral carpets build a framework lacking a distinct internal zonation since they only grow in areas without pronounced gradients. In the northern Red Sea they show a gradual change with depth from *Porites* (pillarstone) to faviid (mixstone) dominance.

The initiation of frameworks was governed by bottom topography (reefs on steep slopes and highs, coral carpets in flat areas). According to environmental conditions, different coral communities produce different framework and growth fabric types. In step with framework growth the environment is modified. The modified environment in turn modifies the coral communities. Thus an environment–organism–environment feedback loop exists.

Coral reefs, both fossil and modern, have long been recognized as systems of intricate environment–organism interaction (Rosen 1975; Longman 1981; Frost 1981; Hopley 1982; Done 1982, 1983; Perrin *et al.* 1995; Riegl & Piller 1997; Wood 1999) that are largely brought about by the corals' ability to build a solid reef structure (Insalaco 1998). The ecological structure of the reef and its imprint in the geological record is strongly dependent on a combination of environment, species availability and ecology, both for initiation and as shaping factors during its growth (Longman 1981; Leinfelder 1997; Guozhong 1998). Once a solid reef structure is established, this in turn modifies the environment. As the reef structure grows and modifies its own environment, constituent coral communities and accretion rates change (Montaggioni & Faure 1997; Smith *et al.* 1998) creating an organism–environment feedback.

The composition of coral communities, which in the fossil record is recorded by its growth fabric (Insalaco 1998), is influenced not only by physico-chemical factors in the water column, but also by basin geometry and bottom topography which govern the availability of space for the development of reefs or non-reef building communities (Hopley 1982; Done 1982; Kleypas 1996; van Woesik & Done 1997; Insalaco 1998).

In the northern Red Sea, coral reef development follows mainly tectonically generated topographic highs and the mostly steep continental margin (Strasser *et al.* 1992; Gvirtzman 1994; Piller & Pervesler 1989). However, in shallow shelf areas, extensive framework-building coral communities exist in addition to reefs (Piller & Pervesler 1989; Riegl & Piller 1997, 1999). We call these communities 'coral carpets' in accordance with Reiss & Hottinger (1984). The systems 'reef' and 'coral carpet' differ both in their ecological and frame-building response to environmental factors, as well as in their influence on the environment and their representation in the geological record (Riegl & Piller 1999). However, they should not be seen as mutually exclusive systems but rather as different stages into which frame-building coral communities can develop according to environmental constraints.

In this paper we provide a model for how organismic response to the physical environment translates into different types of reef and non-reef frameworks which in turn change their

From: INSALACO, E., SKELTON, P. W. & PALMER, T. J. (eds) 2000. *Carbonate Platform Systems: components and interactions*. Geological Society, London, Special Publications, **178**, 71–88. 0305–8719/00/$15.00

environments. We examine: (1) the different coral framework types in the northern Red Sea; (2) the ability of coral carpets to build frameworks (3) the expected lithological representation in the fossil record; (4) the interaction of the environment with the frame-building coral communities; (5) the feedback of the organisms to the environment via different frame-building capacity.

Material and methods

Study area

Coastal and offshore sites, representing most coral habitats available in the northern Red Sea, were investigated in the Gulfs of Aqaba and Suez as well as the Egyptian Red Sea (Fig. 1). The quantitative sampling sites were located in the Straits of Gubal and the Hurghada area. Additionally, qualitative observations were made in the Gulf of Aqaba north to Eilat in Israel, to Ain Sukhna in the Gulf of Suez and in the main Red Sea basin south to Ras Banas (Egypt).

Geomorphological features in the northern Red Sea region are mainly controlled by tectonism and salt diapirism (Dullo & Montaggioni 1998). Since the Oligocene, the sedimentary evolution of the Red Sea basin has been tectonically controlled, as evidenced by the orientation of major fault structures and the orientation and shape of the reefs which frequently follow and are determined by such structures. Salt diapirism, which again frequently follows the major tectonic lineaments, is another factor creating highs suitable for extensive coral settlement (Orszag-Sperber *et al.* 1998). Purser *et al.* (1998) showed how the tilting of fault blocks influenced carbonate and siliciclastic sedimentation, where fault-line depressions funnel silicilastics while carbonates develop on top of, or on the seaward sides of, structural highs (mostly tilt blocks or diapiric structures). These processes also provide the structural diversification into highs with reefs and moderately deep (<50 m) shelves settled by coral carpets as discussed in this paper.

Meteorologically, the region enjoys stable conditions which result in stable oceanographic conditions over most of the year. Dominant wind and swell direction for about 80% of the year is from the northwest with an average speed of 10 knots (Roberts 1985). In winter, eastward-travelling depressions can cause changes in wind direction to SE or E (Edwards 1987). It is therefore possible to talk about mostly exposed (windward, N-facing) and mostly sheltered (leeward, S-facing) reef sides (see windrose on Fig. 1).

Terminology

For the purpose of this study, coral carpets were defined as laterally more or less continuous veneers of coral framework following the existing sea-floor morphology. They do not create a distinct three-dimensionality and are therefore ecologically relatively uniform (Fig. 2). Riegl & Piller (1997) used the term 'coral carpet' in a broader sense for all low-relief coral communities in Safaga Bay, irrespective of framework-building potential. Riegl & Piller (1999) defined only communities with framework-building potential as carpets. From a geological perspective, coral carpets form biostromes (Cumings 1932; Kershaw 1994).

Reefs were defined as distinctly three-dimensional structures producing a stronger ecological differentiation of animal and plant communities than coral carpets. Our definitions are compatible with and expand that of Wainwright (1965) quoted in Stoddart (1969), namely that in '. . .structural coral reefs, corals are actively contributing by skeletal accumulation to the topographic development of the reef. . .'. The definitions of Rosen (1990), which is a condensation from several other definitions – '. . . organic framework, raised relief, wave resistance, photic zone restriction and tropical (or warm water) distribution. . .' – and Longman (1981, p. 10) – '. . .biologically influenced buildup of carbonate sediment which affected deposition in adjacent areas . . . and stood topographically higher than surrounding sediments during deposition. . .' also support our claim that carpets are indeed different from reefs, mainly because of the absence of clear relief since they may frequently be at almost the same level of the surrounding sediment (see also Wood (1999): '. . .a discrete carbonate structure . . . that develops topographical relief upon the sea floor. . .').

Fagerstrom (1987, p. 13) remarks that '. . .among Holocene photic and aphotic zone reefs there is an unbroken size gradation from isolated solitary corals, to weakly colonial, to strongly colonial of various sizes, to clusters of large, massive coralla, to coral knolls, knoll reefs, and patch reefs with entire reef systems' where 'reefs rarely occur in isolation . . . they occur in variously spaced clusters; each cluster is commonly called a "reef system" or if the cluster is enormous . . . it may be called a "reef province". Reefs plus reef systems (provinces) and the adjacent (external) sediments constitute a "reef complex". . .' (Fagerstrom 1987, p. 7). It

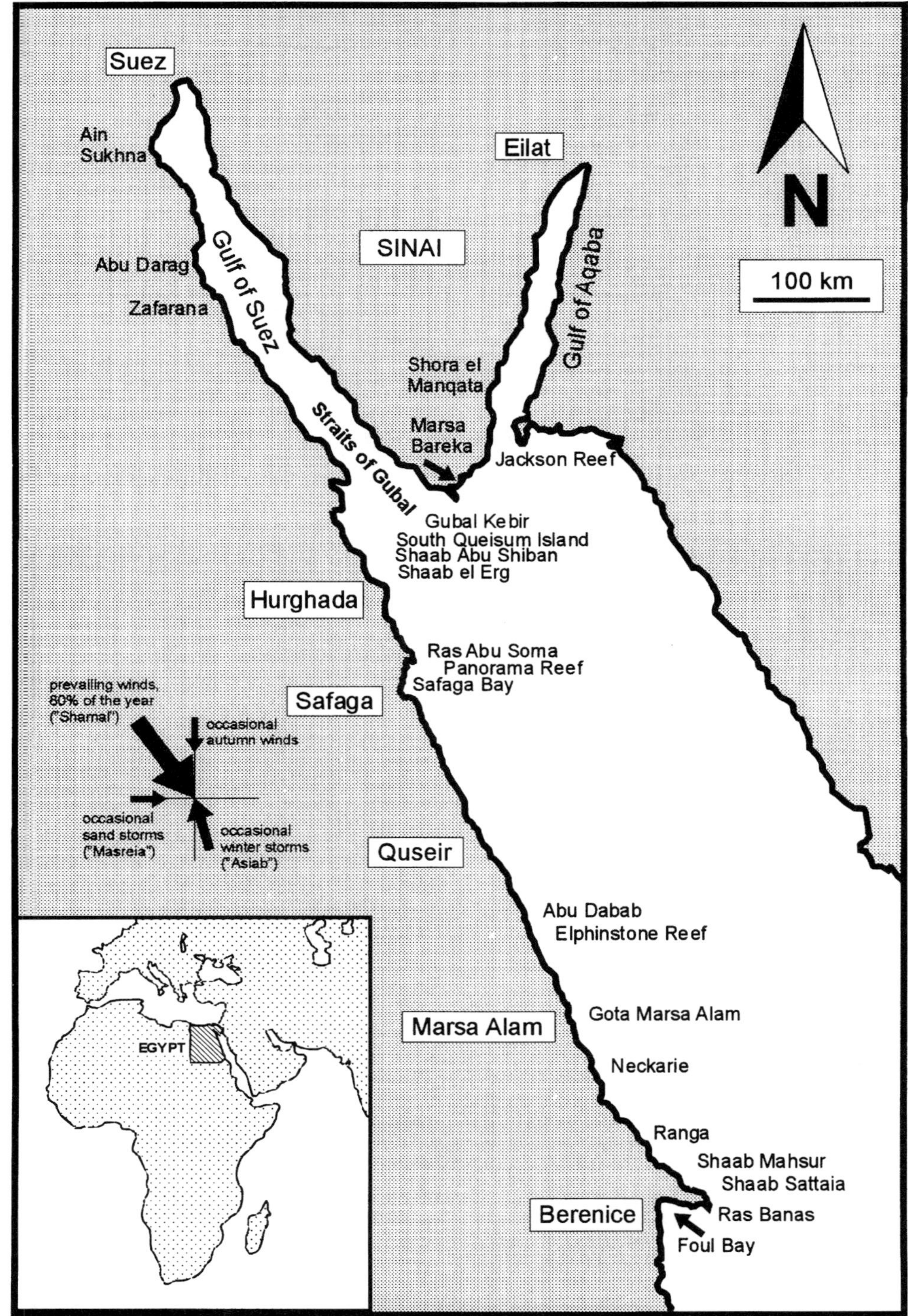

Fig. 1. Location map indicating the study area (Egyptian Red Sea) and names mentioned in the text. Coral carpets are tied to shallow shelf areas. In the Gulf of Suez they replace fringing reefs north of Zafarana. In the Gulf of Aqaba they occupy the sloping seafloor in the fore reef areas.

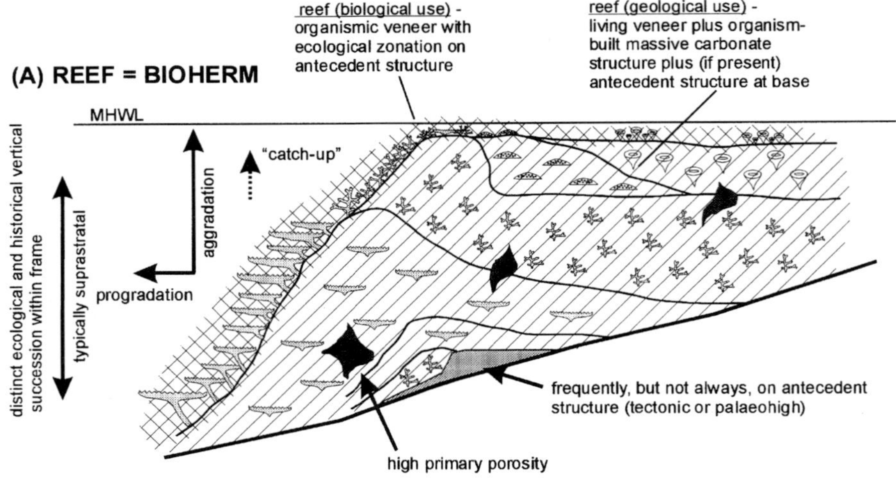

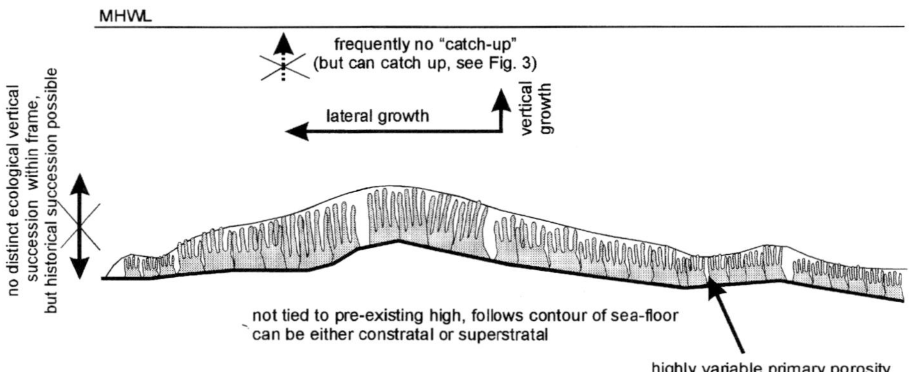

Fig. 2. Definition of the terms 'reef' (**A**) and 'coral carpet' (**B**) as used in this paper, illustrated by the cases of a fringing reef and a *Porites* carpet. (A) is modified and adapted from Montaggioni & Faure (1997) from Mauritius. The ecological succession within the reef framework is based on their research. The community succession in the carpet framework (B) is known in the upper part; it is unknown, however, whether a different initiation community existed at the base.

is important to keep this variability in mind when following our discussion of reefs and carpets. The reef complex in our case defines the totality of all reefs and carpets which are at least in close proximity if not ecologically connected. This indicates the plasticity of the system, which may allow any transition between the two systems and the resultant lithologies. However, in most observed cases they remain structurally distinct.

The term 'community' is used in this paper *sensu* Kidwell & Bosence (1991, p. 118) '. . .to denote recurrent groups of living organisms, Recent or ancient'. This differs from the term coral community used by Geister (1983, p. 178) as 'incoherent growths composed of only a single generation of framebuilders. . .'. The definition of Wainwright (1965) and Stoddart (1969) – '. . .a coral community . . . is an assemblage of organisms growing on a substratum other than of their own production in shallow, tropical seas. . .' – is not specific as to whether their 'coral communities' are frame-building at all. Our present definition also differs from the terminology used by Riegl & Piller (1997, p. 144) who reserved the term 'community' for quantitative descriptions of entities of ecologically well defined and distinct groups of organisms. In their study of Pleistocene coral communities on Grand Cayman, Hunter & Jones (1996) follow Whittaker (1975) and Kauffmann & Scott (1976) in defining communities as '. . .species that lived together and interacted with each other and their environment. . .'. This is also comparable to the definition of communities given by Fagerstrom (1987, p. 154) as '. . .consisting of numerous interacting species, having a broadly predictable composition, and occupying a broadly predictable habitat. . .'. It is unfortunate that the term 'community' is introduced into the ecological as well as the geological literature. In order to avoid confusion, the term 'coral community' *sensu* Wainwright (1965) and Geister (1983) should best be referred to as 'non-framework coral community'.

Drawing attention to further uses of the term 'reef', it is also necessary to differentiate between the use of the term as acceptable to ecologists studying the distribution of biota along gradients and to geologists looking for actual carbonate buildup. In an ecological sense, even a thin veneer of corals covering a fossil reef will exhibit differentiation into different communities along gradients (usually most notably depth and hydrodynamic exposure), and look like a real reef in a geological sense, as defined above (see also Fig. 3). Therefore any structure covered by coral communities exhibiting such an ecological differentiation tends to be called 'reef' in the biological literature, regardless of whether it would be called a reef by geologists. This has, however, nothing to do with the differentiation made by Dunham (1970) and James (1983) into stratigraphic reef (several superimposed bioherms) and the ecologic reef ('. . .rigid, wave-resistant structure generally formed during one specific period of time'), yet another conflicting use of the term.

The term 'framework' is used according to Fagerstrom (1987, p. 5) as '. . .the mass of large, colonial or gregarious, intergrown skeletal organisms in general growth position. . .', which corresponds well with that of Rosen (1990). Framework lithologies were defined according to Insalaco (1998). Although Insalaco (1998) advocates the use of the term 'growth fabric' rather than framework, we still use it extensively in this paper because working on Recent structures, we have the luxury of being able to watch living frameworks as they form and can recognize all the critical components. In order to give our findings palaeontological relevance, we also discuss which growth fabrics (*sensu* Insalaco 1998) could be formed. This interpretation is based on a quantitative evaluation of the Recent coral communities and experience with the local Pleistocene and Miocene. Primary porosity was evaluated visually: large pores or cavern systems are easily visible in Recent frameworks.

Quantitative methods

We used continuous intercept recording line transects of 10 m length placed parallel to the depth contour and taken in 1 m depth increments. The ideal transect length had been established by previous studies in the area (Riegl & Velimirov 1994). Along these line transects, the intercepts of all underlying coral species, benthic invertebrates and macro-algae were recorded to the nearest centimetre. Also the type of substratum, which was classified as either sand, limestone or rubble, was recorded.

The dataset was explored by means of agglomerative hierarchical cluster analyses using the Bray-Curtis similarity measure or Euclidean distance as distance measure and group average method of grouping (Digby & Kempton 1987; James & McCulloch 1990). Cluster analysis was used because it has advantages for delineating groups in a very distinct community setting (Field *et al.* 1982; Kenkel & Orloci 1986) as established from previous experience in the area (Riegl & Velimirov 1994; Riegl & Piller 1997) and with similar datasets. When the statistically obtained groupings were compared to the situation in the field and found to coincide they were used to describe community patterns. Analyses were performed using the PRIMER and SPSS statistical software.

In order to gain information about the spatial distribution of coral carpets, it was necessary to map the distribution of coral associations in a sample area, in our case northern Safaga Bay. After the different communities and framework types were obtained by statistical analysis, we visually examined other localities throughout the study area and assigned the encountered communities to those quantitatively described. The results are published in Riegl & Piller (1997) and were reused for this paper.

Since we were unable to obtain permission for either drilling or seismic surveys, we had to limit our evaluation of framework thickness in reefs and coral carpets to visual observations along natural or artificial (resulting from dynamite damage) scars and cracks. Therefore reliability of framework thickness estimation is problematic. Thicknesses given in this paper represent minimum estimates and we are presently unable to provide serious data either on framework growth rates or carbonate accumulation rates.

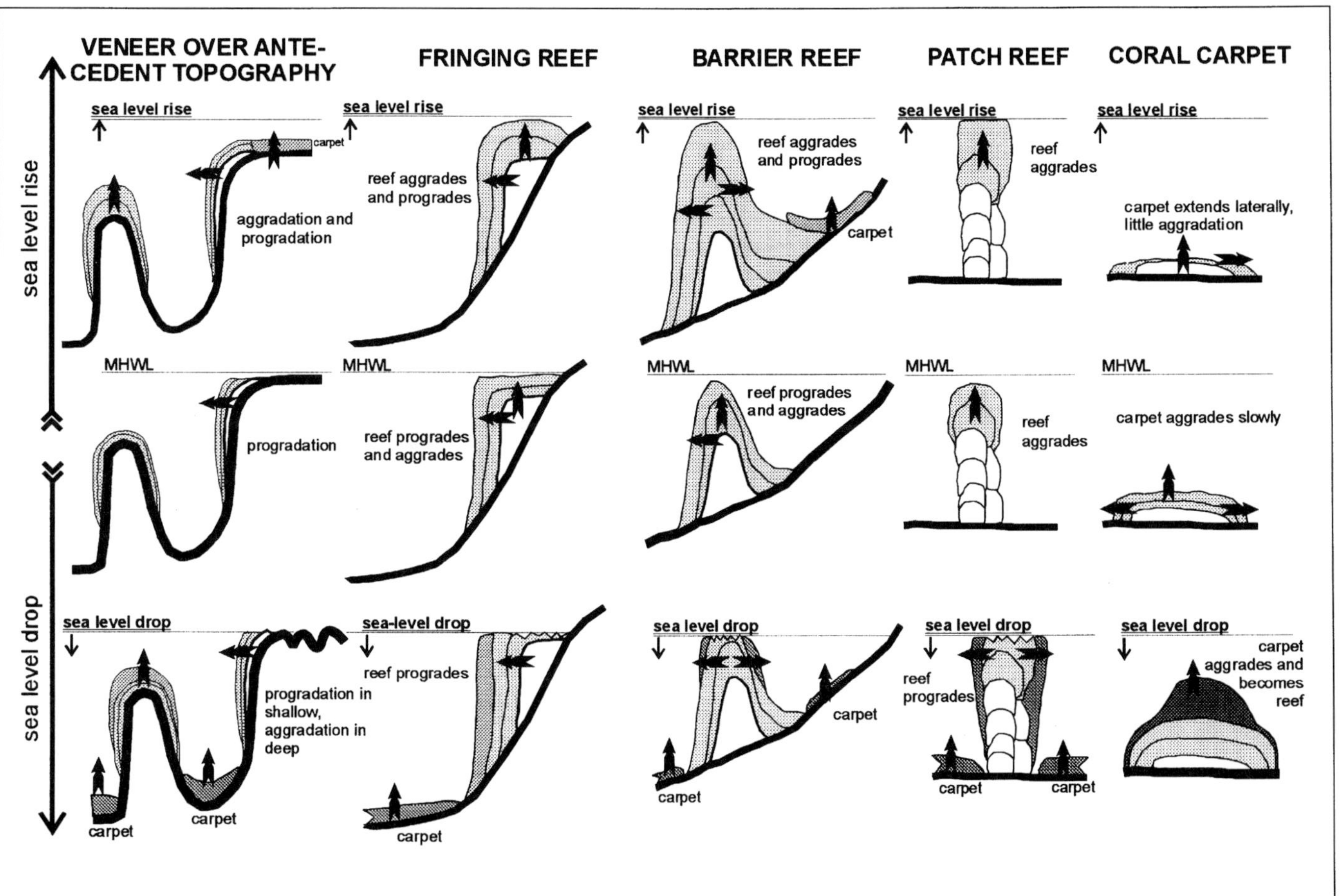

Fig. 3. Conceptual models illustrating several possibilities of coral reef and carpet growth, with spatial relationships and inferred differences in growth patterns. The growth geometries and sequences were inspired by the literature (Longman 1981; James & Macintyre 1985; Liu *et al.* 1998; Camoin *et al.* 1998; Schlager 1992, 1998) and are not based on actual data obtained during this study. Only some mechanisms of reef or carpet growth (simple aggradation and progradation) are illustrated while others, like backstepping or downstepping, were ignored because the added complexity would have done little to clarify the point.

Results

Spatial distribution of coral frameworks

The study area included several shelf areas (e.g. offshore Hurghada and Gulf of Suez) and shallow bays (Safaga Bay, northern Foul Bay) where extensive coral carpets occurred (Fig. 1). Safaga Bay was used as our central model for distribution of framework types in relation to bottom topography. Figure 4 shows how coral carpets occupy mainly the areas of flat, medium deep sea floor (5–30 m) while reefs are primarily tied to topographical highs, such as the steep mainland coastline at Ras Abu Soma and the Tubya islands. Patch reefs are found on the Tubya-Gamul ridge. From the map (Fig. 4) we calculated that reef frameworks occupy approximately 1.5 km^2 within northern Safaga Bay, while coral carpets occupied about 16.6 km^2 and scleractinian non-frameworks (patches and non-framework coral communities) occupied 20.3

Fig. 4. Distribution of coral framework types in northern Safaga Bay, Red Sea, Egypt (modified after Riegl & Piller 1999).

km^2. The *Acropora*-dominated patches were included in this calculation with the non-frameworks, since they do not form a continuous framework. In terms of total surface area, carpet frameworks cover approximately ten times more space than reef frameworks in northern Safaga Bay.

The cluster analysis of 150 line transects allowed the differentiation of several coral communities that could be interpreted as reef communities, coral carpets and non-framework coral communities.

Reefs

The present and previous studies (Riegl & Velimirov 1994; Riegl & Piller 1997, 1999) showed a clear ecological subdivision of the coral assemblages on reefs along a depth and hydrodynamic gradient. Table 1 gives an overview of the biological characteristics of scleractinian reef communities that can be allocated to framework types. These community types are illustrated in Fig. 5 A–C. A special reef type is represented by the coral patches *sensu* Piller & Pervesler (1989), which are also described by Riegl & Piller (1997, 1999). They are only a few metres in diameter and do not show clear ecological zonation. They grow in a hydrodynamically exposed position at less than 20 m depth and are dominated by *Acropora* and a diverse assemblage of faviids. Typical species are tabular and open arborescent. The patches are not included in Table 1. The relationships of these communities to water energy, irradiance and turbidity (inshore–offshore) gradients are outlined in Fig. 6. The lithological representation of these communities' growth fabrics once fossilized would be (according to Insalaco 1998) as follows: *Acropora* community, continuous rigid mixstone; *Porites* community, continuous rigid domestone to dense rigid pillarstone; *Millepora* community, continuous dense rigid mixstone. All would be clearly superstratal with well developed primary porosity, caused by ledges and overgrowths.

Coral carpets

Coral carpets have a stronger lateral than vertical growth component and follow the underlying sea-floor morphology (Figs. 2, 3). Maximum apparent framework thickness above the surrounding sea bed was between 8 and 11 m. Framework density is laterally variable depending on community type. Against their periphery, carpets tend to thin out and/or disintegrate into isolated patches. Coral carpets were represented by four types in the Red Sea proper (*Porites* carpet, *Porites (Synaraea)* carpet, faviid carpet, platy scleractinian carpet); (Fig. 6), two types in the Gulf of Suez (*Stylophora* carpet in the north, faviid carpet in the south) and one type in the Gulf of Aqaba (faviid carpet). The growth fabric of these frameworks is more variable than in reefal frameworks but always remains superstratal. However, the primary porosity is more variable than in the reefal growth fabrics: low in the platy growth fabric type, high in the columnar (pillarstone) type.

The *Porites* carpet (Fig. 5D) was typically found between 5 and 15 m depth except in the Gulfs of Aqaba and Suez. In the Red Sea, it increased in areal extent and framework thickness towards the south. It was dominated by *P. columnaris*, *P. lutea* and *P. (Synaraea) rus*. These corals have columnar growth forms (except on reef slopes where *P. lutea* is often massive). The *Porites* carpet is an area of vigorous framework building. The thickest *Porites* framework in the Hurghada/Safaga area was 2 to 3 m thick and increased in the Abu Dabab Group to >8 m (Fig. 1). The framework was characterized by numerous caverns between the stick-like *Porites* columns, where live tissue and coral growth were restricted to the tips. The dead parts of the *Porites* columns were settled by other corals, particularly encrusting and platy species (*Montipora* spp., *Echinophyllia aspera*, *Mycedium elephantotus*, *Echinopora lamellosa*, *Turbinaria mesenterina*) and small, corymbose *Acropora* (*A. valida*, *A. granulosa*, *A. humilis*). Faviids and alcyoniids were rare. Xeniids were common, although not as much as in other deep associations, like the faviid carpet (see below). At 20–25 m depth, *Porites lutea* and *P. columnaris* were replaced by *P. (Synaraea) rus.* This situation was frequently encountered on reefs in southern Egypt (Shaab Mahsur, Gota Marsa Alam). At greater depth the *Porites* carpet was replaced by a faviid carpet. In low light conditions, the *P. (S.) rus* carpet was not developed and the faviid carpet was directly adjacent to a *P. lutea* or *P. columnaris* carpet (Shaab Abu Shiban, Shaab el Erg, Ranga, Neckarie). Typical live coral cover of the substratum was between 60 and 90%. The lithological representation (Insalaco 1998) of this framework would be a superstratal dense rigid pillarstone with high growth fabric continuity. It is characterized by high primary porosity.

The highly diverse faviid carpet (Fig. 5E) was the most widely distributed coral assemblage at depths of 10 m or more (lower limit *c.* 30–45 m) throughout the Gulf of Aqaba and the Red Sea. The coral-built framework was on average 50–100 cm thick and had an irregular surface topography. The carpet was dissected by fissures,

Table 1. *Ecological zonation (indicator species), average living coral cover values and growth fabric type on northern Red Sea reefs*

		Exposed			Semi-exposed			Sheltered		
Position	Depth (m)	Indicator species	Average cover (%)	Growth fabric type	Indicator species	Average cover (%)	Growth fabric type	Indicator species	Average cover (%)	Growth fabric type
Reef crest	0–1	*Pocillopora verrucosa*, *Acropora gemmifera*, *Stylophora mordax*	43±18	Mixstone	*Stylophora pistillata*, *Acropora secale*, Faviidae	48±10	Mixstone	*Stylophora pistillata*, Faviidae, *Porites lutea*	24±17	Domestone
Reef edge	1–3	*Acropora hyacinthus* group	54±11	Platestone	*Millepora dichotoma*	58±16	Mixstone	*Porites lutea*	68±30	Domestone
Reef slope	3–15	Various *Acropora*, diverse without clear dominance	59±18	Mixstone or platestone	*Millepora dichotoma* and various massives	56±14	Mixstone	*Porites lutea* and various massive species	85±29	Domestone
Slope base	15–25	Tabular *Acropora* (*A. clathrata*, *A. divaricata*)	58±25	Pillarstone	*Acropora hemprichi*	29±6	Pillarstone	Tabular *Acropora* (*A. clathrata*, *A. divaricata*)	26±11	Pillarstone
Carpet	5–35	Diverse carpet (faviid carpet with *Acropora*)	28±3	Mixstone	Diverse carpet (usually faviid carpet)	28±3	Mixstone	*Porites* carpet and/or faviid carpet	85±15	Pillarstone, Mixstone
Deep areas	35–50	Platy scleractinia (*Turbinaria mesenterina*, *Echinophyllia aspera*, *Leptoseris papyracea*)		Sheetstone	Platy scleractinia *Turbinaria mesenterina*, *Echinophyllia aspera*, *Leptoseris papyracea*		Sheetstone	Platy scleractinia *Turbinaria mesenterina*, *Echinophyllia aspera*, *Leptoseris papyracea*		Sheetstone

Species named are the visually dominant species within the reef zone. Modified from Riegl & Velimirov (1994), Riegl & Piller (1997, 1999). Depths are approximate and can vary according to exposure and reef morphology. Depths indicated for coral carpets overlap with those for reefs since they can occur in different, independent, settings

Fig. 5. Illustration of framework types in the northern Red Sea. (**A**) Windward *Acropora* reef at Tubya kebir, Safaga Bay. Species are *A. gemmifera* and *A. digitifera* on the shallowest parts, and mainly *A. polystoma* on the slope with prominent *A. hyacinthus* tables. The framework produced in areas of pure *Acropora* dominance would be a continuous superstratal rigid platestone. (**B**) Leeward *Porites* reef framework at Abu Dabab, southern Egypt. The dominant corals are *Porites lutea*. The framework produced would be a continuous superstratal rigid domestone to dense rigid pillarstone. (**C**) Semi-exposed *Millepora* reef framework at Tubya al-Hamra (Safaga Bay), 4 m depth, showing *M. dichotoma* dominance and a small *Goniastrea retiformis* in the foreground. The framework produced would be a continuous dense rigid mixstone. (**D**) *Porites* carpet framework at Gota Marsa Alam, southern Egypt, 10 m depth. Dominant corals are *P. lutea* and *P. columnaris.* The framework produced is a superstratal dense rigid pillarstone. (**E**) Faviid carpet framework at Marsa Muqebla, 15 m, Gulf of Aqaba. The dominant framebuilders are faviids (mainly *Goniastrea* spp.) with interspersed *Acropora* and *Stylophora*. The framework produced would be a superstratal rigid non-uniform mixstone. (**F**) *Stylophora* carpet framework, Abu Darag, Gulf of Suez. The framework produced would be a superstratal sparse pillarstone.

caves and gullies which were the habitat for a rich semi-cryptic to cryptic community. Visually dominant corals were *Goniastrea*, mostly *G. pectinata, G. retiformis* and *Platygyra lamellina*, as well as various *Favia*, *Favites*, *Leptastrea*, *Cyphastrea*, *Pavona*, *Echinopora*, *Acropora granulosa* and the rare *A. squarrosa*. The most frequent tabular species was *A. pharaonis* but

Coral reef communities

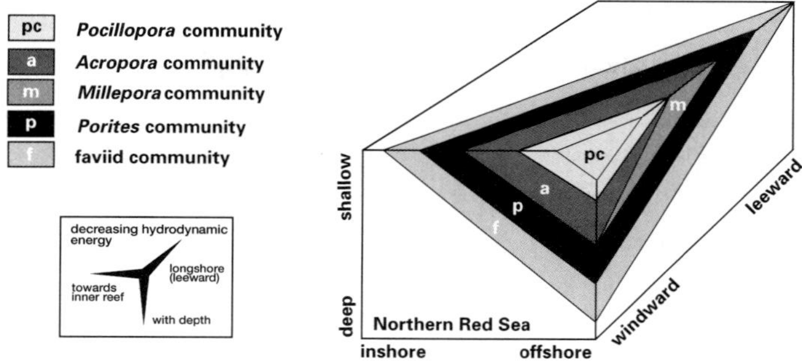

Coral carpet communities

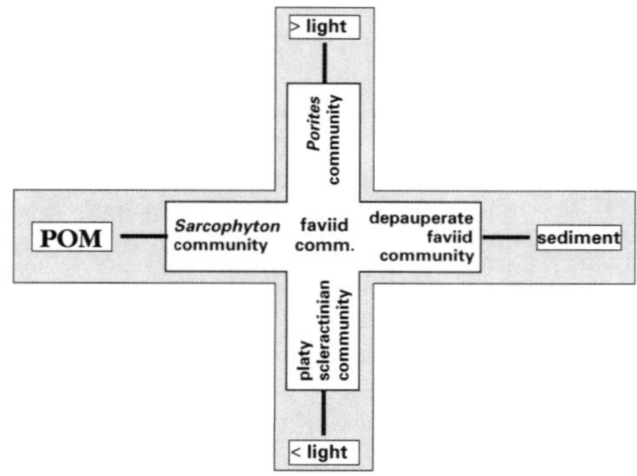

Fig. 6. Schematic representation of the distribution of northern Red Sea coral assemblages on coral reefs and carpet frameworks. Adapted from Riegl & Piller (1997).

large tabular colonies were missing. Fungiids were common. Platy corals (*Turbinaria mesenterina*, *Echinophyllia aspera*, *Mycedium elephantotus*) increased towards the lower depth limit, where they formed a specific community: the platy scleractinian assemblage of Riegl & Piller (1997, 1999). Soft corals (*Sarcophyton* and xeniids) made up an important percentage (up to 50%) of all colonies. Total live coral cover of all substrata including sand was typically between 20 and 30%. The lithological representation of the regular faviid carpet framework according to Insalaco (1998) would be a superstratal rigid non-uniform mixstone. The platy scleractinian assemblage would be a continuous to discontinuous rigid to loose (if growing on reef-base unconsolidated sediment) sheetstone. Both growth fabrics have low primary porosity.

The *Stylophora* carpet (Fig. 5F) was only found north of Ras Zafarana, in the Gulf of Suez in the nearshore area. Largely monospecific thickets consisted almost entirely of *Stylophora* cf. *pistillata* (or another yet undescribed *Stylophora*) with some individual massive corals (*Platygyra, Favia, Porites*). The alcyonacean soft coral *Litophyton* sp. was also common. The skeletons of adjacent *Stylophora* colonies fused, and the entire carpet thus formed an open, continuous framework. Framework thickness reached approximately 1 m and carpets grew to the low-water mark, becoming exposed to air at spring low tide. Coral skeletons made up the entire framework, which was very porous; frequently, however, only the colonies on the landward and seaward side were alive, while the central framework was made up almost exclusively of dead corals. The expected lithology according to Insalaco (1998) would be a sparse to dense superstratal pillarstone, which could be either loose or rigid.

Non-frameworks

Two communities with less than 30% total coral cover that did not build a framework, but may be incipient carpets, were found.

A *Stylophora–Acropora* community was observed in shallow (>5 m) sandy areas, where a substantial amount of rocky substratum was available and individual coral heads or small coral patches settled. The visually dominant coral was *Stylophora pistillata*, which formed dense bushy colonies of up to 30 cm. *Acropora robusta, A. tenuis, A. pharaonis* and *A. anthocercis* formed large corymbose or tabular colonies. The bases of these colonies had dense xeniid growth. Characteristic massive corals were *Platygyra lamellina* and *Porites lutea* and *P. solida*, which formed big microatolls of up to several metres across. Besides xeniids, the alcyoniids *Litophyton arboreum, Sarcophyton* spp. and *Lobophytum* cf. *venustum* were common. This community did not form a framework but consisted of widely spaced colonies. Live coral cover varied between 5 and 20%. The lithological representation of this community would be rudstone or floatstone.

The soft coral communities were *Sarcophyton*-dominated off Hurghada and in Safaga Bay between 10 and 30 m depth, and *Lobophytum*-dominated at South Queisum island in the Straits of Gubal in the same depth range. The most typical Scleractinia were *Siderastrea savignyana, Astraeopora myriophthalma* and fungiids. *Porites* and *Acropora* were rare. A mixed *Lobophytum/Sarcophyton* community occurred in Foul Bay between 1 and 10 m depth. Xeniids were widespread and dominated in water depths greater than 20 m. Although numerous Scleractinia occurred, they were usually small and no framework building took place in either community. Live coral cover was between 10 and 20% in the *Sarcophyton* and *Lobophytum* community and up to 60% in the xeniid community. It would be difficult to recognize this community in the fossil record since the visually dominant species in the Recent do not fossilize well. It would be recognizable at best as a shelly hardground.

Discussion

Our study shows that several types of framework and non-framework coral communities exist in the northern Red Sea in response to bottom topography and oceanographic factors (Figs. 5, 6). Several levels of feedback between environment and organisms can be inferred. Oceanographic and topographic factors shape the ecological properties of coral communities (Longman 1981; Dupraz 1999), for example species composition and successional stage. This in turn defines and controls frame-building potential since different coral communities build different frames (Figs. 4, 8), a situation also described from other reef areas (Kleypas 1996; van Woesik & Done 1997). The frameworks can create environmental conditions that can influence the constituent coral communities. For example, windward reef communities, by growing into the waves and thus creating a sheltered side in their lee, cause a differentiation into windward and leeward reef communities with concomitant differences in accretion speed (Figs. 5, 6; Smith *et al.* 1998).

Laterally, reef and coral carpet frameworks as well as non-framework communities can grade into each other within the same reef complex (definition of Henson (1950) and Ladd (1977): '. . .the entire structure – surface reef, lagoon deposits, and off-reef deposits. . .'). Conventionally, three main types of reef are found in nearshore settings (oceanic reefs are not taken into consideration by our model) – patch, fringing and barrier reefs (Geister 1983; Fagerstrom 1987; Tucker & Wright 1990; Wood 1999) – which have true superstratal frameworks (*sensu* Gili *et al.* 1995; Insalaco 1998). Additionally there exist coral carpets (which in the studied case are also superstratal) and veneers of corals over antecedent structures (Fig. 3). From a biological viewpoint, concerning the functioning of their benthos, the latter are very similar to reefs (and are frequently treated as such, especially in the biological literature). From a geological viewpoint they differ from a 'true' reef since they only cover an already existing structure without adding substantially to it.

Reef frameworks frequently, but not exclusively, develop in areas of pre-existing topographic highs, where it is easier for the reef to initiate, 'catch up' and 'keep up' (*sensu* Neumann & Macintyre 1985). This leads to the deposition of the biohermal, ecologically structured, massive, often lenticular bodies. By the process of catching up and growing to the surface, the environmental setting, most notably the light availability and hydrodynamic regime, is changed by the reef structure itself (Smith *et al.* 1998). This leads to the observed differentiation of facies (Montaggioni & Faure 1997; Webster *et al.* 1998) (Fig. 2). Once these systems are initiated, their further development is strongly dependent, among other factors, on sea-level behaviour (Hopley 1982; Tucker & Wright 1990; Schlager 1992, 1998; Pomar *et al.* 1996; Pomar & Ward 1999). When sea-level remains relatively stable, all systems will aggrade to fill accommodation space and then prograde (Schlager 1992; Pomar 1991).

Carpet frameworks operate in a more indeterminate setting: many grow too deep to reach sea level; however, some may be able to catch up in the future and some initiate in very shallow water (e.g. *Stylophora* carpet). Therefore coral carpets should not be seen *a priori* as 'give-up reefs' nor as 'incipient reefs' (van Woesik & Done 1997; Kleypas 1996), since many may not develop into reefs at all. If they grow deep, their position may be caused by delayed initiation or indeed by slower vertical accretion than reefs (due to lower light levels in deep water; Fig. 3, middle row). If sea level rises, all systems will have to aggrade in order to keep up. Once that is achieved, progradation will continue. Rising sea levels will disadvantage the coral carpets in deeper areas, since the additional water column will decrease available light and slow growth rates further. We can assume decreased vertical accretion rates, but probably maintained lateral expansion. Some coral carpets could initiate in areas that had previously been too shallow or too harsh an environment (in our model, for example, lagoonal areas between the barrier reef and the shoreline; Fig. 3, lower row). The coral veneer covering the top of an antecedent structure could also take the appearance of a coral carpet, if no or little ecological differentiation is visible laterally and vertically throughout its frame (Fig. 3). Falling sea level will cause subaerial exposure and erosion of the shallow reef areas and progradation of the reef and veneer. The coral carpets, now in shallower water, will receive more light and will therefore aggrade more quickly. Some carpets may develop into reefs if they catch up to the surface. The process of catching up and the resultant differences in environment along the structure will lead to ecological differentiation of the constituent communities resulting internally in different, ecologically distinct facies. According to our definition, the carpet will have changed into a reef (see definitions in Fig. 2). Several new carpets may be initiated in areas which were previously too deep for corals, but are now within their tolerance limits (Fig. 3, lower row).

Changing water depth does not simply cause accelerated or reduced carpet growth, but also has the potential to change the composition of the constituent coral community (Fig. 6). Carpets can develop an internal sequence of communities by two processes. Firstly, repeated changes in environment, for example by sea-level fluctuations, could lead to layers of facies dominated by different species. However, these would typically be closely related facies. In deep carpets these would be different deeper-water facies, in shallow carpets different shallow-water facies. These zonations are not ecological, resulting from environmental gradients within a continuous community, but a historical succession ('discontinuous communities') reflecting different environments. Secondly, and in clear contrast to the first possibility, by catching up to the surface a continuous ecological gradient of facies (deep to shallow) will be developed.

The carpets described here have some affinity with the incipient reefs and non-frame-building coral communities described from the southern Great Barrier Reef by van Woesik & Done (1997) and Kleypas (1996). The setting is similar but in Australia these systems were switched off (*sensu* Buddemeier & Hopley 1988) by turbidity and tidal range. The Red Sea carpets described possibly represent a similar system in a more benign environmental setting where the carpets are switched on and framework production was able to proceed. Even though frame-building activity in carpet areas may be delayed, they are areas of important carbonate production and accretion, additional to reefs. In the study area, carpet frameworks cover more space than reef frameworks and we believe that they are as important as the latter in the carbonate budget (see Fig. 4 for spatial distribution). Davies & Peerdeman (1998) describe similar systems of laterally extensive areas of coral and algal dominance forming a coral–algal rubble bank, and suggest that these form the subreef facies of the Great Barrier Reef. This facies could be interpreted as a thin coral carpet without strong framebuilding (Davies & Peerdeman 1998, figure 11) and could further demonstrate the importance of carpets in early stages of reef building. However, there is not necessarily a gradation from one system into the other, as shown by the numerous fossil biostromes that never developed into reefs; (Piller 1981; Kuss 1983; Piller *et al.* 1996; Kershaw 1994; Nose 1995; Insalaco 1996).

Coral carpets have a stronger lateral than vertical component. This gives rise to the biostromal nature and explains the absence of strong internal ecological gradients. Gradients in coral assemblages within carpet frameworks are weak and driven primarily by irradiance and sedimentation/turbidity (Fig. 6). We believe that ecological dynamics provide a stable coral community state in deeper coral carpets (*Porites* and faviid carpet) that is not conducive to rapid 'catch up'. In the study area, deep coral carpets grow in areas virtually free of disturbances (i.e. below storm wave base, no terrigenous influx,

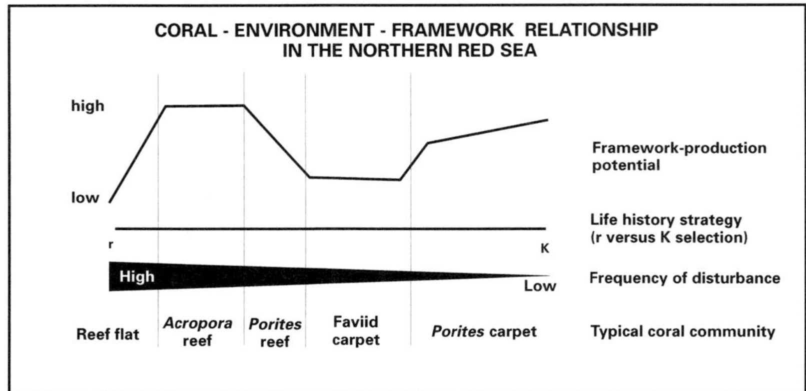

Fig. 7. Schematic relationship between biological characteristics of coral communities and coral frameworks in the northern Red Sea.

minimum sedimentation stress). Since these disturbances are major factors influencing successional stages of coral communities, their absence favours K-selected, near-climax communites (Pianka 1970; Potts *et al.* 1985; Done & Potts 1992). For example, in *Porites* carpets the dominant corals (*Porites lutea*) have a growth rate of 6.17 mm a^{-1} (Gulf of Aqaba) to 6.42 mm a^{-1} (southern Egypt; Heiss 1994, p. 79) which is about a factor of ten slower than in *Acropora* (100–200 mm a^{-1}, Heiss 1994, p. 109). However, owing to the reef's closer proximity to the surface it is subject to more frequent disturbances (e.g. hydrodynamic, temperature anomalies, UV irradiation; Done & Potts 1992; Riegl & Velimirov 1994; Riegl & Piller 1997). The assemblages of the *Acropora* reef framework are more r-selected with a higher turnover rate but also a high frame-building potential due to fast growth rates (Fig. 7). The disturbances, however, have the potential to reduce the actual frame accumulation by breakage and downslope export of skeletons subsequent to coral death (Hughes 1999; Riegl & Piller 1999). This is not the case in the generally deeper and more protected carpets, where due to the flat morphology and the high primary porosity (especially in *Porites* carpets) most produced sediment and fragments are more likely to be retained.

Table 1 and Fig. 5 show that the frameworks described here made up by specific, environmentally controlled coral communities can be described in terms of the descriptive nomenclature and classification of growth fabrics by Insalaco (1998). It should be possible to use the environmental model presented here for the development of specific growth fabrics and apply it to younger Neogene deposits for palaeoecological analyses (Perrin *et al.* 1995). Some lack of clarity remains concerning the term 'platestone' and whether the windward *Acropora* communities develop into platestones or mixstones. Kan *et al.* (1995) and Webster *et al.* (1998) illustrate cases where tabular *Acropora* communities apparently form platestones. We believe that parts of Red Sea *Acropora* reef slopes (especially the reef edges and upper reef slopes in certain wave exposure) can form platestones *sensu* Insalaco (1998). In the reefs investigated for this study, however, we are hesitant to call the entire reef slope a platestone facies.

A similar downslope sequence of domestone to pillarstone to platestone facies as observed in sheltered Recent Red Sea reefs could be interpreted based on the descriptions from the Upper Miocene reefs of Mallorca (Perrin *et al.* 1995; Pomar *et al.* 1996) where the palaeoslope is still preserved, and the Alicante-Elche basin (Calvet *et al.* 1996). Alternations of *Porites* pillarstone and faviid domestone facies are also found in Tortonian and Messinian patch reefs in southern Spain (Martin *et al.* 1989; Braga *et al.* 1990; Esteban *et al.* 1996), which, however, grew in a different environment.

The coral reef/carpet system – and the resulting growth fabrics – provide us with evidence for environment–organism–environment feedback on several hierarchical levels (Fig. 8). The central unit for our model is the coral community since it builds the carbonate framework that alters its own environment. The largest-scale factors are geological processes triggering oceanographic change (Longman 1981; James & Macintyre 1995; Insalaco 1998; Wood 1999). Plate movements, sea-level changes and changes in current patterns or sea temperature cause

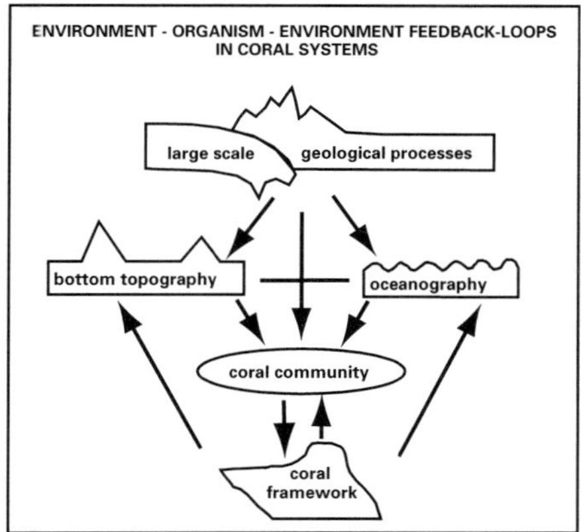

Fig. 8. Suggested interactions between geological and biological processes in coral framework production (inspired by Longman 1981).

restriction or expansion in the biota available for framework construction (Edinger & Risk 1994; Kershaw 1998) as seen in the differences between local faunas in the tropical Atlantic and Indo-Pacific (Veron 1995). Furthermore, geological processes influence bottom topography which in turn influences coral frameworks (Longman 1981; Leinfelder 1994, 1997). Our study area shows how different modes of faulting led to environments favouring different framework types. Block faulting (Roberts 1985; Purser & Bosence 1998) has led to wide shelf areas with coral carpet frameworks in the Gulf of Suez and the adjacent northern Red Sea area. Strike/slip faulting has led to steeper basin margins which favour reef growth rather than coral carpets, e.g. in the Gulf of Aqaba.

On both small and large scales, bottom topography and basin geometry influence oceanography by influencing currents and waves. Physical and chemical factors coupled with climatic influences shape coral communities (Rosen 1975; Glynn 1993; Leinfelder 1994; Perrin *et al.* 1995; Riegl & Piller 1997). The structure of the coral community is critical for the type of framework produced (van Woesik & Done 1997). Many studies show that in the Indo-Pacific, most vigorous coral growth and reef accretion are concentrated in shallow areas, particularly in *Acropora* communities (Braithwaite 1982; Montaggioni & Faure 1997; Kan *et al.* 1995; Webster *et al.* 1998). These are the typical reef framework 'catch-up' and 'keep-up' settings where accretion is limited by accommodation space. In addition to this, our study shows that significant framework accretion potential also exists in shallow shelf areas (10–40 m depth in the study area). There, coral carpet frameworks are produced which in most areas do not catch up, or have not yet caught up, to sea level.

Conclusion

- Coral framework initiation and growth is governed by oceanography and bottom topography.
- Two different coral framework types exist in the study area – coral reefs and coral carpets – which are built by different coral communities.
- Coral reefs and all coral carpets described here (faviid carpet and *Porites* carpet) are superstratal.
- Growth fabrics are: platestone to mixstone facies (windward *Acropora* community); mixstone facies (semi-exposed *Millepora* facies and faviid carpet); domestone facies (leeward *Porites* community); pillarstone facies (*Porits* carpet, *Stylophora* carpet).
- In the northern Red Sea, reefs grow on steep slopes and highs, coral carpets in areas with little topographic differentiation.
- Coral community successional stage and life history attributes differ between reefs and carpet. The highest growth potential is on shallow reefs where coral communities are fast growing, r-selected and have a high turnover. Coral carpets are slow-growing, K-selected systems.
- Reefs grow towards the surface and thus modify their own environment (light availability, hydrodynamic exposure) which feeds back into their coral communities (windward–leeward, reef edge–reef slope). Coral carpets grow mainly laterally in deeper water and interfere much less with their environment. They build a highly structured hard substratum that increases settlement space and has a high sediment retention potential.

We gratefully acknowledge support by the Austrian Science Foundation through grants P5877, P7507-GEO, P10715-GEO, P13165-GEO, Hochschuljubiläumsstiftung der Stadt Wien, EU (Gulf of Aqaba protectorates development project) and USAID (Promotion of sustainable tourism development project). We thank K. E. Luke, A. M. Mansour, M. Rasser, D. Smith, C. Yanni, M. Zuschin for their help during field work and fruitful discussions. We thank B. Rosen and E. Insalaco for discussions and suggestions that greatly improved the manuscript's quality.

References

BRAGA, J. C., MARTIN, J. M. & ALCALA, B. 1990. Corals in coarse-terrigenous sedimentary environments

(Upper Tortonian, Granada Basin, southern Spain). *Sedimentary Geology,* **66**, 135–150.

BRAITHWAITE, C. J. R. 1982. Patterns of accretion of reefs in the Sudanese Red Sea. *Marine Geology*, **46**, 297–325.

BUDDEMEIER, R. W. & HOPLEY, D. 1988. Turn-ons and turn-offs: causes and mechanisms of the initiation and termination of coral reef growth. *Proceedings of the 6th International Coral Reef Symposium, Townsville,* **1**, 253–261.

CALVET, F., ZAMARREÑO, I. & VALLES, D. 1996. Late Miocene reefs of the Alicante-Elche Basin, southeast Spain. *In*: FRANSEEN, E. K., ESTEBAN, M., WARD, W. C., ROUCHY, J.-M. (eds) *Models for carbonate stratigraphy from Miocene reef complexes of Mediterranean regions*. SEPM, Concepts in Sedimentology and Palaeontology, **5**, 177–190.

CAMOIN, G. F., ARNAUD-VANNEAU, A., BERGERSEN, D. D., ENOS, P. & EBREN, Ph. 1998. Development and demise of mid-oceanic carbonate platforms, Wodejebato Guyot (NW Pacific). *In:* CAMOIN, G. F. & DAVIES, P. J. (eds) *Reefs and Carbonate Platforms in the Pacific and Indian Ocean*. International Association of Sedimentologists, Special Publications, **25**, 39–67.

CUMINGS, E. R. 1932. Reefs or bioherms? *Geological Society of America Bulletin*, **43**, 331–352.

DAVIES, P. J. & PEERDEMAN, F. M. 1998. The origin of the Great Barrier Reef – the impact of Leg 133 drilling. *In:* CAMOIN, G. F. & DAVIES, P. J. (eds) *Reefs and Carbonate Platforms in the Pacific and Indian Ocean*. International Association of Sedimentologists, Special Publication, **25**, 23–38.

DIGBY, P. E. & KEMPTON, R. A. 1987. *Multivariate Analysis of Ecological Communities*. Chapman and Hall, London.

DONE, T. J. 1982. Patterns in the distribution of coral communities across the central Great Barrier Reef. *Coral Reefs*, **1**, 95–107.

—— 1983. Coral zonation: its nature and significance *In*: BARNES, D. J. (ed.) *Perspectives on Coral Reefs*. Clouson, Australia, 107–147.

—— & POTTS, D. C. 1992. Influences of habitat and natural disturbances on contributions of massive *Porites* corals to reef communities. *Marine Biology*, **114**, 479–493.

DULLO, W. C. & MONTAGGIONI, L. 1998. Modern Red Sea coral reefs: a review of their morphologies and zonation. *In*: PURSER, B. H. & BOSENCE, D. W. J. (eds) *Sedimentation and Tectonics in Rift Basins. Red Sea – Gulf of Aden.* Chapman & Hall, London, 583–593.

DUNHAM, R. J. 1970. Stratigraphic reefs versus ecologic reefs. *AAPG, Bulletin*, **54**, 1931–1931.

DUPRAZ, D. 1999. Paleontologie, paleoecologie at evolution des facies recifaux de l'Oxfordien Moyen-Superieur (Jura suisse et Francais). *GeoFocus*, **2**.

EDINGER, E. N. & RISK, M. J. 1994. Oligocene-Miocene extinction and geographic restriction of Caribbean corals: roles of turbidity, temperature, and nutrients. *Palaios*, **9**, 576–598.

EDWARDS, F. J. 1987. Climate and oceanography. *In*: EDWARDS, F. J. & HEAD, S. M. (eds) *Red Sea – Key Environments*. Pergamon Press, Oxford, 45–69.

ESTEBAN, M., BRAGA, J. C., MARTIN, J. M. & SANTISTEBAN, C. D. 1996. Western Mediterranean reef complexes. *Society for Sedimentary Geology, Concepts in Sedimentology and Paleontology*, **5**, 55–72.

FAGERSTROM, J. A. 1987. *The Evolution of Reef Communities*. Wiley Interscience, New York.

FIELD, J. G., CLARKE, K. R. & WARWICK, R. M. 1982. A practical strategy for analysing multispecies distribution patterns. *Marine Ecology Progress Series*, **8**, 37–52.

FROST, S. H. 1981. Oligocene reef coral biofacies of the Vicentin, Northeast Italy. *In*: TOOMEY, D. F. (ed.) *European Fossil Reef Models*. Society of Economic Paleontologists and Mineralogists, Special Publications, **30**, 483–539.

GEISTER, J. 1983. Holocene West Indian coral reefs: geomorphology, ecology and facies. *Facies*, **9**, 173–284.

GILI, E., MASSE, J.-P. & SKELTON, P. W. 1995. Rudists as gregarious sediment-dwellers, not reef-builders, on Cretatious carbonate platforms. *Palaeogeography, Palaeoclimatology, Palaeoecology*, **118**, 245–267.

GLYNN, P. 1993. Coral bleaching: ecological perspectives. *Coral Reefs*, **12**, 1–18.

GUOZHONG, W. 1998. Tectonic and monsoonal controls on coral atolls in the South China Sea. *In:* CAMOIN, G. F. & DAVIES, P. J. (eds) *Reefs and Carbonate Platforms in the Pacific and Indian Ocean*. International Association of Sedimentologists, Special Publications, **25**, 237–248.

GVIRTZMAN, G. 1994. Fluctuations of sea level during the past 400000 years: the record of Sinai, Egypt (northern Red Sea). *Coral Reefs*, **13**, 203–214.

HEISS, G. A. 1994. *Coral reefs in the Red Sea: growth, production, and stable isotopes*. GEOMAR Report, **32**.

HENSON, F. R. S. 1950. Cretaceous and Tertiary reef formations and associated sediments in the Middle East. *AAPG Bulletin,* **34**, 215–238.

HOPLEY, D. 1982. *The Geomorphology of the Great Barrier Reef: Quarternary Development of Coral Reefs*. Wiley-Interscience New York.

HUNTER, I. G. & JONES, B. 1996. Coral associations of the Pleistocene Ironshore Formation on Grand Cayman. *Coral Reefs*, **15**, 249–267.

HUGHES, T. P. 1999. Off-reef transport of coral fragments at Lizard Island, Australia. *Marine Geology*, **157**, 1–6.

INSALACO, E. 1996. Upper Jurassic microsolenid biostromes of northern and central Europe: facies and depositional environment. *Palaeogeography, Palaeoclimatology, Palaeoecology*, **121**, 169–194.

—— 1998. The descriptive nomenclature and classification of growth fabrics in fossil scleractinian reefs. *Sedimentary Geology*, **118**, 159–186.

JAMES, F. C. 1983. Reef. *In*: SCHOLLE, P. A., BEBOUT, D. G. & MOORE, C. H. (eds) *Carbonate Depositional Environments. AAPG Memoir*, **33**, 345–462.

—— & MCCULLOCH, C. E. 1990. Multivariate analysis in ecology and systematics: Panacea or Pandora's box? *Annual Review of Ecology and Systematics*, **21**, 129–166.

JAMES, N. P. & MACINTYRE, I. G. 1985. Carbonate

depositional environments. Modern and Ancient. Part 1: Reefs, zonation, depositional facies, diagenesis. *Colorado School of Mines Quarterly*, **80**, 1–70.

KAN, H., HORI, N., NAKASHIMA, Y. & ICHIKAWA, K. 1995. The evolution of narrow reef flats at high-latitude in the Ryukyu islands. *Coral Reefs*, **14**, 123–130.

KAUFMANN, E. G. & SCOTT, R. W. 1976. Basic concepts of community ecology and paleoecology. *In:* SCOTT, R. W. & WEST, R. R (eds) *Structure and Classification of Paleocommunities*. Dowden, Hutchinson and Ross, Stroudsburg, 1–28.

KENKEL, N. C. & ORLOCI, L. 1986. Applying metric and non-metric multidimensional scaling to some ecological studies: some new results. *Ecology*, **67**, 919–928.

KERSHAW, S. 1994. Classification and geological significance of biostromes. *Facies*, **31**, 81–92.

—— 1998. The applications of stromatoporoid palaeobiology in palaeoenvironmental analysis. *Palaeontology*, **41**, 509–544.

KIDWELL, S. M. & BOSENCE, D. W. J. 1991. Taphonomy and time-averaging of marine shelly faunas. *In*: ALLISON, P. A. & BRIGGS, D. E. G. (eds) *Releasing the Data Locked in the Fossil Record*. Topics in Geobiology, **9**, Plenum Press, New York and London, 115–209.

KLEYPAS, J. A. 1996. Coral reef development under naturally turbid conditions: fringing reefs near Broad Sound, Australia. *Coral Reefs*, **15**, 153–167.

KUSS, J. 1983. Faziesentwicklung in proximalen Intraplatform-Becken: Sedimentation, Palökologie und Geochemie der Kössener Schichten (Ober-Trias, Nördliche Kalkalpen). *Fazies*, **9**, 61–172.

LADD, H. S. 1977. Types of coral reefs and their distribution. *In*: JONES, O. J. & ENDEAN, R. (eds) *Biology and Geology of Coral Reefs*. Academic, London, 1–19.

LEINFELDER, R. R. 1994. Distribution of Jurassic reef types: a mirror of structural and environmental changes during breakup of Pangea. *Canadian Society of Petroleum Geologists*, **17**, 677–700.

—— 1997. Coral reefs and carbonate platforms within a siliciclastic setting: general aspects and examples from the late Jurassic of Portugal. *Proceedings of the 8th International Coral Reef Symposium, Panama,* **2**, 1737–1742.

LIU, K., PIGRAM, C. J., PATERSON, L. & KENDALL, C. G. ST. C. 1998. Computer simulation of a Cainozoic carbonate platform, Marion plateau, north-east Australia. *In*: CAMOIN, G. F. & DAVIES, P. J. (eds) *Reefs and Carbonate Platforms in the Pacific and Indian Ocean*. International Association of Sedimentologists, Special Publication, **25**, 145–161.

LONGMAN, M. W. 1981. A process approach to recognizing facies of reef complexes. *In*: TOOMEY, D. F. (ed.) *European Fossil Reef Models.* Society of Economic Paleontologists and Mineralogists, Special Publications, **30**, 9–40.

MARTIN, J. M., BRAGA, J. C. & RIVAS, P. 1989. Coral successions in Upper Tortonian reefs in SE Spain. *Lethaia*, **22**, 271–286.

MONTAGGIONI, L. F. & FAURE, G. 1997. Response of reef coral communities to sea-level rise: a Holocene model from Mauritius (Western Indian Ocean). *Sedimentology*, **44**, 1053–1070.

NEUMANN, A. C. & MACINTYRE, I. G. 1985. Reef response to sea-level rise: keep-up, catch-up, or give-up. *Proceedings of the 5th International Coral Reef Symposium, Tahiti*, **3**, 105–110.

NOSE, M. 1995. Vergleichende Faziesanalyse und Palökologie korallenreicher Verflachungsabfolgen des iberischen Oberjura. *Profil*, **8**, 1–237.

ORSZAG-SPERBER, F., PURSER, B. H., RIOUAL, M. & PLAZIAT, J. C. 1998. Post-Miocene sedimentation and rift dynamics in the southern Gulf of Suez and northern Red Sea. *In*: PURSER, B. H., & BOSENCE, D. W. J. *Sedimentation and Tectonics in Rift Basins – Red Sea-Gulf of Aden*. Chapman & Hall, London, 239–270

PERRIN, C., BOSENCE, D. & ROSEN, B. 1995. Quantitative approaches to paleozonation and paleobathymetry of corals and coralline algae in Cenozoic reefs. *In*: BOSENCE, D. W. J. & ALLISON, P. A. (eds) *Marine Paleoenvironmental Analysis from Fossils*. Geological Society, London, Special Publications, **83**, 181–229.

PIANKA, E. R. 1970. On r- and K-selection. *American Naturalist*, **104**, 592–597.

PILLER, W. E. 1981. The Steinplatte reef complex, part of an Upper Triassic carbonate platform near Salzburg, Austria. *In*: TOOMEY, D. F. (ed.) *European Fossil Reef Models.* Society of Economic Paleontologists and Mineralogists, Special Publications, **30**, 261–290.

—— & PERVESLER, P. 1989. The northern Bay of Safaga (Red Sea, Egypt): an actuopalaeontological approach I. Topography and bottom facies. *Beiträge zur Paläontologie von Österreich*, **15**, 103–147.

——, DECKER, K. & HAAS, M. 1996. Sedimentologie und Beckendynamik des Wiener Beckens. *Sediment 96, 11. Sedimentologentreffen, Excursion Guide, Geologische Bundesanstalt Wien.*

POMAR, L. 1991. Reef geometries, erosion surfaces and high frequency sea-level changes, upper Miocene reef complex, Mallorca, Spain. *Sedimentology*, **38**, 243–270.

—— & WARD, W. C. 1999. Reservoir-scale heterogeneity in depositional packages and diagenetic patterns on a reef-rimmed platform, Upper Miocene, Mallorca, Spain. *AAPG Bulletin*, **83**, 1759–1773.

——, —— & GREEN, D. G. 1996. Upper Miocene reef complex of the Llucmajor area, Mallorca, Spain. Concepts in Sedimentology and Paleontology, **5**, Society for Sedimentary Geology, 191–225.

POTTS, D. C., DONE, T. J., ISDALE, P. J. & FISK D. A. 1985. Dominance of a coral community by the genus *Porites* (Scleractinia). *Marine Ecology Progress Series*, **23**, 79–84.

PURSER, B. H. & BOSENCE, D. W. J. 1998. *Sedimentation and Tectonics in Rift Basins – Red Sea-Gulf of Aden*. Chapman & Hall, London.

——, BARRIER, P., MONTENAT, C., ORSZAG-SPERBER, F., OTT D'ESTEVOU, P., PLAZIAT, J.-C. & PHILOBBOS, E. 1998. Carbonate and siliciclastic sedimentation in an active tectonic setting: Miocene of the north-western Red Sea rift, Egypt. *In*: PURSER, B. H. & BOSENCE, D. W. J. *Sedimentation and Tectonics in Rift Basins – Red Sea-Gulf of Aden*. Chapman & Hall, London, 239–270.

REISS, Z. & HOTTINGER, L. 1984. *The Gulf of Aqaba. Ecological Micropaleontology*. Springer, Berlin.

RIEGL, B. & PILLER, W. E. 1997. Distribution and environmental control of coral assemblages in northern Safaga Bay (Red Sea, Egypt). *Facies*, **36**, 141–162.

—— & —— 1999. Coral frameworks revisisted: reefs and coral carpets in the northern Red Sea. *Coral Reefs*, **18**, 241–253.

—— & VELIMIROV, B. 1994. The structure of coral communities at Hurghada in the northern Red Sea. *PSZN Marine Ecology*, **15**, 213–231.

ROBERTS, H. H. 1985. Carbonate platforms forming in a strong tidal current setting: southern Gulf of Suez. *Proceedings of the 5th International Coral Reef Congress, Tahiti*, **3**, 335–341.

ROSEN, B. R. 1975. The distribution of reef corals. *Reports of the Underwater Association (New Series)*, **1**, 1–16.

—— 1990. Reefs and carbonate build-ups. *In*: BRIGGS, D. E. G. & CROWTHER, P. R. (eds) *Palaeobiology – a Synthesis*. Blackwell, Oxford, 341–346.

SCHLAGER, W. 1992. *Sedimentology and Sequence Stratigraphy of Reefs and Carbonate Platforms*. Continuing Education Course Note Series, **34**, AAPG.

—— 1998. Exposure, drowning and sequence boundaries on carbonate platforms. *In:* CAMOIN, G. F. & DAVIES, P. J. (eds) *Reefs and carbonate platforms in the Pacific and Indian Ocean*. International Association of Sedimentologists, Special Publications, **25**, 3–21.

SMITH, B. T., FRANKEL, W. & JELL, J. S. 1998. Lagoonal sedimentation and reef development on Heron Reef, southern Great Barrier Reef Province. *In:* CAMOIN, G. F. & DAVIES, P. J. (eds) *Reefs and Carbonate Platforms in the Pacific and Indian Ocean*. International Association of Sedimentologists, Special Publications, **25**, 281–294.

STODDART, D. R. 1969. Ecology and morphology of recent coral reefs. *Biological Review*, **44**, 433–493.

STRASSER, A., STROHMENGER, C., DAVAUD, E. & BACH, A. 1992. Sequential evolution and diagenesis of Pleistocene coral reefs (South Sinai, Egypt). *Sedimentary Geology*, **78**, 59–79

TUCKER, M. E. & WRIGHT, V. B. 1990.*Carbonate Sedimentology*. Blackwell, Oxford.

VAN WOESIK, R. & DONE, T. J. 1997. Coral communities and reef growth in the southern Great Barrier Reef. *Coral Reefs*, **16**, 103–115.

VERON, J. E. N. 1995. *Corals in Space and Time: Biogeography and Evolution of the Scleractinia*. Comstock, Cornell.

WAINWRIGHT, S. A. 1965. Reef communities visited by the Israel South Red Sea Expedition, 1962. Israel South Red Sea Expedition 1962. Reports No.9. *Bulletin of the Sea Fisheries Research Station Israel*, **38**, 40–53.

WHITTAKER, R. H. 1975. *Communities and Ecosystems*, second edition. Macmillan, New York.

WEBSTER, J. M., DAVIES, P. J. & KONISHI, K. 1998. Model of fringing reef development in response to progressive sea level fall over the last 7000 years – (Kikai-jima, Ryukyu Islands, Japan). *Coral Reefs*, **17**, 289–308.

WOOD, R. 1999. *Reef Evolution*. Oxford University, Oxford.

Diversity, growth forms and taphonomy: key factors controlling the fabric of coralline algae dominated shelf carbonates

JAMES H. NEBELSICK[1] & DAVIDE BASSI[2]

[1] *Institute of Geology and Palaeontology, University of Tübingen, Sigwartstrasse 10, D – 72076 Tübingen, Germany (e-mail: nebelsick@uni-tuebingen.de)*

[2] *Dipartimento di Scienze Geologiche e Paleontologiche, Universita' di Ferrara, Corso Ercole I d'Este 32, I – 44100 Ferrara, Italy*

Abstract: The fabric of biogenic carbonate sediments can be differentiated (1) with respect to diversity of constituent components, (2) using features pertaining to their growth forms and (3) upon consideration of taphonomic aspects. These not only determine limestone fabrics, but also form the basis for facies differentiation and palaeoecological interpretation.

This study is based on coralline algae dominated Lower Oligocene shelf carbonates from northern Slovenia from which seven facies (nummulitic, bivalve, foraminiferal–coralline algal, coralline algal, coralline algal–coral, coral and grainstone) are differentiated. The role of diversity, growth forms and taphonomy of coralline algae in each facies is discussed. Nine species from seven genera of coralline algae including geniculates and non-geniculates were recognized. Numerous different growth forms ranging from crusts, protuberances, lamellae to arborescent types are present. A wide range of taphonomic features including disarticulation, encrustation, fragmentation and abrasion can be observed.

The determination of diversity is dependent on taxonomic identification using preserved diagnostic characters relevant to palaeontological and botanical systematics. Growth-form determination in thin section is influenced by orientation and sectioning effects. The taphonomy of red algae is highly dependent on initial growth form and the specific environment in which they are found. A number of taphonomic processes described in Recent environments (e.g. disease, shallow grazing) cannot be ascertained in fossil material, while others are readily observable. Some taphonomic processes are detrimental (e.g. fragmentation, abrasion) to the preservation and recognition of vegetative and growth-form features, while others have positive effects (e.g. encrustation).

The fabric of biogenic limestones is largely controlled by the diversity of constituent components, their skeletal morphologies and their taphonomic pathways (Fig. 1). Coralline algae dominated shelf carbonates are especially suitable for fabric analysis as the contribution of these three factors can be readily recognized and studied in thin section. In this paper, coralline algae dominated limestones from the Early Oligocene of Slovenia are used as a case study to show how the diversity, growth forms and taphonomy of a dominating biogenic component impart a primary control on the fabric of limestones. This study is enhanced by the excellent preservation of a diverse algal flora (Bassi *et al.* 2000; Bassi & Nebelsick 2000) as well as the presence of a number of different facies allowing the study of changes along ecological gradients to be made.

Of special importance concerning red algae are the following: (1) the determination of diversity by applying modern taxonomic concepts which are highly dependent on the recognition of key features used in modern palaeontological as well as neophycological systematics; (2) the possibility of using growth-form analysis as designated for Recent representatives, although the orientation of thin sections and sectioning effects have to be taken into account; and (3) the constraints of recognizing complex taphonomic pathways.

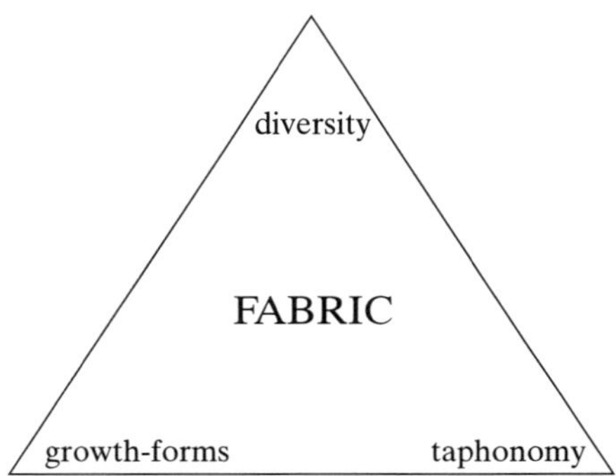

Fig. 1. Factors controlling the fabric of coralline algae dominated limestones.

From: INSALACO, E., SKELTON, P. W. & PALMER, T. J. (eds) 2000. *Carbonate Platform Systems: components and interactions*. Geological Society, London, Special Publications, **178**, 89–107. 0305–8719/00/$15.00

Diversity

Taxonomic concepts

Taxonomic concepts concerning living species of coralline red algae (Corallinales, Rhodophyta) have undergone important changes since the 1960s following a number of major studies, many of which are cited in Johansen (1981), Woelkerling (1988, 1996), Irvine & Chamberlain (1994) and Woelkerling & Lamy (1998). Detailed observations of the many different characters of Recent coralline species such as vegetative anatomy (vegetative structure) and growth forms (external appearance or habit) have shown that the stability of these characters has never been fully assessed. Many of these characters vary too much to be taxonomically reliable and, in fact, for most species few or no data are available (Woelkerling 1985, 1996; Penrose 1992; Verheij 1993; Woelkerling & Harvey 1993; Wilks & Woelkerling 1995; Townsend *et al.* 1995). As a consequence, several of the more common genera found in the older literature have been relegated to synonymy. Braga *et al.* (1993) have shown that these developments have significant implications for the taxonomy of fossil coralline red algae.

The vegetative characters of fossil non-geniculate coralline genera used in palaeophycology have traditionally been different from those recognized for Recent coralline algae (Poignant 1979*a*,*b*, 1984). More recent works have shown, however, that taxonomic criteria currently used by phycologists at generic and subfamily levels can also be identified in fossil material (Bosence 1991; Braga *et al.* 1993; Iryu & Matsuda 1994; Basso 1995*a*,*b*; Bassi 1995, 1998; Braga & Aguirre 1995, 1998; Aguirre *et al.* 1996; Basso *et al.* 1997; Rasser & Piller 1997, 1999). Many fossil taxa have now to be reassigned to different genera (e.g. Braga *et al.* 1993; Aguirre *et al.* 1996; Aguirre & Braga 1998; Bassi 1998; Rasser & Piller 1999). This reiterates the need to (1) reassess the precise circumscription of fossil taxon (avoiding the use of taxa with no preserved type material) and (2) re-examine type collections and/or material from type localities.

The classification of present-day geniculate coralline algal genera is based largely on genicular cell characteristics including the position and nature of conceptacles, growth habits and branching types (e.g. Johansen & Silva 1978; Johansen 1981; Woelkerling 1988; Riosmena-Rodríguez & Siqueiros-Beltrones 1996; Womersley & Johansen 1996*a*,*b*,*c*). The assignment of fossil species within geniculate genera must, however, be treated with caution, especially in the absence of reassessments with respect to current taxonomic concepts.

Diversity gradients

Taxonomic differences along environmental gradients have been shown for Recent environments, especially in tropical reef settings (e.g. Wray 1979; Minnery *et al.* 1985; Adey 1986; Minnery 1990; Martindale 1992; Iryu *et al.* 1995; Perrin *et al.* 1995). These have described depth-dependent coralline algal associations with a shallow-water mastophoroid dominated assemblage and deeper-water melobesioid dominated assemblage. The presence of corallines in cryptic environments can reflect this pattern as shown by Martindale (1992). Data on the occurrence and abundance of corallines with respect to depth and distance from shore given by Littler (1973*a*,*b*) for Oahu, Hawaii, show differences in the taxa present, but little change in overall diversity. A deeper water survey (below 20 m) undertaken by Minnery *et al.* (1985) on the Flower Garden Banks of the Gulf of Mexico showed a general decrease of generic diversity with depth.

Growth forms

Coralline algae show a wide range of growth forms ranging from parasitic or partly endophytic, to smooth or knobby crusts of varying thickness, to rooted upright growing geniculates. The variation of growth forms within species has been often noted and related to differences in hydrodynamic energy and substrate morphology (e.g. Johnson 1961; Cabioch 1969; Adey & Adey 1973; Bosence 1976; Dethier 1994). Variation of coralline algal growth forms across hydrodynamic gradients in reefs has been shown by Martindale (1992) and Gherardi & Bosence (1999). The influence of physical environmental parameters on growth forms of coralline algae has been especially noted for algal ridges (e.g. Laborel 1961; Pérès & Picard 1958; Ginsburg & Schroeder 1973; Adey & Burke 1975; Adey 1978; Bosence & Pedley 1982; Bosence 1983*a*) and rhodoliths.

Rhodoliths consisting of free-living, nodular aggregates of non-geniculate corallines can be produced whenever crustose or branching corallines either settle on non-cohesive particulate substrate or are detached from existing hard substrates (e.g. Adey & McKibbin 1970; Bosellini & Ginsburg 1971; Adey & MacIntyre 1973; Bosence 1976, 1983*b*, *c*, 1991; Montaggioni 1979). They form in diverse environments on continental shelves around the world and have been described in tropical reefs (Scoffin *et al.* 1985; Bosence 1985*b*), temperate fjords and bays (Bosence 1976) and polar regions (Freiwald 1993, 1995), from the lowest intertidal to depths over 200 m (see also Reid & MacIntyre 1988; Littler *et al.* 1991; Iryu *et al.* 1995; Steller & Foster 1995; Piller & Rasser 1996; Basso 1998; Marrack 1999). The form and branching density has often been correlated with environmental parameters especially hydrodynamic gradients in both Recent and fossil environments.

Differences in growth forms have been widely used to delimit and identify genera, species and infraspecific taxa of non-geniculate coralline algae (e.g. Lemoine 1911, 1939; Adey *et al.* 1982). Woelkerling *et al.* (1993) determined the range of growth forms present amongst non-geniculate coralline algae and proposed an intergrading network with ten focal points. These include: unconsolidated, encrusting, warty, lumpy, fruticose, discoid, layered, foliose, ribbon-like and arborescent. These growth forms have been applied to fossil coralline algae in thin section analysis (Bassi 1998; Rasser & Piller 1999).

There are two important effects that can influence our perception of the growth forms of coralline algal thalli in thin section: (1) the orientation of the thalli in the sediment, and (2) sectioning effects. Fossil remains usually lie with the longest axis parallel to the sediment surface: in the case of a flattened, sheet-like thallus, thin sections cut perpendicular to the bedding plane (as in the present study) will result in a preponderance of cross-sections. Sectioning effects can lead to the preferential recognition of certain growth forms. Sporadic occurrences of protuberances on a relatively flat thallus surface, for example, have a low chance of being included within a section plane. The chances of recognizing more fully developed lumpy and fruticose growth forms is also reduced. This effect resembles that when sectioning geometric bodies (for example a cone) with an oblique section resulting in a reduced perception of the true height or morphology of the object. The recognition of other growth forms such as encrusting and lamellate types should, however, not be seriously affected.

Taphonomy

Although taphonomic aspects have been included in past investigations, the taphonomy of coralline algae as such has rarely been specifically addressed or studied in detail. Coralline algae, as all potential fossils, are necessarily subject to a wide variety of necrolytic, biostratinomic and diagenetic processes, all affecting their preservation potentials. Necrolysis such as death by disease (for example the coralline lethal orange disease (CLOD) described by Littler & Littler (1995, 1997)) has little chance of being discerned in the fossil record. Biostratinomy and diagenesis, however, have a high potential of being recognized and can potentially be useful in discerning palaeoecological parameters and gradients.

Disarticulation

Disarticulation is a process obviously pertaining to geniculate coralline algae in which the intergenicula are connected by decalcified genicula which decompose after death. The in situ preservation of articulated intergenicula has been observed when the genicular joints are encrusted by coralline algae (Bosence 1984) or foraminifera (Bassi 1998).

Encrustation

While obviously a biological process of eminent importance for coralline algae, encrustation is also an important factor affecting preservation and is thus treated here with respect to its taphonomic connotations. This is especially the case where the process of encrustation protects the encrusted organisms from destruction by abrasion. There is a complex ecology of encrustation involving coralline algae within multispecies frameworks including aspects of space competition and encrusting hierarchies (Steneck 1986, 1997; Keats & Maneveldt 1994; Keats *et al.* 1994). The succession of encrusting frameworks and associated changes in coralline algal growth forms have been noted across hydrodynamic gradients (Gherardi & Bosence 1999). Coralline algae and rhodoliths can be accompanied by a varied epifauna including encrusting foraminifera, hydrozoans, serpulids, bryozoans, brachiopods and other coralline algae (Bosence & Pedley 1982; Bosence 1983*b*, 1984). This can lead to the preservation of those parts of the coralline algae (for example the epithallus) which are not usually preserved through the process of bio-immuration (Voigt 1981).

Bioerosion

Bioerosion plays a major role in the ecology of coralline algae and can be caused by surface grazing and endophytic bioerosion boring. Grazing can be subdivided into shallow denuders (e.g. gastropods, patellids) and deep excavators (e.g. polyplacophores, sea urchins (Lawrence

1975) and fish (van den Hoek *et al.* 1975; Wanders 1977; Brock 1979)). Two processes have been identified to prevent fouling of the coralline algal surface: the promotion of herbivory and epithallial sloughing (Johnson *et al.* 1997; Steneck 1997). The positive feedback of grazing and coralline algal growth through biotic interaction has been a field of intense study including both qualitative and quantitative experimental investigations (e.g. Adey & MacIntyre 1973; Steneck 1997). The relationship goes so far as to include the specific production of proteinaceous, morphogenetic substances which regulate the metamorphosis of grazing gastropods (Morse & Morse 1984). Epithallial sloughing represents a self-cleaning mechanism through the controlled senescence and mortality of epithallial cells, thus preventing overgrowth.

Herbivory is obviously destructive if too deep and intense. Figueiredo (1997) noted that grazing activity can be detrimental to red algal growth. The association of deep herbivores (certain sea-urchins and parrot fish) and coralline algae has lead to the a co-evolutionary scenario with the development of multilayered thalli and sunken conceptacles in coralline algae as a protection against deep herbivores (Steneck & Adey 1976; Steneck 1982, 1986). Bioerosion through endophytic boring is a destructive process by both micro- and macroborers such as clionid sponges, boring bivalves, polychaetes and echinoids (Sebens 1986; Bosence 1983*a*; Kidwell & Bosence 1991).

Fragmentation

Fragmentation in coralline algae is associated especially with the production of maerl, the Breton term for unattached corallines and algal gravels found off the northwest coast of France (Lemoine 1910; Bosence 1983*b*, 1985*a*). This facies is very common on the boreal to subartic shelves in northwestern Europe (Jacquotte 1962; Adey & McKibbin 1970; Adey & Adey 1973; Bosence 1976, 1980, 1983*b*; Freiwald 1995; Basso 1998). The formation of maerl is tied to the fragmentation of rhodolith branches.

Abrasion

Abrasion caused by transport and/or sediment agitation can lead to the destruction of important surface characters such as the epithallus and (raised) conceptacles. Chave (1964) demonstrated the very low durability of coralline algal skeletal material in tumbling experiments. Abrasion can decidedly influence growth forms as described by Bosence (1976) and Testa (1997).

Early diagenesis

The early diagenesis of coralline algae is complex. Alexandersson (1974) describes the diagenetic history within living rhodoliths with the production of internal cements consisting of both aragonite (needles and spherulitic cement) and Mg-calcite (fringes, micrites and intraskeletal microdruses). These are related to a unique saturated or even supersaturated internal microenvironment sustained by the algae, which counters external dissolving processes. This is contrasted to post-mortem alterations which are entirely destructive. Further intricacies of precipitation are described by Bosence (1985*a*, 1991) with the successive growth of different crystal morphologies including a primary layer with needle-shaped crystals within and parallel to the cell wall followed by a secondary layer of bladed to fibrous calcite crystals. Eventually the cells can be infilled by micrometre-sized Mg-calcite crystals which may fill the cell cavity. Coarse, aragonitic botryoidal cements with blade-shaped aragonite crystals can be common as early diagenetic features of coralline algae (Bosence 1991). Micritization can affect the surface of the thalli obscuring or obliterating cell structures (Martindale 1992).

Paleogene succession of the Gornji Grad area

The Gornji Grad area (northern Slovenia) (Fig. 2) represents an important transition zone between the northernmost extension of the Mediterranean Tethys to the south and the developing Paratethys to the north and east. The trangressive Paleogene succession overlies Triassic carbonates and is initiated by the limno-fluviatile Basal Unit consisting of conglomerates, sandstones and marls of varying thickness. This series is followed by the Gornji Grad Beds consisting of 5 to 30 m of marine marls, sandstones and carbonates (of which the facies are described in detail below). These beds trangress above Triassic limestones, Eocene limestones and the Basal Series. The stratigraphical framework of the Gornji Grad beds is given by Drobne *et al.* (1985) who recognized *Nummulites fichteli* and *N. germanicus* as indicating an Early Oligocene age. The carbonates are followed by the marine marls reaching a thickness of *c.* 170 to 270 m (Hemleben 1964) deposited during NP23 to NP24 (Rupelian) (Jelen *et al.* 1980; Bricl & Pavšič 1991). The sedimentary development is capped by a thick succession of the volcanoclastic tuffites.

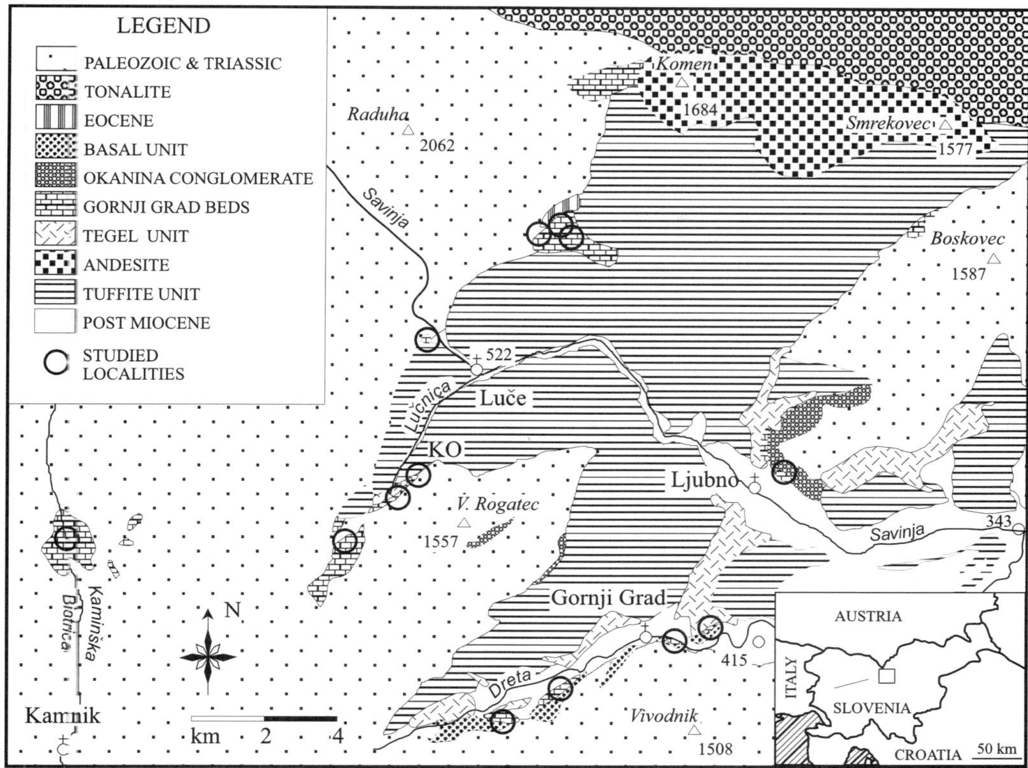

Fig. 2. Study area with lithologies and location of profiles (KO = Korenovec see Fig. 3)

Calcareous algae in the Gornji Grad Beds

The vegetative and reproductive features of the fossil coralline algae generally show excellent preservation allowing the systematics of Recent representatives to be applied at the genus level (Bassi *et al.* 2000; Bassi & Nebelsick 2000). The determination of species was made following an open nomenclature because of the lack of revisions of the type collections of fossil material. In all, 11 species belonging to nine genera were recognized (Table 1): *Lithoporella melobesioides* (Foslie) Foslie 1909, *Neogoniolithon* sp. 1 (Fig. 7a), *Spongites* sp. 1, *Spongites* sp. 2, *Lithothamnion* sp. 1 (Fig. 7b), *Lithothamnion* sp. 2, *Mesophyllum* sp. 1, *Sporolithon* sp. 1, *Subterraniphyllum thomasii* Elliott 1957, *Polystrata alba* (Pfender) Denizot 1968 (Cryptonemiales, Peyssonneliaceae), and the green algae *Cymopolia* sp. 1 (Dasycladales, Dasycladaceae) and *Halimeda* sp. 1. (Syphonales, Halimedaceae). Unidentified geniculate coralline algal fragments are also present. This careful neontological systematic approach allowed taxonomic differences to be recognized with respect to growth forms and taphonomy both within and across facies boundaries.

Carbonate facies of the Gornji Grad Beds

The Gornji Grad Limestones are dominated by coarse, poorly sorted rudstones with wackestone to packstone matrix, Nebelsick *et al.* (2000). Packstones and grainstones are subordinate. The textural classification of Embry & Klovan (1972) is used allowing for a general component and matrix description. Microfacies distinction was based on 500 point counts on 80 thin sections cut perpendicular to bedding. This allows changes in component distributions to be followed within sections (Fig. 3). Seven microfacies have been distinguished in a detailed study which will be published elsewhere: nummulitic, bivalve, foraminiferal–coralline algal, coralline algal, coralline algal–coral, coral and grainstone facies. The dominance of coralline algae, their

Table 1. *Coralline algal taxonomy and growth forms of the Lower Oligocene Gornji Grad Beds*

	Encrusting		Protuberances			Lamellae		
Species	Type 1	Type 2	Warty	Lumpy	Fruticose	Layered	Foliose	Arborescent
Lithop. melobesioides		×						
Neogoniolithon sp. 1	×	×				×		
Spongites sp. 1		×	×	×				
Spongites sp. 2			×	×		○		
Lithothamnion sp. 1		×	×					
Lithothamnion sp. 2		×	○				×	
Mesophyllum sp. 1					×			
Geniculate corall. *s. l.*								×
Subterr. thomasi					×			
Sporolithon sp. 1		○	×	×				
Polystrata alba		○					×	

Lithop. = *Lithoporella*; *Subterr.* = *Subterraniphyllum*; corall. *s. l.* = corallines *sensu lato*; × = common; ○ = rare; no entry = absent

excellent preservation and the variation of facies in which they occur make these limestones ideal for the study of the relationships between their diversity, growth forms and taphonomy (compiled in Table 2).

Facies description

The nummulitic facies is characterized by packstones dominated by nummulites and terrigenous components up to 1 cm in diameter. The dominating nummulites are small and highly fragmented and abraded. Small benthic foraminifera are dominated by miliolids and rotaliids. Bryozoans are common; bivalves and rare barnacles are present. The coralline algae are rare, very highly fragmented and remain unidentified.

The rudstones of the bivalve facies are characterized by the total dominance of highly fragmented and abraded oysters and pectinid bivalves along with common isolated corals as well as miliolids and textulariids. Unidentified coralline algae are present in moderate amounts.

The foraminiferal–coralline algal facies represents a fine packstone and is dominated by small benthic foraminifera especially miliolids followed by textulariids. Highly fragmented specimens of coralline algae can also be common consisting of *Lithothamnion* sp. 1 and *Mesophyllum* sp. 1. The coralline algal growth forms present show warty and lumpy protuberances; encrusting types and fruticose protuberances can also occur.

The coralline algal facies consisting of rudstones and bindstones with wacke- to packstone matrix is totally dominated by diverse coralline algae containing all non-geniculate taxa found in the Gornji Grad Beds. The most characteristic species is *Neogoniolithon* sp. 1 consisting of a thick crust, 500 μm in thickness, and up to several centimetres long. *Lithoporella melobesioides*, *Spongites* sp. 2, *Lithothamnion* sp. 1, *Mesophyllum* sp. 1 and *Sporolithon* sp. 1 are common. *Spongites* sp. 1, *Lithothamnion* sp. 2 and *Polystrata alba* are also present. A correspondingly wide variety of growth forms occurs including encrusting types, warty, lumpy and fruticose protuberances as well as layered and foliose lamellate forms. Thick algal encrustation sequences can be observed and the components are generally well preserved; rotaliid small benthic foraminifera are common.

The coralline algal–coral facies consists of rudstones with wackestone to packstone matrix dominated by coralline algae and corals. A diverse coralline algal flora is present with common *Lithoporella melobesioides*, *Spongites* sp. 1 and sp. 2, *Lithothamnion* sp. 1, *Sporolithon* sp. 1 and *Subterraniphyllum thomasii*. *Neogoniolithon* sp. 1, *Lithothamnion* sp. 2, *Mesophyllum* sp. 1 and *Polystrata alba* as well as geniculates are also present. A wide variety of growth forms occur with encrusting types, protuberances (warty, lumpy and fruticose) and lamellate (layered, foliose) forms. The encrusting foraminifera *Haddonia heissigi* is common as are small articulate brachiopods. Thick encrustation sequences are again present especially around coral fragments.

Corals totally dominate the rudstones with wackestone to packstone matrix of the coral facies, and are often heavily encrusted. Coralline algae are subordinate with a fauna which is less diverse than the coralline algal and coralline algal–coral facies: *Lithoporella melobesioides*, *Spongites* sp. 1 and sp. 2 are

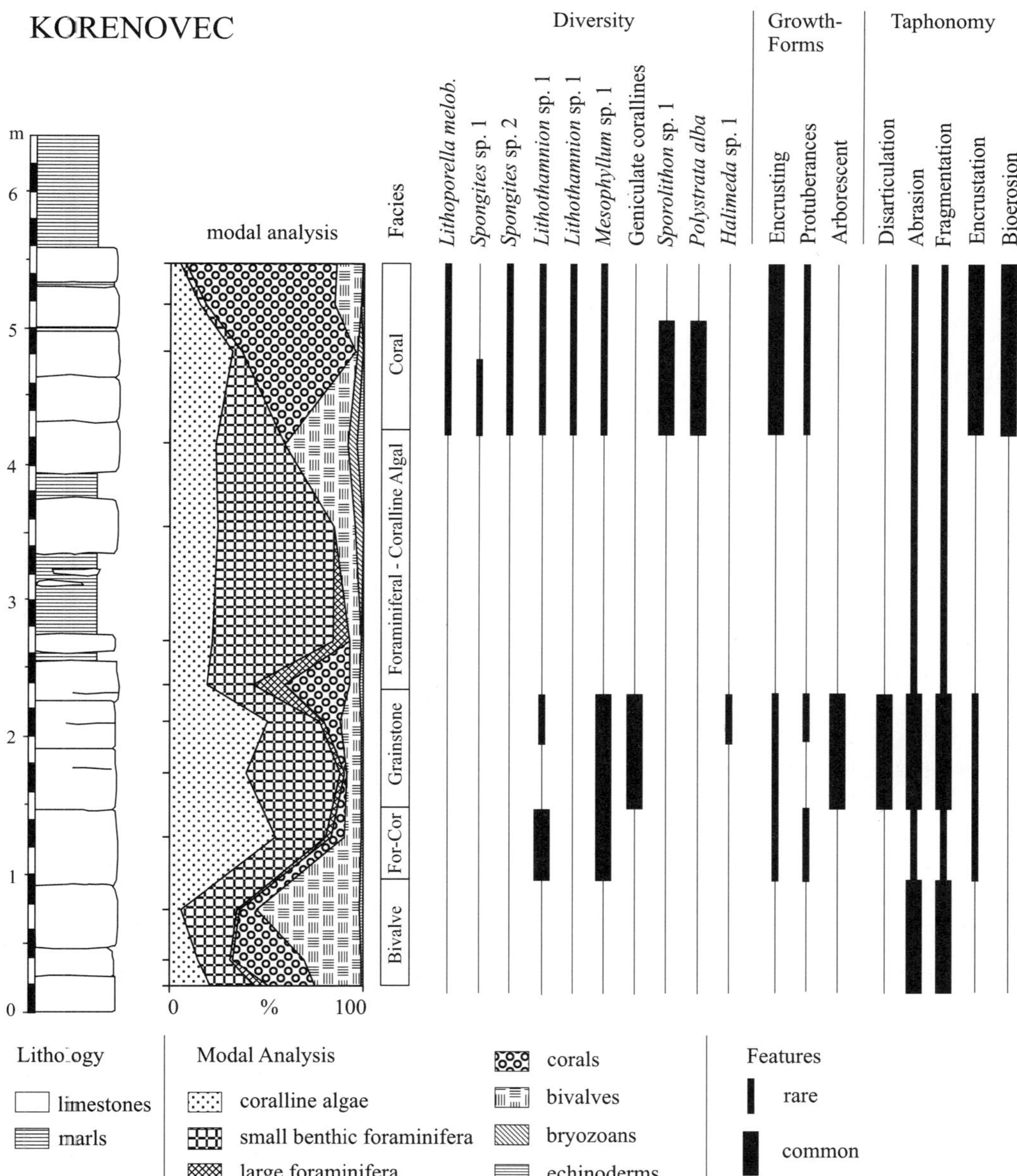

Fig. 3. Profile (locality Korenovec) with modal analysis of biogenic components as well as qualitative assessment of diversity, growth forms and taphonomy of coralline red algae within the section. Coralline algae are present throughout the section (For-Cor = foraminiferal–coralline algal facies; *melob.* = *melobesioides*).

common; *Neogoniolithon* sp. 1, *Lithothamnion* sp. 1, *Mesophyllum* sp. 1, *Sporolithon* sp. 1 and *Polystrata alba* also occur. Growth forms are restricted to encrusting types and protuberances (warty, lumpy and fruticose).

The grainstone facies is rarely present and is the only facies showing intergranular cements. It is dominated by relatively small, highly fragmented, highly abraded coralline algal remnants with a subordinate amount of miliolid foraminifera and corals. The coralline algal flora is represented by *Lithothamnion* sp. 2, and *Mesophyllum* sp. 1 showing fruticose growth forms as well as disarticulated intergenicula of geniculate

Table 2. *Comparison of facies, coralline algal taxonomy, growth forms and taphonomy from the Carbonates of the Gornji Grad Beds*

Facies		Nummulitic	Bivalve	Foraminiferal–coralline algal	Coralline algal	Coralline algal–coral	Coral	Grainstone
Taxonomy								
Corallinaceae	*Lithoporella melob.*				×	×	×	
	Neogoniolithon sp. 1				×	O	O	
	Spongites sp. 1				O	×	×	
	Spongites sp. 2				×	×	×	O
	Lithothamnion sp. 1			×	×	×	O	
	Lithothamnion sp. 2				O	O		×
	Mesophyllum sp. 1			×	×	O	O	×
	Geniculate corallines					O		×
	Subterr. Thomasi					×		
	Corallinaceae indet.	×	×	×	×	×	×	×
Sporolithaceae	*Sporolithon* sp. 1				×	×	O	
Peyssonneliaceae	*Polystrata alba*				O	O	O	
Dasycladaceae	*Cymopolia* sp. 1							O
Halimedaceae	*Halimeda* sp. 1							O
Growth-Forms								
Encrusting	Type 1				×	O	O	
	Type 2		O	O	×	×	×	
Protuberances	Warty		O	O	×	×	×	
	Lumpy			×	×	×	O	
	Fruticose		O	×	O	O	O	×
Lamellae	Layered				×	O		
	Foliose				×	O		
Arborescent						O		×
Taphonomy								
	Disarticulation							×
	Encrustation		O		×	×	×	
	Bioerosion				×	×	×	
	Fragmentation	×	×	O	O	O	O	×
	Abrasion	×	×	O	O	O	O	×
	Diagenesis	×	×	×	×	×	×	×

× = common; O = rare, no entry = absent

coralline algae; *Spongites* sp. 2 is also present. The facies also contains the green algae *Cymopolia* sp. 1 and *Halimeda* sp. 1.

General facies interpretation

Carbonate facies distributions within the Gornji Grad Beds have been interpreted to be dependent on water energy (turbulence), the nature of the substrate, sedimentation rate and light (Nebelsick *et al.* 2000). Rapidly changing substrate conditions result from interrupted carbonate development in a nearshore, mixed carbonate–siliciclastic setting with a high input of both coarse and fine clastic material. The nummulitic and bivalve facies represent high water turbulence and low substrate stability. Colonization by coralline algae is promoted by decreasing turbulence and increasing substrate stability. The increased presence of corals representing relatively stable secondary hardgrounds allows intense encrustation. The grainstone facies represent higher energy environments. The carbonate facies were all deposited within the photic zone.

Coralline algae in the Gornji Grad Beds

Diversity

The facies can be divided into three groups with regard to coralline algal diversity: (1) the nummulitic and bivalve facies containing only non-determinable fragmented material; (2) the foraminiferal–coralline algal and grainstone facies with two or three non-geniculate taxa which are dominated by melobesioids; and (3) the coralline algal, coralline algal–coral, and coral facies with six or seven non-geniculate taxa. This latter group shows a mixed mastophoroid/melobesioid occurrence. The coral facies is dominated by two no-geniculate genera. The fact that the general distribution of mastophoroids and melobesioids reported from Recent well lit, oligotrophic environments (e.g. Wray 1979; Minnery *et al.* 1985; Adey 1986; Minnery 1990; Martindale 1992; Iryu *et al.* 1995; Perrin *et al.* 1995) is not followed may be because of the generally turbid, shallow-water setting of the Gornji Grad Beds.

Growth forms

Coralline algal growth forms as defined by Woelkerling *et al.* (1993) can be applied with some modifications to the coralline algae in the Gornji Grad Beds (Table 1, Fig. 4) with: (1) two types of encrusting growth on (a) soft muddy

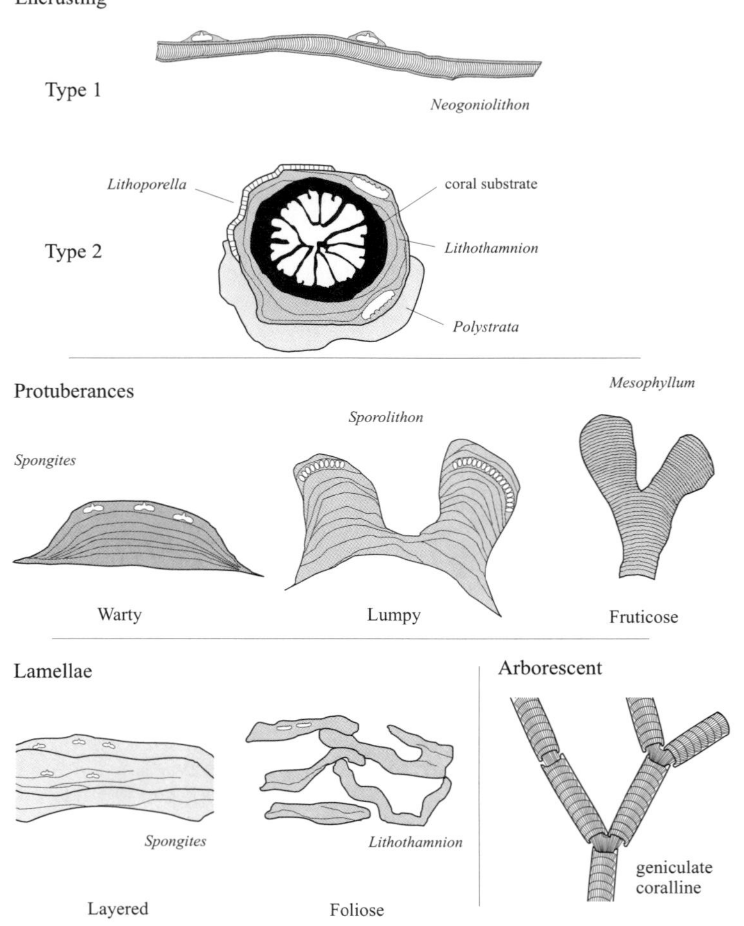

Fig. 4. Growth forms of coralline algae in the Lower Oligocene Gornji Grad Beds.

and (b) hard biogenic substrates; (2) three types of protuberances including warty, lumpy and fruticose; (3) layered and foliose lamellae; as well as (4) the arborescent growth form reconstructed for geniculate coralline algae. The unconsolidated, discoid and ribbon-like growth forms (Woelkerling *et al.* 1993) were not recognized. There is a considerable variation among growth forms at both the genus and species levels with obvious transitions between them. It is important to note that neither rhodoliths (rare) nor maerl (not present), on which the numerous deductions concerning growth forms and the taphonomy of coralline algae have been made, play a significant role in the Gornji Grad Beds.

The encrusting growth is very common and consists of two different forms depending on the type of encrusted substrate: Type 1, is found in association with soft substrates; and Type 2 encrusting hard substrates, in this case the skeletons of other biogenic components.

Type 1 is most typically formed by *Neogoniolithon* which constructs relatively thick crusts, about 500 μm in thickness, and can reach several centimetres in length (Figs 4, 7a). These are found lying parallel to the bedding plane and can form a typical crustose fabric. Sometimes there are encrusting foraminifera (acervulinids) on top of the crusts. In some cases, thin layers of nondescript sparry cements are present between the lower surfaces of the thalli and the sediment. These probably correspond to either primary voids or decayed material subsequently filled by the cements. The thalli show little abrasion and fragmentation. There are two possible interpretations concerning the origin of these crusts: (1) they are formed from transported thalli of primary encrusters of soft material (e.g. algae, seagrass) which has subsequently decomposed; and (2) they have formed as a result of primary growth on an initial hard substrate which has subsequently reached out over the sediment surface (compare Basso 1995*a*,*b*). The latter interpretation is favoured here because of the orientation of the thalli in the sediment as well as general facies interpretation of the coralline algal facies in which they dominate.

The Type 2 encrusters on biogenic substrates are very common among numerous different coralline algal taxa (see Table 1). It is common in the coralline algal, coralline algal–coral and coral facies, and present in the foraminiferal–coralline algal and bivalve facies (Table 2). These can form multilayered sequences (Fig. 7c) including various encrusting foraminifera (*Haddonia*, gypsinids, acervulinids, planorbulinids), unilaminar bryozoans as well as other coralline algae. The core of these encrustation sequences often consists of coral branches (Figs 4, 7b). Continuous growths around the coral branches suggest the encrustation of living colonies. The encrustation of morphologically differentiated surfaces can result in irregular forms. The genera showing this growth form include *Lithothamnion*, *Spongites*, *Lithoporella* and *Polystrata*. This encruster type is, in principle, the growth form that dominates free-living rhodoliths which are also constructed by a multilayer succession of encrusting thalli.

Protuberances are present in all facies except for the nummulitic facies and occur in three forms (Figs 4, 7e,f) between which transitions may be present. They are frequently associated with the presence of conceptacles as often seen for the lumpy growth form of *Sporolithon*. Both warty and lumpy growth forms occur in *Spongites*, *Lithothamnion* and *Sporolithon*. The warty growth form shows the same distribution as the Type 2 encrusters. The lumpy growth forms dominate in the facies characterized by coralline algae, and are present in the coral and bivalve facies. The fruticose growth form is restricted to rare occurrences of *Mesophyllum* and possible *Subterraniphyllum* and is common in the foraminiferal–coralline algal as well as grainstone facies.

Lamellae (Figs 4, 7d) occur in two main types: densely layered as in *Spongites*, and the more loosely arranged layers of the foliose growth shown by *Lithothamnion* sp. 2 and *Polystrata*. Both are common in the coralline algal facies and present in the coralline algal–coral facies. An arborescent growth form (Fig. 4) can be implied for disarticulated intergenicula of unidentified geniculate coralline algae and is therefore common in the grainstone facies and present in the coralline algal–coral facies.

Taphonomy

The observed taphonomic features (Fig. 5) show a characteristic distribution with respect to the different facies (Table 2). Some taphonomic features can be directly observed, some inferred, while others are impossible or difficult to distinguish from one another or be preserved as such in the fossil record.

Disarticulation can be inferred for the geniculate coralline algae and is thus most commonly observed in the grainstone facies where geniculates occur. The non-calcified genicula connecting the intergenicula represent inherent points of weakness of the skeleton which subsequently disarticulate after death and decay. Neither an in

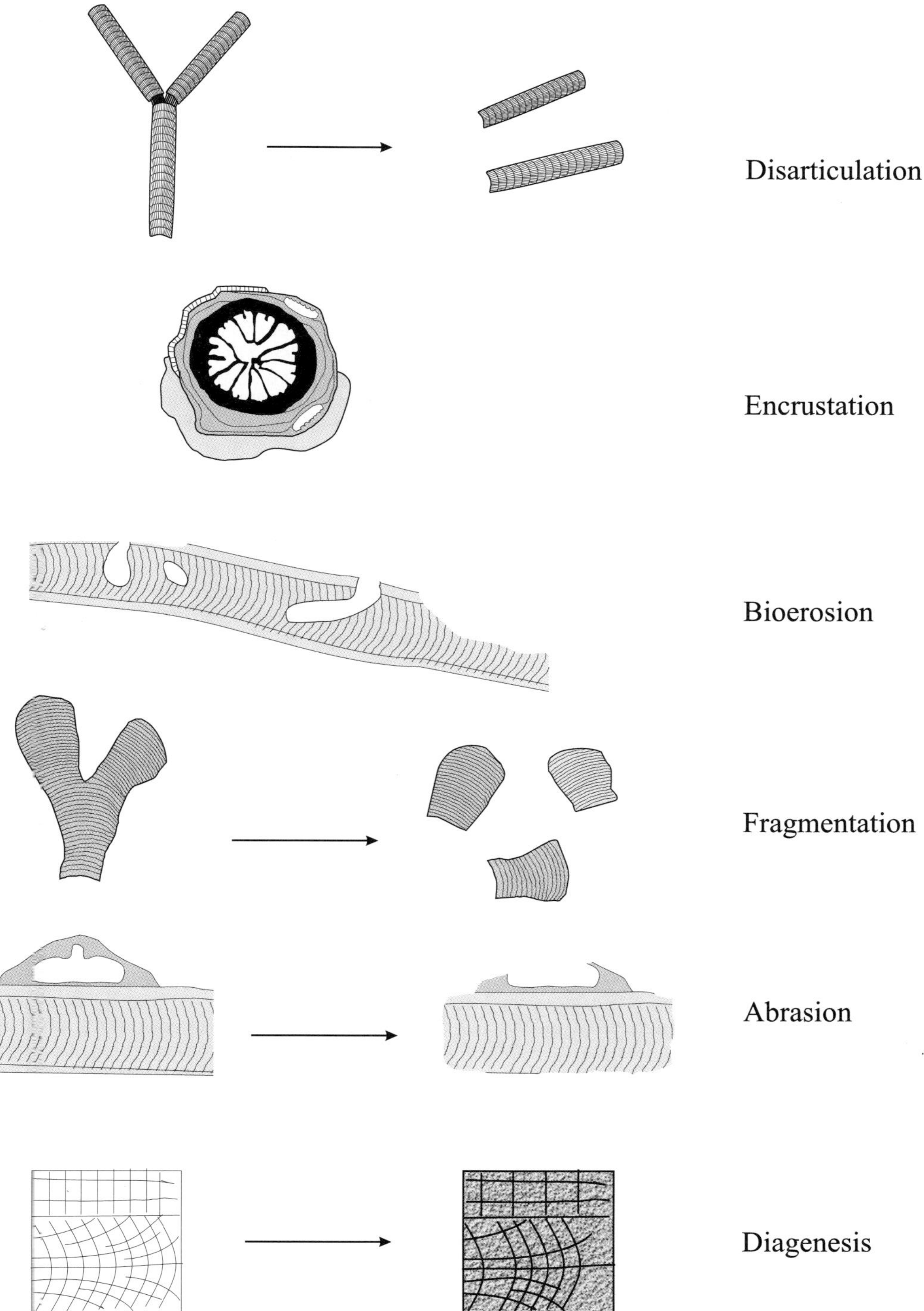

Fig. 5. Taphonomic features of coralline algae in the Lower Oligocene Gornji Grad Beds.

situ preservation of intergenicula nor preservation of articulation by encrustation was observed.

Encrustation can be readily observed and is very common in all the coralline algae dominated facies. Complex encrustation sequences can involve not only several different coralline algal taxa (Fig. 7c), but also encrusting foraminifera and bryozoans. It is especially important as a taphonomic factor because encrustation can ensure the preservation of the encrusted organisms. Coralline algae with thin thalli such as the unistratose palisade cell layers of *Lithoporella*, are thus most commonly found within encrustation sequences. Growth forms such as protuberances can also be preserved within encrusting sequences.

Bioerosion is obvious in the study material in the presence of endolithic borings which can be readily recognized as mud-filled, rounded holes that often continue through the coralline algal crusts into the underlying biogenic (often coral) substrate. It is most common in the coralline algal, coralline algal–coral and coral facies. Bioerosion caused by herbivory, although of eminent importance for the biology of coralline algae (see Adey & MacIntyre 1973; Steneck 1997) is impossible to distinguish from abrasion (see below).

Fragmentation (Fig. 5) is common resulting in isolated protuberances and fruticose fragments, as well as thallial segments of crusts. It is most common in the nummulitic, bivalve and grainstone facies where higher levels of water turbulence are envisioned. Fragmentation present in the other facies is thought to originate primarily from biological activity as these are interpreted to represent quiet water conditions. The determination of both taxonomy and growth forms of the coralline algae is highly dependent on the degree of fragmentation as demonstrated in the nummulitic and bivalve facies in which only unidentified, highly fragmented specimens of coralline algae were found.

Abrasion shows a similar distribution to fragmentation (Table 2) and is especially evident in the grainstone facies leading to the production of rounded components (Fig. 7h). This is especially true for intergenicula of geniculate coralline algae, which after disarticulation potentially represent relatively robust cylinders. Abrasion caused by transport and sediment agitation can remove important characters used for taxonomic identification.

Diagenesis obviously affects all the study material, though it is difficult to investigate considering the dominance of muddy matrix in most facies. Aragonitic skeletons are leached which affects the thalli of *Polystrata*. Micritization, leading to a 'masking' of characters such as cell wall features, is especially observed in the grainstone facies. All conceptacles are filled with nondescript sparite; pressure solution is also observed (Fig. 7g).

Synthesis of diversity, growth forms and taphonomy

In the following, a synthesis for the controlling factors diversity, growth forms and taphonomy of coralline algae is given for the individual facies. A basic question is to what extent the preserved algal flora reflects the once living flora, i.e. how severely do taphonomic processes mask 'true' diversities. This is obviously tied to the nature and severity of taphonomic processes. It is important to note that, even in well preserved flora, not all fragments can be identified with respect to higher taxonomic levels or primary growth forms. A comparison of facies to the taxonomic diversity, growth forms and taphonomic features of the algal flora of the Gornji Grad Beds is compiled in Table 2. This is based on all studied samples. An example from a single profile is shown in Fig. 3. A schematic reconstruction for three facies comparing the interpreted biofacies with the situation after biostratinomic processes (taphofacies) is shown in Fig. 6.

The nummulitic and bivalve facies show the lowest diversity with respect to taxa with only unidentified coralline fragments present. Both of these facies represent shallow-water, higher energy environments resulting in high rates of fragmentation and abrasion. This reduces the possibility of identifying coralline algal taxa and so diminishing the prospects of recognizing the primary algal diversity. The nummulitic facies is not considered to be conducive for a rich algal flora. We thus consider the paucity of algae in this facies to reflect the original floral diversity. In the bivalve facies, however, a number of different growth forms can be recognized. This facies also contains larger biogenic particles including bivalve shells and corals, which could act as substrates for coralline growth. The low algal diversity observed in the bivalve facies may thus represent a taphonomic artefact.

The foraminiferal–coralline algal facies is characterized by a low diversity coralline algal flora with lumpy as well as fruticose protuberances. The low diversity of corallines corresponds to a lack of suitable, large particulate substrates. The interpreted lower water energy, with the presence of muddy matrix, allows the preservation of lumpy and fruticose growth forms.

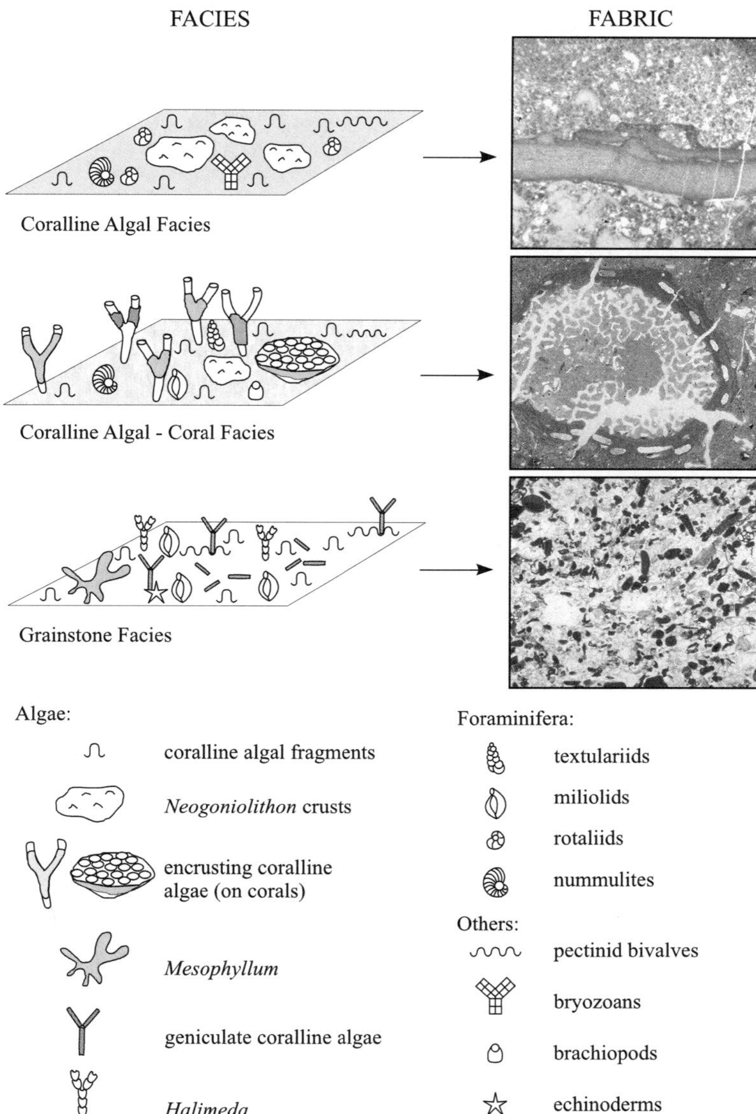

Fig. 6. Schematic representation of the three coralline algae dominated facies within the Gornji Grad Beds comparing the reconstructed facies and resulting limestone fabric in thin section. This represents a gradient from lower (top) to higher (bottom) energy environments which is accompanied by a corresponding change of dominant algal taxa, growth-form morphologies, and style and intensity of taphonomic processes.

A higher algal diversity is found in the coralline algal facies (Fig. 6). Six of the nine non-geniculate taxa observed in this facies are common (Table 2). Genera of both mastophoroids and melobesioides are present. The most characteristic growth form is the Type 1 encrusting thalli of the genus *Neogoniolithon*. This facies also shows the highest presence of layered and foliose lamellae. The well preserved nature of the components and the growth forms support the interpretation of this facies as a quiet, muddy environment.

The coralline algal–coral facies (Fig. 6) shows the highest coralline algal diversity in the study area with ten recorded genera, six of them common. There is no clear dominance of either mastophoroid or melobesioid taxa. A correspondingly wide variation of growth forms is present with the exception of the arborescent growth of geniculate corallines. Biogenic components, especially corals, support multilayered crusts which can include the algal taxa *Lithothamnion*, *Spongites*, *Lithoporella* and *Polystrata* as well as encrusting foraminifera. Fragmentation of the coralline algae can occur, but is not common. The high diversity of corallines is correlated with the diversity of available substrates with both soft and hard substrates present.

The coralline algal flora of the coral facies is slightly less diverse with eight non-geniculate taxa, only three of them common. The growth forms are dominated by thin encrustations around corals as well as warty protuberances. The mastophoroid taxa *Lithoporella* and *Spongites* dominate; melobesioids are still present. The diminished diversity of coralline algae in the coral facies is correlated with the reduced diversity of available substrates.

The grainstone facies (Fig. 6) shows a less diverse but distinct calcareous algal flora with fruticose non-geniculates, disarticulated arborescent geniculates, and the green algae *Cymopolia* and *Halimeda*. This facies shows the highest degree of fragmentation, abrasion and micritization among the coralline algae dominated facies. The higher energy conditions reconstructed for this facies potentially result in increased fragmentation and abrasion. The rounded, cylindrical intergenicula of geniculate corallines potentially represent stable components capable of withstanding a high degree of transport (compare Riosmena-Rodríguez & Siqueiros-Beltrones 1996). The lack of a larger sized particulate substrate needed for the fixation of geniculate algae may point to transport into this facies. The green algae *Cymopolia* and *Halimeda* which are restricted to this facies are known to flourish within higher energy environments (Hillis 1991; Berger & Kaever 1992).

Conclusions

The study material provides an example of how the fabric of limestone is controlled by the diversity, growth-form morphology and taphonomy of its constituent components. Although this study concentrates on the dominating coralline algal flora, this approach can be applied to other limestones dominated by particulate biogenic skeletal components. Taxonomy, growth forms and taphonomic aspects are all facies dependent, subject to the various controlling ecological factors affecting benthic environments. These features are, however, inherent to the development of carbonate fabrics and thus in the recognition and definition of the facies themselves. Coralline algae are especially useful in such an analysis considering the high potential of recognizing these features in thin sections. They are also highly sensible to environmental changes which can be expected between different shallow-water, carbonate facies.

It is important that the systematics of Recent coralline algal representatives be applied to the fossil counterparts. These can be applied at least to the genus level if vegetative and reproductive features are well enough preserved. The careful systematic approach also allows taxonomic differences to be recognized with respect to growth forms and taphonomy both within and across facies boundaries.

Coralline algal growth forms as defined by Woelkerling *et al.* (1993) can be recognized in thin section analysis of fossil floras as previously shown by Bassi (1998) and Rasser & Piller (1999). The possibility of distinguishing between growth forms can be affected by taphonomic factors such as fragmentation and abrasion. Orientation and sectioning effects also have to be taken into account. The recognition of growth forms can, nonetheless, be important in the characterization and interpretation of coralline algae dominated facies.

Important taphonomic processes known to affect coralline algae from investigations on Recent material are very difficult, if not impossible, to recognize in thin sections. This is especially the case for disease (Littler & Littler 1995, 1997) and shallow grazing which seem to have a major impact on coralline algal ecology and evolution (Adey & MacIntyre 1973; Steneck 1997). Other processes such as bioerosion due to endophytic organisms have potentially good chances of being recorded in the fossil record. Some taphonomic processes can be inferred (e.g. the disarticulation of geniculate corallines), others are obvious such as encrustation and endophytic bioerosion. Taphonomic processes such as abrasion and fragmentation are detrimental to the preservation (and recognition) of taxonomic characters and growth-form features making the assessment of coralline algal floras in highly agitated environments difficult. Other taphonomic processes, however, can have positive effects including encrustation which protects thin algal crusts from destruction through abrasion.

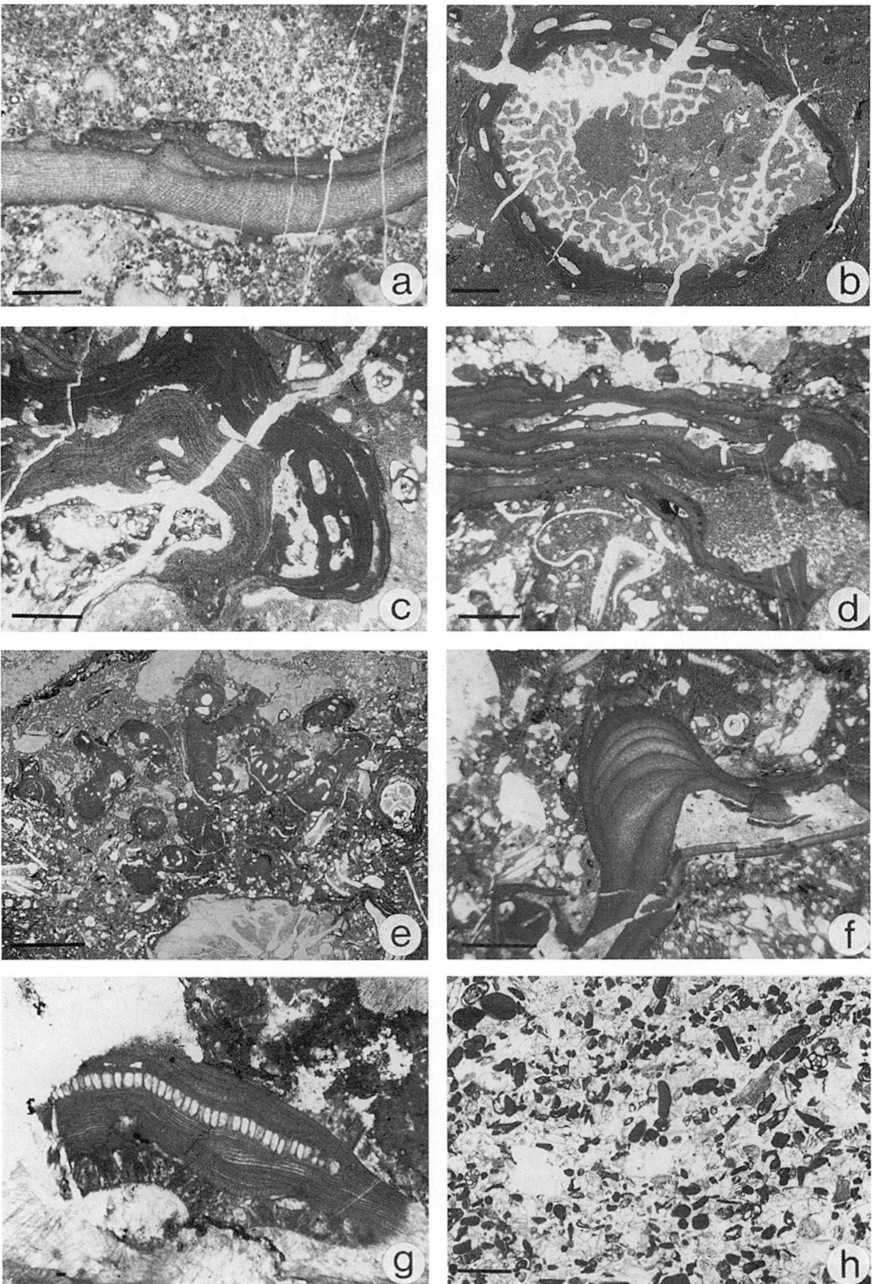

Fig. 7. (**a**) Thick crust of *Neogoniolithon* in the coralline algal facies showing the Type 1 encrusting growth form. Scale bar = 1 mm. Sample SLO94–041. (**b**) Thin crust of *Lithothamnion* sp. 1 on a leached coral substrate showing the Type 2 encrusting growth form. Scale bar = 1 mm. Sample SLO94–021. (**c**) Multilaminar encrustation with at least two encrusting coralline algae (*Spongites* followed by *Lithothamnion*). Scale bar = 1 mm. Sample SLO94–027B. (**d**) Foliose growth form. Scale bar = 1 mm. Sample SLO94–041. (**e**) *Lithothamnion* showing protuberances laden with conceptacles. Scale bar = 4 mm. Sample SLO94–020. (**f**) Single protuberance of an undetermined coralline alga. Scale bar = 1 mm. Sample SLO94–027B. (**g**) Fragment of *Sporolithon* (encrusting the foraminifer *Haddonia*) which has been incised by pressure solution. Scale bar = 0.5 mm. Sample SLO94–004. (**h**) Overview of the grainstone facies with abraded, micritized fragments of disarticulated intergenicula of geniculate coralline algae. Scale bar = 1 mm. Sample SLO94–038.

This study was supported by DFG (German Science Foundation) SFB275 (Sonderforschungsbereich): 'Climate-coupled Processes in Mesozoic and Cainozoic Geological Systems' at the University of Tübingen (Germany), and by MURST ex-40% (C. Loriga) at the University of Ferrara (Italy). We thank K. Drobne (Slovenian Academy of Sciences, Ljubljana, Slovenia), C. Hemleben, V. Mosbrugger, (University of Tübingen) and C. Loriga Broglio (University of Ferrara). We thank M. Rasser (University of Vienna, Austria) and W. Woelkerling (La Trobe University, Australia) for useful discussions and valuable comments on coralline red algae. We are grateful to J. Pavšič as the co-leader of a bilateral project between the University of Tübingen and Ljubljana University. We also thank two reviewers for their useful comments.

References

ADEY, W. H. 1978. Algal ridges of the Caribbean sea and West Indies. *Phycologia*, **17**, 361–367.

—— 1986. Coralline algae as indicators of sea-level. *In*: VAN DE PLASSCHE, O. (ed.) *Sea level Research: A Manual for the Collection and Evaluation of Data*. Free University, Amsterdam, 229–280.

—— & ADEY, P. J. 1973. Studies on the biosystematics and ecology of the epilithic crustose corallines of the British Isles. *British Phycological Journal*, **8**, 343–407.

—— & BURKE, R. 1975. Holocene bioherms of lesser Antilles, geologic control of development. *In*: FROST, S. H., WEISS, M. P. & SAUNDERS, J. B. (eds) *Reefs and Related Carbonates, Ecology and Sedimentology*. Studies in Geology, **4**, AAPG, Tulsa, 67–81.

—— & MACINTYRE, I. G. 1973. Crustose coralline algae: a re-evaluation in the geological sciences. *Geological Society of American Bulletin*, **84**, 883–904.

—— & MCKIBBIN, D. L. 1970. Studies on the maerl species *Phymatolithon calcareum* (Pallas) nov. comb. and *Lithothamnium corallioides* Crouan in the Ria de Vigo. *Botanica Marina*, **13**, 100–106.

——, TOWNSEND, R. A. & BOYKINS, W. T. 1982. The crustose coralline algae (Rhodophyta; Corallinaceae) of the Hawaiian Islands. *Smithsonian Contributions to Marine Sciences*, **15**, 1–74.

AGUIRRE, J. & BRAGA, J. C. 1998. Redescription of Lemoine's (1939) types of coralline algal species from Algeria. *Palaeontology*, **41**, 489–507.

——, —— & PILLER, W. E. 1996. Reassessment of *Palaeothamnium* Conti, 1946 (Corallinales, Rhodophyta). *Review of Palaeobotany and Palynology*, **94**, 1–9.

ALEXANDERSSON, T. 1974. Carbonate cementation in coralline algal nodules in the Skagerrak, North Sea: Biochemical precipitation in undersaturated waters. *Journal of Sedimentary Petrology*, **44**, 7–26.

—— 1977. Carbonate cementation in recent coralline algal constructions. *In*: FLÜGEL, E. (ed.) *Fossil Algae, Recent Results and Developments*. Springer, Berlin, 261–269.

BASSI, D. 1995. Crustose coralline algal pavements from Late Eocene Colli Berici of northern Italy. *Rivista Italiana di Paleontologia e Stratigrafia*, **101**, 81–92.

—— 1998. Coralline red algae (Corallinales, Rhodophyta) from the Upper Eocene Calcare di Nago (Lake Garda, Northern Italy). *Annali dell'Università di Ferrara*, **7**, supplemento, 1–51.

—— & NEBELSICK, J. H. 2000. Calcareous algae from the Lower Oligocene Gornji Grad beds of northern Slovenia. *Rivista Italiana di Paleontologia e Stratigrafia*, **106**, 99–122.

——, —— & WOELKERLING, W. J. 2000. Taxonomic and biostratigraphical re-assessments of *Subterraniphyllum* Elliott (Corallinales, Rhodophyta). *Palaeontology*, **43**, 405–425.

BASSO, D. 1995*a*. Study of living calcareous algae by a paleontological approach: the non-geniculate Corallinaceae (Rhodophyta) of the soft bottoms of the Tyrrhenian Sea (Western Mediterranean). The genera *Phymatolithon* Foslie and *Mesophyllum* Lemoine. *Rivista Italiana di Paleontologia e Stratigrafia*, **100**(4), 575–596.

—— 1995*b*. Living calcareous algae by a paleontological approach: the genus *Lithothamnion* Heydrich nom. cons. from the soft bottoms of the Tyrrhenian Sea (Mediterranean). *Rivista Italiana di Paleontologia e Stratigrafia*, **101**, 349–366.

—— 1998. Deep rhodolith distribution in the Pontian Islands, Italy: a model for the paleoecology of a temperate sea. *Palaeogeography, Palaeoclimatology, Palaeoecology*, **137**, 173–187.

—— FRAVEGA, P. & VANNUCCI, G. 1997. The taxonomy of *Lithothamnium ramosissimum* (Gümbel non Reuss) Conti and *Lithothamnium operculatum* (Conti) Conti (Rhodophyta, Corallinaceae). *Facies*, **37**, 167–182.

BERGER, S. & KAEVER, M. J. 1992. *Dasycladales: An illustrated monograph of a fascinating algal order*. Thieme, Stuttgart.

BOSELLINI, A. & GINSBURG, R. N. 1971. Form and internal structure of Recent algal nodules (rhodolites) from Bermuda. *Journal of Geology*, **79**, 669–682.

BOSENCE, D. W. J. 1976. Ecological studies on two unattached coralline algae from western Ireland. *Palaeontology*, **19**(2), 365–395.

——- 1980. Sedimentary facies, production rates and facies models for Recent coralline algal gravels, Co. Galway, Ireland. *Geological Journal*, **15**, 91–111.

—— 1983*a*. Coralline algal reef framework. *Journal of the Geological Society of London*, **140**, 365–376.

—— 1983*b*. The occurrence and ecology of recent rhodoliths – a review. *In*: PERYT, T. M. (ed.) *Coated Grains*. Springer, Berlin, 225–242.

—— 1983*c*. Description and classification of rhodoliths (rhodoids, rhodolites). *In*: PERYT, T. M. (ed.) *Coated Grains*. Springer, Berlin, 217–224.

—— 1984. Construction and preservation of two modern coralline algal reefs, St Croix, Caribbean. *Paleontology*, **27**, 549–574.

—— 1985*a*. The 'Coralligène' of the Mediterranean – A recent analogue for the Tertiary coralline algal

limestone. *In*: TOOMEY, D. F. & NITECKI, M. H. (eds) *Palaeoalgology: Contemporary Research and Applications*. Springer, Berlin, 216–225.
—— 1985*b*. The morphology and ecology of a mound-building coralline alga (*Neogoniolithon strictum*) from the Florida Keys. *Palaeontology*, **28**, 189–206.
—— 1991. Coralline algae: mineralization, taxonomy and palaeoecology. *In*: RIDING, R. (ed.) *Calcareous Algae and Stromatolites*. Springer, Berlin, 98–113.
—— & PEDLEY, H. M. 1982. Sedimentology and palaeoecology of a Miocene coralline algal biostrome from the Maltese islands. *Palaeogeography, Palaeoclimatology, Palaeoecology*, **38**, 9–43.
BRAGA, J. C. & AGUIRRE, J. 1995. Taxonomy of fossil coralline algal species: Neogene Lithophylloideae (Rhodophyta, Corallinaceae) from southern Spain. *Review of Palaeobotany and Palynology*, **86**, 265–285.
—— & —— 1998. Redescription of Lemoine's (1939) types of coralline algal species from Algeria. *Palaeontology*, **41**(3), 489–507
——, BOSENCE, D. W. J. & STENECK, R. S. 1993. New anatomical characters in fossil coralline algae and their taxonomic implications. *Palaeontology*, **36**, 535–547.
BRICL, B. & PAVŠIČ, J. 1992. Frequency of nannoplankton in Oligocene marine clay in Slovenia. *Razprave IV. raureda SAZU*, **32**, 154–173.
BROCK, R. E. 1979. An experimental study on the effects of grazing by parrotfishes and the role of refuges in the benthic community structure. *Marine Biology*, **51**, 381–388.
CABIOCH, J. 1969. Les fonds de maerl de la Baie de Morlaix et leur peuplement végétal. *Cahiers Biologique Marine*, **9**, 33–55.
CHAVE, K. E. 1964. Skeletal durability and preservation. *In*: IMBRIE, J. & NEWELL, N. (eds) *Approaches to Paleoecology*. Wiley, New York, 377–387.
DETHIER, M. N. 1994. The ecology of intertidal algal crusts: variation within a functional group. *Journal of Experimental Marine Biology and Ecology*, **177**, 37–71.
DROBNE, K., PAVLOVEC, R., DROBNE, F., CIMERMAN, F. & ŠIKIĆ, L. 1985. Some larger foraminifera from the Upper Eocene and the basal Oligocene beds in North Slovenia. *Geoloski glasnik*, **28**, 77–117.
EMBRY, A. F. & KLOVAN, J. E. 1972. Absolute water depth limits of Late Devonian Paleoecological Zones. *Geologische Rundschau*, **61**, 672–686.
FIGUEIREDO, DE O. M. A. 1997. Colonization and growth of crustose coralline algae in Abrolhos, Brazil. *In*: LESSIOS, H. A. & MACINTYRE, I. G. (eds) *Proceedings of the 8th International Coral Reef Symposium*, **1**, 689–694.
FREIWALD, A. 1993. Coralline algal maerl frameworks-islands within the phaeophytic kelp belt. *Facies*, **29**, 133–148.
—— 1995. Sedimentological and biological aspects in the formation of branched rhodoliths in northern Norway. *Beiträge zur Paläontologie*, **20**, 7–19.
GHERARDI, D. F. M. & BOSENCE, D. W. J. 1999. Modelling of the ecological succession of encrusting organisms in Recent coralline-algal frameworks from Atol das Rocas, Brazil. *Palaios*, **14**, 145–158.
GINSBURG, R. N. & SCHROEDER, J. H. 1973. Growth and submarine fossilization of algal cup reefs, Bermuda. *Sedimentology*, **20**, 575–614.
HEMLEBEN, C. 1964. *Geologisch-paläontologische Untersuchungen im Gebiet zwischen Gornji Grad (Oberburg) und Nova Stifta (Neustift) in Nordslowenien/Jugoslawien*. Masters Thesis, University of Munich.
HILLIS, L. 1991. Recent calcified Halimedaceae. *In*: RIDING, R. (ed.) *Calcareous Algae and Stromatolites*. Springer, Berlin, 167–188.
IRVINE, L. M. & CHAMBERLAIN, Y. M. 1994. *Seaweeds of the British Isles. Volume 1 Rhodophyta Part 2B Corallinales, Hildenbrandiales*. HMSO, London.
IRYU, Y. & MATSUDA, S. 1994. Taxonomic studies of the *Neogoniolithon fosliei* complex (Corallinaceae, Rhodophyta) in the Ryukyu Islands. *Transaction and Proceedings of Palaeontological Society of Japan*, New Series, **174**, 426–448.
——, NAKAMORI, T., MATSUDA, S. & ABE, O. 1995. Distribution of marine organisms and its geological significance in the modern reef complex of the Ryukyu Islands. *Sedimentary Geology*, **99**, 243–258.
JACQUOTTE, R. 1962. Etude des fonds de maerl de la Mediterrenee. *Recueil des Travaux de la Station Marine d'Endoume*, **26**, 141–235.
JELEN, M., LAPANJE, V. & PAVŠIČ, J. 1980. Nanoplankton in dinoflagelati iz oligocenskih plasti na Homu pri Radmirju (Nannoplankton and dinoflagellates from the Oligocene beds of Hom). *Geologija*, **23**, 177–188.
JOHANSEN, H. W. 1981. *Coralline Algae, A First Synthesis*. CRC, Boca Raton.
—— & SILVA, P. C. 1978. Janiae and Lithotricheae: two new tribes of articulated Corallinaceae (Rhodophyta). *Phycologia*, **17**(4), 413–417.
JOHNSON, J. H. 1961. *Limestone-building Algae and Algal Limestones*. Colorado School of Mines, Boulder.
——, KNIGHT, M. A. & PUESCHEL, C. M. 1997. Antifouling effects of epithallial shedding in three crustose coralline algae (Rhodophyta, Corallinales) on a coral reef. *Journal of Experimental Marine Biology and Ecology*, **213**, 281–293.
KEATS, D. W. & MANEVELDT, G. 1994. *Leptophytum foveatum* Chamberlain & Keats (Rhodophyta, Corallinales) retaliates against competitive overgrowth by other encrusting algae. *Journal of Experimental Marine Biology and Ecology*, **175**, 243–251.
——, MATTHEWS, I. & MANEVELDT, G. 1994. Competitive relationships in a guild of crustose algae in the eulittoral zone, Cape Province, South Africa. *South African Journal of Botany*, **60**, 108–113.
KIDWELL, S. M. & BOSENCE, D. W. J. 1991. Preservation in time-averaging of marine shelly faunas. *In*: ALLISON, P. A. & BRIGGS, D. E. G. (eds) *Taphonomy: Releasing the Data Locked in the Fossil Record*. Academic, London, 116–209.

LABOREL, J. 1961. La concrétionnment algal 'coralligène' et son importance geomorphologique en Mediterranée. *Recueil des Travaux de la Station Marine d'Endoume*, **37**, 37–60.

LAWRENCE, J. M. 1975. On the relationships between marine plants and sea urchins. *Oceanographic Marine Biological Annual Review*, **13**, 231–286.

LEMOINE, Mme. P. 1910. Repartition et mode de vie du maerl (*Lithothamnium calcareum*) aux environs de Concarneau (Finistère). *Annales de l'Institute de Oceanographie*, **1**, 1–28.

—— 1911. Structure anatomique des Mélobésiées. Application à la classification. *Annales de l'Institute de Océanographie de Monaco*, **2**, 1–213.

—— 1939. Algues calcaires fossiles de l'Algérie. *Matériaux pour la Carte Géologique d'Algerie, serie 1a, Paléontologie*, **9**, 1–128.

LITTLER, M. M 1973*a*. The population and community structure of Hawaiian fringing reef crustose Corallinaceae (Rhodophyta, Cryptonemiales). *Journal of experimental marine biology*, **11**, 103–120.

——- 1973*b*. The distribution, abundance and communities of deepwater Hawaiian crustose Corallinaceae (Rhodophyta, Cryptonemiales). *Pacific Science*, **27**, 281–289.

—— & LITTLER, D. S. 1995. Impact of CLOD pathogen on Pacific coral reefs. *Science,* 267, 1356–1360

—— & —— 1997. Disease-induced mass mortality of crustose coralline algae on coral reefs provides rationale for the conservation of herbivorous fish stocks. *In*: LESSIOS, H. A. & MACINTYRE, I. G. (eds) *Proceeding 8th International Coral Reefs Symposium*, **1**, 719–724.

——, —— & HANISAK, M. 1991. Deep-water rhodolith distribution, productivity, and growth history at sites of formation and subsequent degradation. *Journal of Experimental Marine Biology and Ecology*, **150**, 163–182.

MARRACK, E. C. 1999. The relationship between water motion and living rhodolith beds in the southwestern Gulf of California, Mexico. *Palaios*, **14**, 159–171.

MARTINDALE, W. 1992. Calcified epibionts as palaeoecological tools: examples from the Recent and Pleistocene reefs of Barbados. *Coral Reefs*, **11**, 167–177.

MINNERY, G. A. 1990. Crustose coralline algae from the Garden Flower Banks, Northwestern Gulf of Mexico. Controls on the distribution and growth morphology. *Journal of Sedimentary Petrolology*, **60**, 992–1007.

——, REZAK, R. & BRIGHT, T. J. 1985. Depth zonation and growth form of crustose coralline algae: Flower Garden banks, northwestern Gulf of Mexico. *In*: TOOMEY, D. F. & NITECKI, M. H. (eds) *Paleoalgology: Contemporary Research and Applications*. Springer, Berlin, 237–247.

MONTAGGIONI, L. F. 1979. Environmental significance of rhodolites from the Mascarene reef province western Indian Ocean. *Bulletin des Centres de Recherches Exploration-Production Elf-Aquitaine*, **3**, 713–723.

MORSE, A. N. C. & MORSE, D. E. 1984. Recruitment and metamorphosis of Haliotis larvae induced by molecules uniquely available at the surfaces of crustose red algae. *Journal of Experimental Marine Biology and Ecology*, **75**, 191–215.

NEBELSICK, J. H., BASSI, D. & DROBNE, K. 2000. Microfacies Analysis and palaeoenvironmental interpretation of Lower Oligocene, shallow water carbonates (Gornji grad Beds, Slovenia). *Facies*, **43**, 157–176.

PENROSE, D. 1992. *Neogoniolithon fosliei* (Corallinaceae, Rhodophyta), the type species of *Neogoniolithon*, in southern Australia. *Phycologia*, **31**(3/4), 338–350.

PÉRÈS, J. M. & PICARD, J. 1958. Manual de bionomie benthique de la mer Méditerranée. *Recueil des Travaux de la Station Marine d'Endoume*, **23**, 7–122.

PERRIN, C., BOSENCE, D. & ROSEN, B. 1995. Quantitative approaches to palaeozonation and palaeobathymetry of coral sand coralline algae in Cenozoic reefs. *In*: BOSENCE, D. W. J. & ALLISON, P. A. (eds) *Marine Palaeoenvironmental Analysis from Fossils*. Geological Society, London, Special Publications, **83**, 181–229.

PILLER, W. E. & RASSER, M. 1996. Rhodolith formation induced by reef erosion in the Red Sea, Egypt. *Coral Reefs*, **15**, 191–198.

POIGNANT, A. F. 1979*a*. Les Corallinacées mésozoïques et cénozoiques: Hypothèses phylogénétiques. *Bulletin des Centres de Recherches Exploration-Production Elf-Aquitaine*, **3**(2), 753–755.

——- 1979*b*. Détermination générique des Corallinacées Mésozoïques et Cénozoiques. *Bulletin des Centres de Recherches Exploration-Production Elf-Aquitaine*, **3**(2), 757–765.

—— 1984. La notion de genre chez leas Algues fossiles. A- Les Corallinacées. *Bulletin de la Societé géologique de France*, **16**(4), 603–604.

RASSER, M. & PILLER, W. E. 1997. Depth distribution of calcareous encrusting associations in the northern Red Sea (Safaga, Egypt) and their geological implications. *In*: LESSIOS, H. A. & MACINTYRE, I. G. (eds) *Proceedings of the 8th International Coral Reef Symposium*, **1**, 743–748.

—— & —— 1999. Application of neontological taxonomic concepts to Late Eocene coralline algae (Rhodophyta) of the Austrian Molasse Zone. *Journal of Micropaleontology*, **18**, 67–80.

REID, P. R. & MACINTYRE, I. G. 1988. Foraminiferal–algal nodules from the Eastern Caribbean: growth history and implications on the value of nodules as paleoenvironmental indicators. *Palaios*, **3**, 424–435.

RIOSMENA-RODRÍGUEZ, R. & SIQUEIROS-BELTRONES, D. A. 1996. Taxonomy of the genus *Amphiroa* (Corallinales, Rhodophyta) in the southern Baja California Peninsula, México. *Phycologia*, **35**(2), 135–147.

SCOFFIN, T. P., STODDART, D. R., TUDHOPE, A. W & WOODROFFE, C. 1985. Rhodoliths and coralliths of Muri Lagoon, Rarotonga, Cook Islands. *Coral Reefs*, **4**, 71–80.

SEBENS, K. P. 1986. Spatial relationships among encrusting marine organisms in the New England subtidal zone. *Ecological Monographs*, **56**, 73–96.

STELLER, D. L. & FOSTER, M. S. 1995. Environmental

factors influencing distribution and morphology of rhodoliths in Bahía Concepción, B. C. S., México. *Journal of Esperimental Marine Biology and Ecology*, **194**, 201–212.

STENECK, R. S. 1982. Adaptive trends in branching crustose coralline algae: patterns in space and time. *Geological Society of America,* Abstract with Programs *Bulletin,* **14**, 86.

—— 1986. The ecology of coralline algal crusts: convergent patterns and adaptative strategies. *Annual Review of Ecological Systems*, **17**, 273–303.

—— 1997. Crustose corallines, other algal functional groups, herbivores and sediments: complex interactions along reef productivity gradients. *In*: LESSIOS, H. A. & MACINTYRE, I. G. (eds) *Proceedings of the 8th International Coral Reef Symposium*, **1**, 695–700.

—— & ADEY, W. H. 1976. The role of environment in control of morphology in *Lithophyllum congestum*, a Caribbean algal ridge builder. *Botanica Marina*, **19**, 197–215.

TESTA, V. 1997. Calcareous algae and corals in the inner shelf of Rio Grande do Norte, NE Brazil. *In*: LESSIOS, H. A. & MACINTYRE, I. G. (eds) *Proceeding 8th International Coral Reefs Symposium*, **1**, 737–742.

TOWNSEND, R. A., WOELKERLING, Wm J., HARVEY, A. S. & BOROWITZKA, M. 1995. An account of the red algal genus *Sporolithon* (Sporolithaceae, Corallinales) in Southern Australia. *Australian Systematic Botany*, **8**, 85–121.

VAN DEN HOEK, C., CORTEL-BREEMAN, A. M. & WANDERS, J. B. W. 1975. Algal zonation in the fringing coral reef of Curaçao, Netherlands Antilles, in relation of corals and gorgonians. *Aquatica Botanica*, **1**, 269–308.

VERHEIJ, E. 1993. The genus *Sporolithon* (Sporolithaceae fam. nov., Corallinales, Rhodophyta) from the Spermonde Archipelago, Indonesia. *Phycologia*, **32**(3), 184–196.

VOIGT, E. 1981. Erster fossiler Nachweis des Algen-Genus *Fosliella* HOWE, 1920 (Corallinaceae; Rhodophyceae) in der Maastrichter und Kunrader Kreide (Maastrichtium, Oberkreide). *Facies*, **5**, 265–282.

WANDERS, J. B. W. 1977. The role of benthic algae in the shallow water reef of Curaçao (Netherland Antilles), III: the significance of grazing. *Aquatic Botany*, **3**, 357–390.

WILKS, K. M. & WOELKERLING, W. J. 1995. An account of Southern Australia species of *Lithothamnion* (Corallinaceae, Rhodophyta). *Australian Systematics Botany*, **8**, 549–583.

WOELKERLING, W. J. 1985. Proposal to conserve *Lithothamnion* against *Lithothamnium* (Rhodophyta, Corallinaceae). *Taxon*, **34**, 302–303.

—— 1988. *The Coralline Red Algae: An Analysis of the Genera and Subfamilies of Nongeniculate Corallinaceae*. British Museum (Natural History), Oxford University, Oxford.

—— 1996. Subfamily Lithophylloideae. *In*: WOMERSLEY, H. B. S. (ed.) *The Marine Benthic Flora of Southern Australia. Part IIIB. Gracilariales, Rhodymeniales, Corallinales and Bonnemaisoniales.* Australian Biological Resources Study, Canberra, 214–237.

—— & HARVEY, A. 1993. An account of southern Australia species of *Mesophyllum* (Corallinaceae, Rhodophyta). *Australian Systematics Botany*, **6**, 571–637.

—— & LAMY, D. 1998. *Non-geniculate Coralline Red Algae and the Paris Muséum: Systematics and Scientific History.* Muséum/ADAC, Paris.

——, IRVINE, L. M. & HARVEY, A. S. 1993. Growth-forms in non-geniculate coralline red algae (Corallinales, Rhodophyta). *Australian Systematics Botany*, **6**, 277–293.

WOMERSLEY, H. B. S. & JOHANSEN, H. W. 1996*a*. Subfamily Amphiroideae. *In:* WOMERSLEY, H. B. S. (ed.) *The Marine Benthic Flora of Southern Australia. Part IIIB. Gracilariales, Rhodymeniales, Corallinales and Bonnemaisoniales.* Australian Biological Resources Study, Canberra, 283–288.

—— & —— 1996*b*. Subfamily Corallinoideae. *In:* WOMERSLEY, H. B. S. (ed.) *The Marine Benthic Flora of Southern Australia. Part IIIB. Gracilariales, Rhodymeniales, Corallinales and Bonnemaisoniales.* Canberra, Australian Biological Resources Study, Canberra, 288–317.

—— & —— 1996*c*. Subfamily Metagoniolithoideae. *In*: WOMERSLEY, H. B. S. (ed.) *The Marine Benthic Flora of Southern Australia. Part IIIB. Gracilariales, Rhodymeniales, Corallinales and Bonnemaisoniales.* Australian Biological Resources Study, Canberra, 317–323.

WRAY, J. L. 1979. Paleoenvironmental reconstructions using benthic calcareous algae. *Bulletin des Centres de Recherches Exploration-Production Elf-Aquitaine*, **3**, 873–879.

Factors regulating the development of elevator rudist congregations

EULÀLIA GILI[1] & PETER W. SKELTON[2]

[1]*Departament de Geologia, Universitat Autònoma de Barcelona, Bellaterra 08193, Spain (e-mail: igpaa@blues.uab.es)*

[2]*Department of Earth Sciences, Open University, Milton Keynes, MK7 6AA, UK (e-mail: P. W.Skelton@open.ac.uk)*

Abstract: Upper Cretaceous carbonate platform deposits show a widespread development of distinctive lenticular to tabular bodies of rock (lithosomes) formed by congregations of upright tubular ('elevator') rudist bivalves. Here we discuss the factors that regulated the initiation, consolidation and termination of such lithosomes, based on examples in the Santonian of the southern Central Pyrenees.

In the study area, rudist congregations developed between phases of sediment influx from neighbouring source areas. For the initiation of settlement, sedimentation rate evidently had to be very low or nil. The first rudist settlers both provided more hard substrates for subsequent recruitment, and fuelled the in situ formation of bioclastic sediment, leading to the embedding and consolidation of the congregation (both positive feedbacks to establishment). Thereafter, rudist density correlated with inferred sediment destabilization at the benthic boundary layer, which in turn affected rudist recruitment. Thus, successful rudist larval settlement declined with increasing numerical density of individuals in the congregation – a crucial negative feedback mechanism. The density of the rudist congregations could then have been maintained at more or less the same level through time by this stabilizing process. Finally, development of the rudist congregations ceased with progressive shallowing, usually involving reworking of their upper parts and/or burial by renewed influxes of sediment.

The late Cretaceous carbonate platforms of the Tethyan Realm hosted widespread congregations of elevator rudist bivalves (Gili *et al.* 1995*a*). Many were generated by slender cylindrical rudists such as hippuritids, growing constratally, i.e. with the upright to inclined rudists growing in tandem with the accumulating sediment. Hence, only the growing tips of the shells usually projected (some centimetres) above the sediment surface, and the bulk of the skeletal fabric was embedded in sediment and supported by it. As most of the interstitial sediment tended to be generated within the congregations, through biodegradation of shells, its accumulation and the growth rate of the rudists were perhaps mutually linked (Götz 1999). Characteristically, however, hippuritid congregations did not generate significant topographical relief, and they formed lenticular to tabular bodies of rock (lithosomes) termed 'rudist banks' by Masse & Philip (1981). Several studies have described the stratigraphical, sedimentological and palaeontological aspects of various hippuritid rudist lithosomes (e.g. Philip 1970; Freytet 1973; Bilotte 1985; Höfling 1985; Grosheny & Philip 1989; Gili 1992, 1993; Carannante *et al.* 1995; Skelton *et al.* 1995; Stössel 1997; Stössel & Bernoulli 2000). Much less has been written about the factors, including feedback effects, that controlled the development of rudist congregations, though Skelton *et al.* (1995), Stössel (1997) and Götz (1999) discussed some of these aspects.

This paper discusses the factors that regulated the initiation, consolidation and termination of hippuritid congregations, based on well exposed examples contained in a Santonian carbonate platform succession in the southern Central Pyrenees. An understanding of these factors may help to clarify the controls on the distribution of the rudist lithosomes in the carbonate platform sequences. The platform succession crops out in both flanks of the E–W oriented Sant Corneli anticlinal mountain, near the city of Tremp (42°11′0″ N, 01°00′20″ E). The succession in the northern (Sant Martí de Vilanoveta) section has been studied in detail by Ross (1989), Gili *et al.* (1995*b*,*c*, 1996), Skelton *et al.* (1995), and that in the southern (les Collades de Basturs) section by Gili (1992, 1993).

Stratigraphical setting of the lithosomes

In the eastern part of the Vilanoveta section, laterally extensive tabular hippuritid lithosomes are present in several of the cycles (which are up to 20 m maximum thickness) that make up most of the platform succession. Each lithosome-bearing cycle commences with a more or less developed unit of coral-rich marls and marly

From: INSALACO, E., SKELTON, P. W. & PALMER, T. J. (eds) 2000. *Carbonate Platform Systems: components and interactions*. Geological Society, London, Special Publications, **178,** 109–116. 0305–8719/00/$15.00

limestones, and this is followed by the rudist lithosome, which is then conformably overlain by a blanket of bioclastic floatstone to grainstone (Gili *et al.* 1995*c*; Skelton *et al.* 1995).

In les Collades de Basturs, lenticular and tabular hippuritid lithosomes, together with rudist and coral lithosomes, and bioclastic calcarenites, also constitute an important part of the carbonate platform deposits (Gili 1993). There, the succession consists of three shallowing-upward cycles. Each commences with coraliferous marls and marly limestones, and passes up to a thick rudist-rich calcareous bed. In the east, examples of the latter are largely composed of lenticular lithosomes of hippuritids interbedded with bioclastic deposits. Both kinds of unit are laterally replaced, towards the west, by mixed rudist and coral lithosomes with minor bioclastic deposits, succeeded by tabular hippuritid lithosomes. Each cycle is variably capped by a conformable bioclastic sheet.

Growth fabric of the hippuritid congregations

Both in the Sant Martí de Vilanoveta section and in les Collades de Basturs, the hippuritid congregations are notably paucispecific. Gili (1992) recorded that, in les Collades, only one to three of the total list of gregarious hippuritid species known for the area dominate the congregation at any one time, and Skelton *et al.* (1995) described a hippuritid lithosome, which is for the most part 1.5–2 m thick and at least 0.25 km^2 in area, dominated largely by a single hippuritid species (*Hippurites socialis* Douvillé). Accompanying macrofauna are very scarce and are predominantly other rudists (isolated *Vaccinites* spp., *Plagioptychus paradoxus* Matheron and various radiolitids), small colonial corals and encrusting 'chaetetid' sponges.

At the base of congregations, hippuritids are often in disorder and broken, when installed on bioclastic sands. When the congregation is overlying calcareous sandy marls, by contrast, hippuritid individuals may be seen in growth position and largely clustered in small, discrete, upright bouquets of several individuals. In the main part of congregations, however, hippuritid shells are most commonly grouped in somewhat inclined bunches of a few parallel to subparallel shells. Usually, individuals are attached together in aggregates of two to four (rarely more), and the adhesion area involves only a small part of the shell. A preliminary estimation of density, recorded as the number of individuals per unit area, was done following the unbiased counting rule of Gundersen (1978). Calculations indicated that rudist aggregation may reach densities of ten individuals per 35 cm^2 (about 2900 individuals per m^2) though for the most part densities range from five to seven individuals per 35 cm^2 (about 1400 to 2000 individuals per m^2) (Table 1). The percentage of surface area

Table 1. *Recruitment and change of numerical density of hippuritids from les Collades de Basturs*

Sample	Density (D_n), (individuals per 35 cm^2)	$D_n - D_1$	Young (*1*)	Young (*2*)	Young (*3*)
28.757	10	0 ($n = 2$)	7	–	1(1)
3.413	10	1 ($n = 4$)	13	3	7(2)
s/n	7	−3 ($n = 3$)	–	–	–
28.748	7	2 ($n = 3$)	2	–	–(3)
28.747	6	0 ($n = 3$)	–	–	–
28.758	–	–	3	–	2(4)
28.765	–	–	3	–	2(5)
28.696	5	1 ($n = 4$)	5	6	6(6)

Samples 28.758 and 28.765 are bouquets. Their numerical densities have not been estimated because of their small sizes (in bouquets, density generally decreases as diameters of rudists increase, although new recruits also augment the number of individuals). Data are from serial bed-parallel sections separated vertically by 3 cm. Key to abbreviations: D_1, density in the lowest section of sample; D_n, density in the nth section; Young (*1*), number of young attached to adults; Young (*2*), number of young isolated or attached to other young individuals; Young (*3*), number of young surviving. (1) At least one juvenile survived (another may have done so, but been lost from the sample); (2) at least seven juveniles survived (five died, two were lost from the sample and two appear in the last section, so may not have survived); (3) the two young appear in the last section, so may not have survived; (4) at least two juveniles survived (one of the young is lost from the sample); (5) at least two juveniles survived (one of the young appears in the last section, so may not have survived); (6) at least six juveniles survived (two of the young appear in the last section, so may not have survived). All samples from the Palaeontology collections of the Universidad Autònoma de Barcelona.

covered by matrix sediment roughly correlates with rudist densities (most commonly it is around 20–50%). Although only a small number of samples (eight) have been analysed, it is obvious that definite trends do exist for the packing density and spacing of individuals. Nevertheless, larger numbers of measurements in exposures of greater extent will be necessary in order to discern statistically significant trends.

Two patterns of hippuritid settlement were observed. Both the hippuritid bouquets and the bunches of parallel hippuritid shells show a great number of young individuals, in their earliest stages of development, attached to the flanks of established shells (see Figs 1, 2). In a few bunches, however, some isolated young hippuritids have been observed nestling among adult shells (see, for example, specimen nos 23 and 24 in Fig. 1). Attachment persisted to varying extents in the juvenile and adult stages. Secondary attachment among clustered adults may also be seen (see specimens nos 6 and 9 in Fig. 1, and 5 and 8 in Fig. 2).

Detailed inspection of successive sections across clustered hippuritids also revealed that only in the bouquets is a significant net increase in the number of individuals shown. Although the total number of new rudist settlements can be very high (sometimes approximately equal to the number of adults) in the dense bunches of parallel shells, an insignificant or nil *net* increase in the number of hippuritids was observed in all cases investigated, as a consequence of mortality. However, there is a distinct difference in the pattern of juvenile survival and growth in bunches of hippuritids of low numerical density and those of high numerical density. In the former, the number of hippuritid settlers that survived is very high and roughly equals the number of adult hippuritids that died. In one of the samples (Fig. 1), for example, the initial number of adults was 15, and the total number of juveniles recruited was 11. Of these, ten of the original adults died but at least six of the juveniles survived. In this sample, the local density of specimens increased slightly from five to six individuals per 35 cm^2. In high density bunches (Fig. 2), by contrast, the survivorship of juveniles is extremely low, as is the mortality of adults. In

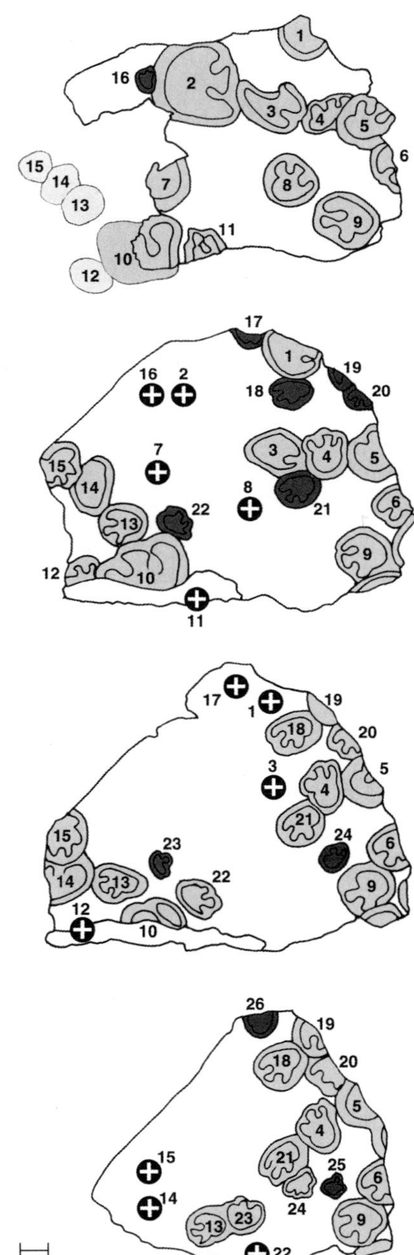

Fig. 1. Ascending series of cross-sections (arranged down page, with basal section at top) of hippuritid cluster from rudist lithosome at les Collades de Basturs. Adults shaded pale grey; new recruits shaded dark grey; positions of individuals that have disappeared from previous section (= mortality) shown as dark circles with crosses. Nos 16–26 are new settlers. Ten of the 15 adult individuals present in the first section died, and at least six juveniles survived. Numerical density increased from five to six individuals per 35 cm^2. Vertical distance between cross-sections 3 cm. (Universitat Autònoma de Barcelona, Palaeontology collection no. 28.757) Scale bar 1 cm.

this example, only one adult died from the 12 present in the beginning, but there was only one survivor among the seven juveniles (another juvenile may have survived, but been lost from the sample), so that the density remained at ten individuals per 35 cm^2.

The bulk of the hippuritid fabric is largely embedded in matrix sediment, which is a distinctive wackestone consisting largely of bioeroded hippuritid debris, together with dark micrite (e.g. Gili 1992, fig. 2; and Skelton *et al.* 1995, fig. 13c, d); peloids are another common, though local, component.

Lithosomes formed by hippuritid congregations may range from small examples, with a few tens of individuals, to major tabular lithosomes containing millions of individuals. The latter are from 1.5 m to a few metres thick and may be hundreds of metres, or more, in lateral extent. Usually, in thick lithosomes, distinct hippuritid congregations are superimposed, separated by discontinuities in the rudist growth.

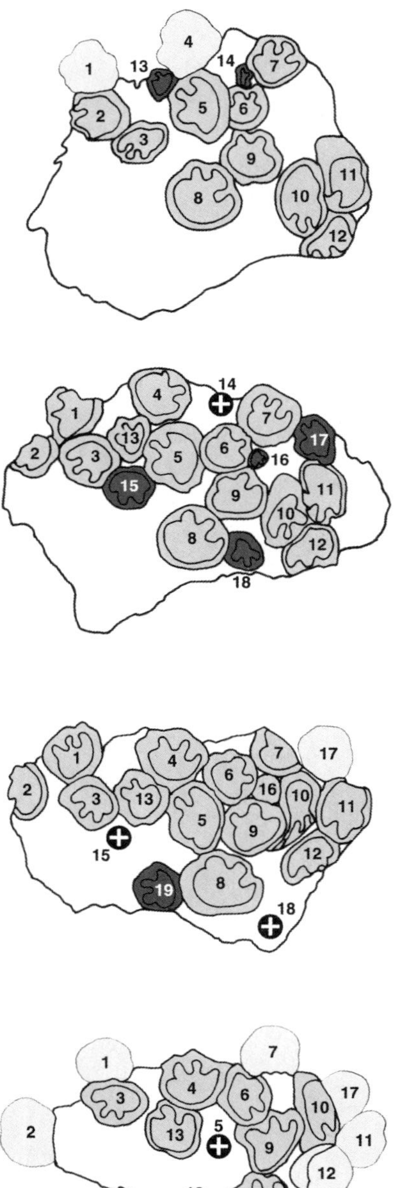

Fig. 2. Ascending series of cross-sections (arranged down page, with basal section at top) of another hippuritid cluster from rudist lithosome at les Collades de Basturs. Adults shaded pale grey; new recruits shaded dark grey; positions of individuals that have disappeared from previous section (= mortality) shown as dark circles with crosses. Nos 13–19 are new settlers. Only one of the 12 individuals present in the first section died, and only one of the juveniles demonstrably survived (though another, no. 17, may have survived but been lost from the sample). Numerical density remained at ten individuals per 35 cm^2. Vertical distance between cross-sections 3 cm. (Universitat Autònoma de Barcelona, Palaeontology collection no. 28.696) Scale bar 1 cm.

Sedimentary context of the hippuritid congregations

Hippuritid lithosomes can be underlain by marly limestones and marls with abundant colonial corals and 'chaetetid' sponges, accompanied towards the top by moderately abundant and diverse rudists (*Vaccinites* being the most conspicuous), or by bioclastic deposits. The latter can be of two kinds. They may consist of a bioclastic packstone containing mainly rudist, echinoid and foraminiferal bioclasts, and encrusting calcareous algae, as in the bioclastic sheet underlying much of Bed 3 of Skelton *et al.* (1995). Or they may comprise a rudist and coral floatstone with wackestone to grainstone matrix, also containing fragments of echinoids, bryozoans, foraminifers and calcareous algae, as in the bioclastic deposits on which hippuritid lithosomes developed in the eastern part of les Collades section.

Overlying the hippuritid lithosomes is usually a blanket of coral and rudist-rich bioclastic floatstone, with packstone to wackestone matrix, passing up to bioclastic packstone and rarer grainstone. The floatstone contains mixed autochthonous–allochthonous concentrations of bioclasts, including poorly sorted, subangular hippuritid, as well as radiolitid, rudist fragments, gastropods, coral and echinoid grains and benthic foraminifers, among others.

Developmental history of the hippuritid congregations

Initiation

For the initiation of a hippuritid congregation, constraints on hippuritid settlement would have to have been removed. Although some hippuritids could withstand influxes of siliciclastic sediment once established (Steuber *et al.* 1998), initial settlement seems to have been sensitive to sustained allochthonous sedimentation. Hence, cessation of allochthonous sediment influx would have been necessary to allow initiation of a congregation. At Vilanoveta, the interruption of continuous sediment fluxes (a mixture of marine and some land-derived material forming the calcarenites of the lower Aramunt Vell Member), has been related by Gili *et al.* (1995*c*) to relative sea-level rise allowing sequestering of the sediment in accommodation space created elsewhere on the platform. Alternatively, Gili (1992) associated the development of hippuritid congregations with periods of calm between sporadic sediment fluxes of bioclastic material driven from more open marine areas, in the eastern (inner) part of the platform in les Collades, where the distribution of both hippuritid lithosomes and bioclastic deposits varies vertically and laterally.

We do not know which biotic constraints may have inhibited hippuritid congregations, but a characteristically sharp change in the biotic diversity – from quite diverse in the beds beneath hippuritid lithosomes, to the usually paucispecific assemblages in the congregations – suggests some form of environmental restriction, with exclusion of other potentially inhibiting taxa. Gili *et al.* (1995*b*) have emphasized the importance of depositionally induced changes in the physical environment, especially those associated with shallowing, in causing such biotic changes. With the exclusion of other taxa, hippuritids were able to occupy most of the available space.

As with the majority of living bivalves, rudists probably depended upon a planktonic larval phase for dispersal. Settlement was on small hard substrates such as individual shells or shell fragments, spread over the sea floor. The scarcity of hard substrates on marly bottoms could account for the arrangement of hippuritids in distinct bouquets at the base of the congregations resting on coraliferous marls, though biochemical attraction of conspecific larvae (as in living oysters) is an additional possibility.

Consolidation

The initial rudist growth facilitated the further settlement of hippuritids. Both living shells and dead shell material derived from earlier-settling individuals provided additional sites for larval attachment and stabilized the sea bed. This positive feedback loop ('taphonomic feedback' of Kidwell & Jablonski 1983) could have triggered the explosive increase in the number of rudist individuals observed in the congregations, and resulted in the in situ production of large amounts of shell debris. The now crowded hippuritids would also have produced large quantities of faeces and pseudofaeces, so providing an additional supply of autochthonous sediment. Sediment trapping by the rudist shells would then have resulted in the constratal growth of both hippuritid shells and substrate.

Slender hippuritid shells are often preserved inclined about 30–45° from the vertical, with their porous upper valves facing towards the inferred main downstream direction. This orientation appears primary from the consistently bed-parallel orientation of tabulae seen in the lower valves, unlike those in evidently toppled shells (Gili 1992; Skelton *et al.* 1995). The upper valve pores have been interpreted as allowing the ingress of feeding currents (Skelton 1976), and an experimental study of hippuritid hydrodynamics (Gili & LaBarbera 1998) found that individuals inclined downstream would thus have filtered a mixture of water from the mainstream flow and that eddying up from the sediment–water interface. The latter supply would probably have been enriched in bottom-derived bacteria and detrital organic particles. These preliminary findings suggest that the downstream-inclined postures could have been advantageous in waters of low or fluctuating nutrient levels. The gregarious growth of the hippuritid shells would have further enhanced the effect, because of the increased eddy-induced disturbance of the surface sediment.

Nevertheless, for the inclined posture to have supplemented feeding in this way, some spacing between hippuritid shells in the congregations would have been necessary. In hippuritid lithosomes overlying coral-rich marls, the numerical density of shells in the congregations was at first augmented over a period of several spawns, until an apparently optimal level was reached. Earlier-settling hippuritids grouped in distinct bouquets of a few individuals, and served as settlement sites for succeeding spawns. As the availability of suitable substrates increased, hippuritids spread over the sea bottom, forming the groups of more or less closely growing parallel shells seen in the main parts of the congregations. In the lithosomes developed on bioclastic calcarenites, by contrast, the optimal numerical density of shells developed more or less immediately from the bases of the units. From the data discussed above (see 'Growth fabric of the hippuritid congregations'), it can be deduced that, once established, the level of numerical density in the main congregation was maintained at more or less the same level through time. If density increased, the rate of survival and consequent growth of juveniles in the congregation declined. Since the rate of adult mortality was apparently insensitive to variation in the numerical density, it can be argued that the relative success of settlement must have regulated the density. Therefore, there must have been a negative feedback loop between numerical density and the survivorship of settlers (Fig. 3).

Eckman & Nowell (1984) showed that, in marine environments, flow disturbance around shells protruding above the sea bed can significantly affect particle movement. Moreover, Eckman *et al.* (1981) had previously demonstrated, in laboratory flume experiments, that destabilization of beds by arrays of tubes was more pronounced at higher densities. Hence, as more rudist larvae colonized the congregation, increasing its numerical density, destabilization of sediment at the benthic boundary layer would have become more pronounced. This would have led to increasing mortality or migration among larvae, until an equilibrium density was achieved. If the density of the hippuritid congregation had fallen, destabilization of sediment at the boundary layer would have decreased, leading to decreased larval mortality, and thus allowing the density of the hippuritid congregation to return to its equilibrium value. These feedback effects on recruitment mean that hippuritid congregations could have been effectively self-regulating (Fig. 3). Such effects would not explain the mortality of established adults (e.g. in Fig. 1), which must therefore reflect normal turnover (yielding space for new recruits) in a population more or less at equilibrium.

Assuming that larval dispersal was unlimited across the top of the carbonate platform and over the hippuritid congregations, those larvae

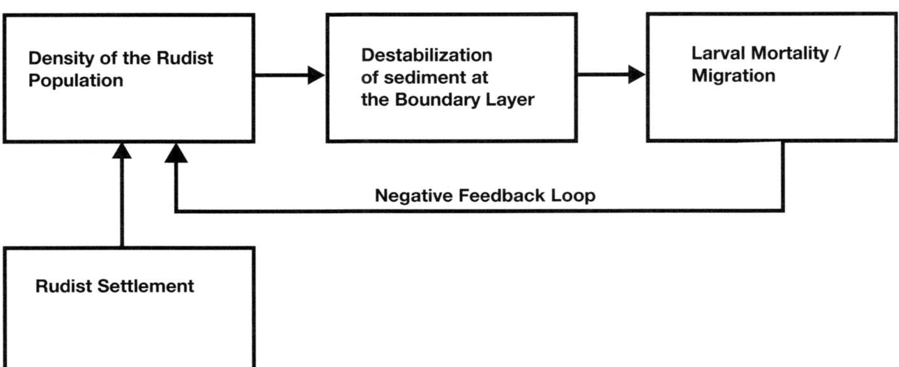

Fig. 3. Model for maintaining the numerical density of a hippuritid congregation at more or less the same level through time, by means of a negative feedback loop. If an excess of hippuritid larvae colonized the congregation, the density of the rudist congregation would have temporarily increased, leading to increasing destabilization of sediment at the boundary layer. This would have resulted in higher mortality or migration among larvae, so reducing the density to its initial level. Correspondingly, if the density of the rudist congregation fell, destabilization of sediment at the boundary layer would also have fallen, leading to decreased larval mortality, so allowing the density of the hippuritid congregation to increase again towards its initial value.

settling towards the centre of the congregation may have been less likely to survive as crowding there led to sediment disturbance. In contrast, larvae settling on the margins of the congregation may have been relatively free of such crowded conditions. Such variation could explain the widespread tabular configuration of rudist lithosomes on the platform.

The main difficulty in reconstructing the internal organization of the hippuritid congregations that we have investigated arises from the limited extent of bedding surfaces suitable for study. More measurements and observations of other rudist formations will be required to explore and test the model we are proposing.

The history of consolidation and ultimate preservation of an individual congregation appears to be related to the accommodation space made available by relative sea-level rise (cf. Stössel 1997). Some additional accommodation space could also have been generated by the compaction of underlying sediments, even when sea level remained static. However, the contribution of this last factor was probably small during the development of the congregations themselves.

Occasionally, congregations suffered severe physical disturbance (especially through the impact of storms; Skelton *et al.* 1995). On the assumption that such slender hippuritids had high reproductive rates, the large supply of larvae would have promoted recolonization of the disturbed substrate, starting the process over again.

Termination

To some extent, the growth of the congregations would have contributed to relative shallowing of the platform top, together with associated bioclastic deposits. But since the congregations did not themselves generate significant localized relief on the platform top, they were not physical obstacles to occasional storm surges and lateral sediment migration. Hence their upper surfaces were prone to reworking and/or burial by allochthonous sediment swept across the platform top. A blanket of bioclastic deposits, frequently associated with evidence for shallower conditions, is often seen covering the rudist lithosomes, especially in the northern part of the study area (Skelton *et al.* 1995). The bioclastic deposits were derived from reworking of in situ hippuritid growth (yielding floatstones), combined with the delivery of allochthonous shell sand and fragments from adjacent areas. Erosional reworking could precede or follow upon the lateral migration of the latter.

Discussion and conclusions

In many ways, the hippuritid lithosomes we have investigated are similar to those described from the Languedoc in southern France by Freytet (1973), from La Cadière d'Azur Formation of SE France by Grosheny & Philip (1989) and from Sardinia, Italy, by Carannante *et al.* (1995). They are also similar in their community structure, geometry and stratigraphical characteristics, to those in the Montagna della Maiella in the central Apennines, Italy, illustrated by Stössel & Bernoulli (2000), though they are distinct in their internal organization. In the 'dense hippuritid lithosomes' of Stössel & Bernoulli, hippuritids are more densely packed and, unlike the ones described in this paper, are oriented vertically. Such differences may to some degree reflect differing autecologies of the rudist taxa themselves, for, even in our own study area, specimens of the more broadly cylindrical *Vaccinites* tend towards a vertical life position, as was also found in the Turkish examples described by Steuber *et al.* (1998).

The initiation and survival of hippuritid congregations in the Sant Corneli carbonate platform appears to have been strongly controlled by the interruption of allochthonous sediment supply, within a zone both favourable for hippuritid growth and unavailable to other potentially inhibiting taxa. The interruption of continuous sediment fluxes can be related to the formation of accommodation space elsewhere on the platform, resulting from a relative sea-level rise. Once established, hippuritid congregations first rapidly expanded, with positive taphonomic feedbacks, but then appear to have reached an equilibrium condition, largely regulated by negative feedbacks on larval recruitment. Finally, the demise of the congregations can be attributed to a decrease in accommodation space over the platform top, often accompanied by blanketing of the lithosomes by renewed influxes of allochthonous sediment.

The distribution of these hippuritid lithosomes was thus controlled by the particular sea-level history of the platform, determining when the sea bottom was available for hippuritid colonization and how long hippuritid congregations had for development.

We are most grateful to the referees, S. Götz and H. Jones, for their pertinent and constructive criticisms, as well as M. La Barbera, for discussions concerning the destabilization of sediment surfaces by flows around hippuritid shells, although we alone accept responsibility for any errors in our explanations. We are also grateful to A. Casanelles for his skilful preparation of the figures. This paper is a contribution by

E. G. to Spanish DGES project no. PB97–0135-C02–02.

References

BILOTTE, M. 1985. Le Crétacé supérieur des plates-formes est-Pyrénéennes. *Strata*, **5**(2), 1–438.

CARANNANTE, G., CHERCHI, A. & SIMONE, L. 1995. Chlorozoan versus foramol lithofacies in Upper Cretaceous rudist limestones. *In*: PHILIP, J. & SKELTON, P. W. (eds) *Palaeoenvironmental models for the benthic associations of Tethyan Cretaceous carbonate platforms*. Special issue of *Palaeogeography, Palaeoclimatology, Palaeoecology*, **119**, 137–154.

ECKMAN, J. E. & NOWELL, A. R.M. 1984. Boundary skin friction and sediment transport about an animal-tube mimic. *Sedimentology*, **31**, 851–862.

——, —— & JUMARS, P. A. 1981. Sediment destabilization by animal tubes. *Journal of Marine Research*, **39**, 361–374.

FREYTET, P. 1973. Edifices recifaux developpés dans un environement detritique, exemple des biostromes a Hippurites (Rudistes) du Senonien inférieur du sillon languedocien (region du Narbonne, sud de la France). *Palaeogeography, Palaeoclimatology, Palaeoecology*, **13**, 65–76.

GILI, E. 1992. Palaeoecological significance of rudist constructions: a case study from les Collades de Basturs (Upper Cretaceous, south-central Pyrenees). *Geologica Romana*, **28**, 319–325.

—— 1993. Facies and geometry of les Collades de Basturs carbonate platform, Upper Cretaceous, South Central Pyrenees. *In*: SIMO, A., SCOTT, R. W. & MASSE J. P. (eds) *Cretaceous Carbonate Platforms*. AAPG, Memoir **56**, 343–352.

—— & LABARBERA, M. 1998. Hydrodynamic behaviour of hippuritid rudist shells: ecological consequences. *Geobios*, M. S. **22**, 137–145.

——, MASSE, J.-P. & SKELTON, P. W. 1995*a*. Rudists as gregarious sediment-dwellers, not reef-builders, on Cretaceous carbonate platforms. *Palaeogeography, Palaeoclimatology, Palaeoecology*, **118**, 245–267.

——, SKELTON, P. W., VICENS, E. & OBRADOR, A. 1995*b*. Corals to rudists – an environmentally induced assemblage sequence. *In*: PHILIP, J. & SKELTON, P. W. (eds) *Palaeoenvironmental models for the benthic associations of Tethyan Cretaceous carbonate platforms*. Special issue of *Palaeogeography, Palaeoclimatology, Palaeoecology*, **119**, 127–136.

——, VICENS, E., OBRADOR, A., SKELTON, P. W. & LÓPEZ, G. 1995*c*. Sequence stratigraphy of the Upper Sant Corneli Platform (Santonian), Southern Central Pyrenees. *Géologie Méditerranéenne*, **21**, 73–76 (issued for 1994).

——, ——, ——, ——. & —— 1996. Las formaciones de rudistas de la plataforma de Sant Corneli (Cretácico superior, unidad central surpirenaica). *Revista Española de Paleontología*, No. Extraordinario, 172–181.

GÖTZ, S. 1999. Elevator rudist mudsticker communities: A self-regulating sedimentary buffer-system? *In*: HÖFLING, R. & STEUBER, T. (eds) *Fifth International Congress on Rudists Abstracts and Field Trip Guides. Erlanger Geologische Abhandlungen*, Sonderband **3**, 23–24.

GROSHENY, D. & PHILIP, J. 1989. Dynamique biosédimentaire de bancs à rudistes dans un environnement péridéltaïque: la formation de La Cadière d'Azur (Santonien, SE France). *Bulletin de la Société Géologique de France*, **6**(8), 1253–1269.

GUNDERSEN, H. J.G. 1978. Estimators of the number of objects per area unbiased by edge effects. *Microscopica Acta*, **81**, 107–117.

HÖFLING, R. 1985. Faziesverteilung und Fossilvergesellschaftungen im karbonatischen Flachwasser-Milieu der alpinen Oberkreide (Gosau-Formation). *Münchner Geowissenschaftliche Abhandlungen*, (A) **3**, 1–241.

KIDWELL, S. M. & JABLONSKI, D. 1983. Taphonomic feedback: ecological consequences of shell accumulation. *In:* TEVESZ, M. J. S. & MCCALL, P. L. (eds) *Biotic Interactions in Recent and Fossil Benthic Communities*. Plenum, New York, 195–248.

MASSE, J.-P. & PHILIP, J. 1981. Cretaceous coral-rudist buildups of France. *In*: TOOMEY, D. F. (ed.) *European Fossil Reef Models*. Society of Economic Paleontologists and Mineralogists, Special Publications, **30**, 399–426.

PHILIP, J. 1970. *Les formations calcaires a Rudistes du Crétacé supérieur provençal et rhodanien*. Thèse Sciences, Université de Provence, Marseille.

ROSS, D. J. 1989. *Facies analysis and diagenesis of Tethyan rudist reefs complexes*. PhD Thesis, University of Wales, Cardiff.

SKELTON, P. W. 1976. Functional morphology of the Hippuritidae. *Lethaia*, **9**, 83–100.

——, GILI, E., VICENS, E. & OBRADOR, A. 1995. The growth fabric of gregarious rudist elevators (hippuritids) in a Santonian carbonate platform in the southern Central Pyrenees. *In*: PHILIP, J. & SKELTON, P. W. (eds) *Palaeoenvironmental models for the benthic associations of Tethyan Cretaceous carbonate platforms*. Special issue of *Palaeogeography, Palaeoclimatology, Palaeoecology*, **119**, 107–126.

STEUBER, T., YILMAZ, C. & LÖSER, H. 1998. Growth rates of early Campanian rudists in a siliciclastic-calcareous setting (Pontid Mts., North-Central Turkey). *Geobios, Mémoire spécial*, **22**, 385–401.

STÖSSEL, I. 1997. *Rudists and platform evolution: the Upper Cretaceous Maiella platform margin, Abruzzi, Italy*. PhD Thesis, Geology Institute, Zürich.

—— & Bernoulli, D. 2000. Rudist lithosome development on the Maiella Carbonate Platform Margin. *This volume*.

El Niño–Southern Oscillation mass mortalities of reef corals: a model of high temperature marine extinctions?

PETER W. GLYNN

Division of Marine Biology and Fisheries, Rosenstiel School of Marine and Atmospheric Science, University of Miami, 4600 Rickenbacker Causeway, Miami, Florida 33149–1098, USA (e-mail: pglynn@rsmas.miami.edu)

Abstract: Protracted high sea temperature anomalies accompanying El Niño-Southern Oscillation (ENSO) events have caused reef-building coral bleaching (loss of zooxanthellae) and mortality in all major coral reef biogeographic regions during the past two decades. Coral reef degradation in the eastern tropical Pacific has resulted from reductions in live coral cover, declines in coral species population abundances, local to regional scale extinctions, disruption of predator/prey spatial relations and relative abundances, bioerosion of reef frameworks, and low coral recruitment. None of the coral species that have suffered regional extinctions has reappeared after 15 years. Intense external and internal bioerosion by fishes, echinoids, lithophagine bivalves and clionid sponges has occurred on reefs affected by the 1982/83 El Niño coral bleaching event, and 1000–5000 year old reef framework accumulations in the Galápagos Islands have been completely eroded and reduced to gravel and sand. Because tropical zooxanthellate reef species are more vulnerable to rising (2–3°C) than falling (8–10°C) temperatures, greenhouse conditions may be more critical in limiting reef growth than icehouse conditions. ENSO warming episodes elicit physiological stress responses resulting in widespread mass coral mortality, leaving scant traces relating to causation. Signals that may help to identify past ENSO disturbances are: (a) temperature-related oxygen isotopic signatures, (b) skeletal stress bands and growth discontinuities, (c) coral debris in beach storm deposits, (d) increases in coral clastics resulting from intensified bioerosion and (e) the skeletal elements of bioeroders. Because this disturbance is the most pronounced and widespread of any known natural perturbation, and may increase markedly in scope with projected global warming predictions, it is considered a likely agent of future and possibly some ancient bioevents.

Marine temperature extremes have been invoked to explain several mass extinction events of tropical invertebrate faunas. Extinctions of warm water taxa occurred during the Late Eocene–Oligocene global/ocean cooling event, spanning a 10 million year interval (Prothero 1994). Stanley (1984) has argued that periods of low temperature stress were responsible for some marine extinctions during the Neogene. He and others have favoured a hypothesis of climatic cooling as the cause of mass bivalve extinctions in the western Atlantic during Pliocene and Pleistocene times (Stanley 1984, 1986; Stanley & Campbell 1981). Strombinid gastropod turnover in tropical America has also been correlated, at least in part, with declining temperatures associated with the intensification of northern hemispheric glaciation at the end of the Pliocene (Jackson *et al* 1996). However, Budd *et al.* (1996) point out that many of the extinctions and originations of Plio-Pleistocene Caribbean reef corals appear to have occurred long before (from 4 to 1 Ma) initiation of the most extreme northern hemisphere glaciations (0.7 Ma).

It is documented extensively that coral reefs demonstrate maximum development and geographic extent during periods of global warming and high sea-level stands (Fagerstrom 1987; Stanley 1992, 1997; Copper 1994). However, relatively few workers have proposed the possibility that extreme high temperatures may cause extinctions or displacement of coral faunas (Fagerstrom 1987; Wood 1999). The logical appeal of such a hypothesis is that subtropical and tropical species generally live precariously close to their upper temperature tolerance limits and slight increases in temperature cause the death of these species (Moore 1972; Johannes 1975; Jokiel and Coles 1990). Schindler (1990) suggested that El Niño-like disturbances (e.g. ocean stratification and anomalously high sea-surface temperatures) could possibly have been one of the causes of the Kellwasser bioevent near the Frasnian/Famennian boundary (Upper Devonian). A later Palaeozoic marine extinction event, attributed to unusually high temperatures during the Late Permian (Changxingian), severely affected such tropical taxa as Fusulinacea, reef-building algae and calcisponges, and rugose and tabulate corals (Waterhouse 1973; Dickens 1984). A global warming-marine anoxia model has also been

From: INSALACO, E., SKELTON, P. W. & PALMER, T. J. (eds) 2000. *Carbonate Platform Systems: components and interactions*. Geological Society, London, Special Publications, **178,** 117–133. 0305–8719/00/$15.00

entertained recently as a likely extinction mechanism during the latest Permian extinction (Hallam & Wignall 1997). It has been suggested that coral–algal reefs were displaced from core tropical environments by rudist bivalves in the Early Cretaceous in the hypothetical Supertethys, an equatorial zone of high temperature and salinity (Kauffman & Johnson 1988, 1997). Although this idea has been challenged (Gili *et al.* 1995), it is perhaps the best documented example of the exclusion of scleractinian corals from a thermally high, stressful equatorial zone during a period of enhanced greenhouse conditions.

This paper demonstrates the variety of El Niño–Southern Oscillation (ENSO)-related disturbances that cause severe reductions of zooxanthellate coral populations, local to regional scale extinctions and reef-framework destruction, and suggests how such perturbations may affect taphonomic processes and thus the recognition of sea warming events in the fossil record. Eastern Pacific examples will be emphasized with occasional reference to other tropical regions. Additional topics that will be addressed briefly are the known life history characteristics of extinction-prone species and the scale of ENSO disturbances in modern tropical seas. While this discussion will focus on elevated sea temperature as the dominant agent of disturbance, other contributing factors that may exacerbate warming mortalities, such as increases in ultraviolet radiation, sedimentation, eutrophication and epizootics, are also noted. Although not considered here, other critical factors that could severely limit coral reef calcification are (1) the theoretical decrease in aragonite saturation state in tropical oceanic waters in response to increasing global concentrations of anthropogenic CO_2 (Smith & Buddemeier 1992; Kleypas *et al.* 1999*a*,*b*) and (2) shifts in the Mg/Ca ratio of sea water that would favour the secretion of either calcitic or aragonitic skeletons (Stanley & Hardie 1998). Additionally, over-fishing and outbreaks of the sea star corallivore *Acanthaster* are also causing significant coral loss in many regions. Although no global extinctions of reef species have yet been documented, several species with very small populations are threatened, and the deterioration of Holocene coral reefs on a global scale has been widely recognized, with some workers concerned that we are possibly entering a period of severe coral reef degradation and extinction (Williams & Bunkley-Williams 1990; Smith & Buddemeier 1992; Chadwick-Furman 1996; Grigg & Kirkeland 1997; Kauffman & Johnson 1997; Wilkinson 1998). Even though many contemporary disturbances are dominantly anthropogenically driven, certain physical stressors on remote reefs, such as elevated temperatures, may mimic natural events and thus offer insight to the causes of ancient disturbances (Ricklefs *et al.* 1990).

ENSO-associated stressors and effects

A variety of stressors, having immediate effects during ENSO events or more delayed effects from years to decades later, have been observed to degrade coral reefs (Table 1). Prolonged sea warming causing coral bleaching (the loss of endosymbiotic algae or zooxanthellae and/or decreases in the per-cell concentration of photosynthetic pigments) can result in significant and widespread coral mortality, which in the eastern Pacific has ranged from the Gulf of California (México) to coastal Ecuador and the Galápagos Islands during the El Niño events of the 1980s and 1990s. High temperature stress (2–3°C increases) has also impacted coral reefs in other parts of the Pacific Ocean, and more recently in nearly all tropical regions during the 1997/98 ENSO (Wilkinson 1998). Sudden extreme low temperature exposures (8–10°C decreases), sometimes occurring during La Niña events, have also been implicated in coral bleaching and mortality on local scales (Glynn & D'Croz 1990). Ultraviolet radiation (UVR) usually penetrates deeper in the water column in areas that experience doldrum-like conditions and high water clarity during ENSO events in the Bahamas (Gleason & Wellington 1993). Several studies suggest that high temperature and UVR stressors interact to cause coral bleaching and mortality (Coffroth *et al.* 1990; Lesser *et al.* 1990; Glynn *et al.* 1993; Fitt & Warner 1995; Brown 1997; Rowan *et al.* 1997). These two stressors, acting alone or in combination, are responsible for the majority of natural, global-scale coral reef disturbance events.

More localized ENSO-associated disturbances include storm damage, subaerial reef exposures, sedimentation, nutrient pulses and dinoflagellate blooms. Mechanical damage to corals and reef structures caused by storms is mainly a result of altered storm paths during ENSO, i.e. the exposure of reefs to storms that normally lie outside of storm tracts, and not necessarily to an increase in the frequency of violent storms. In the Galápagos Islands, large storm waves and the reversal of prevailing seas caused the dislodgement of corals and reef frameworks, scouring and the deposition of coral debris onto nearby beaches. Fluctuations in sea level accompanying ENSO events have

Table 1. *Observed immediate and long-term ENSO disturbances affecting reef-building corals and coral reef structures*

Stressors	Impacts*	Location	Source
Temperature			
high	bleaching, mortality (EN)	eastern Pacific, Indonesia, French Polynesia, Japan, Tokelau Islands	Glynn 1990*a*, Brown 1987
		Several regions globally	Wilkinson 1998
low	bleaching, mortality (LN)	Panamá	Glynn & D'Croz 1990
		East Kalimantan	Wilkinson 1998
UV radiation	bleaching, mortality (EN)	Caribbean	Gleason & Wellington 1993
Storm damage	scouring, dislodgment of corals, burial by sediments (EN)	Galápagos Islands	Robinson 1985, pers. obs.
		Tuamotu Archipelago	Laboute 1985, Harmelin-Vivien & Laboute 1986
Subaerial exposure	bleaching, mortality (LN)	Panamá	Eakin *et al.* 1989
	reduced circulation (EN)	Tokelau Islands	Glynn 1984
Nutrients			
high	overgrowth of corals by filamentous and macroalgae (LN)	Panamá	pers. obs.
Dinoflagellate blooms	mortality of corals, other invertebrates and fishes (LN)	Costa Rica, Panamá	Guzmán *et al.* 1990
Calcification	diminished in corals (EN)	Colombia	Prahl 1986
		equatorial eastern Pacific	Wellington & Dunbar 1995
		Panamá	Eakin 1996
Bioerosion	higher rates following (EN)	equatorial eastern Pacific	Glynn 1988, Scott *et al.* 1988, Eakin 1996, Reaka-Kudla *et al.* 1996
Coral sexual reproduction	reduced/absent (EN); lower recruitment (post-EN)	equatorial eastern Pacific	Glynn *et al.* 1994, 1996
		Florida Keys	Szmant & Gassman 1990
Predator concentration	accelerated mortality of surviving corals (post-EN)	Panamá	Glynn 1985*a*,*b*
Predator–prey spatial relationships	disruption of prey refugia (post-EN)	Panamá	Glynn 1985*b*

*Impact occurring during El Niño (EN) and La Niña (LN) phases of ENSO event.

caused sudden reef exposures and coral mortality in the South Pacific (Tokelau Islands) during the 1982/83 El Niño event (Glynn 1984) and similar disturbances in the eastern Pacific (Panamá) during 1988/89 La Niña activity (Eakin *et al.* 1989). Both nutrient pulses and dinoflagellate blooms are responses to shoaling nutriclines in the eastern Pacific following El Niño events. Coral mortality occurs when macroalgae, responding to high nutrient availability, proliferate over reef substrates. Large coral patches in Chiriquí, Panamá, have been completely overgrown and killed by *Caulerpa sertularioides* (S. G. Gmelin) Howe and *Caulerpa racemosa* (Forskål) J. Agardh, species with chemical deterrents that are not commonly grazed by herbivorous fishes. In mid-1985, dinoflagellate blooms in Costa Rica and Panamá caused significant coral mortality of pocilloporid corals, and reef-associated crustaceans, gastropods and fishes (Guzmán *et al.* 1990). Mortality was most likely a result of a combination of (a) toxicity, (b) oxygen depletion and (c) smothering by mucus produced during the dinoflagellate blooms. Coral colony calcification is greatly diminished or ceases altogether during bleaching events (Wellington & Dunbar 1995). Coral reef calcification also declines, a result of reduced coral cover caused by bleaching-induced mortality (Eakin 1996).

Bioerosion of coral colonies and reef frameworks continues or even accelerates after bleaching-induced mortality. The population abundances of certain internal bioeroders, such as lithophagine bivalve molluscs, may become severely reduced (Scott *et al.* 1988); however, most taxa of internal (e.g. clionid sponges, polychaetous annelids) and external (e.g. echinoids, herbivorous fish grazers) bioeroders are unaffected by warming events and may even undergo increases in abundance. Population outbreaks of *Diadema mexicanum* (A. Agassiz) on some Panamanian coral reefs, increasing from pre- to post-ENSO densities of 2–4 to 60–100 individuals per m^2 respectively, are largely responsible for the loss of up to 0.5 m of vertical frameworks over a 16 year period. *Eucidaris galapagensis* (Döderlein) has demonstrated mean population increases of from 5 to 30 individuals per m^2 following the 1982/83 ENSO bleaching event on coral reefs in the Galápagos Islands (Floreana Island). This increase was due to the redistribution of sea urchins around dead coral reef frameworks, and not to a population outbreak (Glynn 1988). As a consequence, intensified external bioerosion amounted to 15–27 kg $CaCO_3$ m^{-2} a^{-1} which toppled coral colonies (Glynn & Colgan 1992) and was largely responsible for the total loss of reef frameworks between 0.6 and 5 m in vertical thickness since the disturbance.

While seasonally high temperatures generally stimulate sexual activity in most reef-building corals, gonads either regress or fail to develop altogether during periods of elevated temperatures above the seasonal mean. Field and laboratory studies have shown a mean monthly threshold temperature elevation of 1.5°C above the seasonal norm for one species (unpublished results). Further, recruit densities were nil following two severe ENSO events (1982/83, 1997/98). Changes in the relative abundances of corals and corallivores in 1982–83 led to higher rates of predation of surviving corals by corallivores that were initially unaffected by the warming episode. Also, changes in the spatial relations of certain coral species that provided protection to other corals resulted in different patterns of mortality. Massive and foliose corals preferred by the corallivore *Acanthaster planci* (Linnaeus), but protected by barriers of branching corals before the bleaching mortality event, were subsequently exposed and became vulnerable to attack. This period of altered predation, however, was short-lived because the population densities of the predominant gastropod and sea star predators of corals began to decline shortly after 1983.

Causal verification

Compared with palaeontological analyses, the identification of agents of mortality on extant corals is considerably less obscure. Nonetheless, the unequivocal distinction between present-day anthropogenic and natural stressors is far from straightforward (Smith & Buddemeier 1992; Chadwick-Furman 1996; Brown 1997). Elevated sea temperatures and solar radiation, acting alone or together, have been identified as the principal causative agents of coral reef bleaching and mortality in numerous recent studies (e.g., Lesser *et al.* 1990; Glynn 1993, 1996; Goreau & Hayes 1994; Hoegh-Guldberg & Salvat 1995; Schick *et al.* 1996; Brown 1997). Since tropical marine organisms live close to their upper thermal and radiation tolerance limits (Moore 1972; Johannes 1975; Jokiel 1980), relatively modest increases above maximum seasonal norms can result in stress and mortality. The bulk of current field and laboratory studies has demonstrated the primacy of elevated water temperature and solar radiation, particularly UVR, in regional and global scale coral reef disturbances.

Field studies have demonstrated that reef temperatures exceeding 2–3°C above long-term local seasonal maxima can initiate bleaching (Glynn 1990*a*; Goreau & Hayes 1994; Gleeson & Strong 1995; Podestá & Glynn 1997). Even lower positive temperature anomalies (1–2°C) can have comparable negative effects if sustained for several weeks to a few months (Glynn & D'Croz 1990; Goreau *et al.* 1993). Controlled laboratory experiments support the hypothesis that slightly elevated water temperatures are responsible for numerous bleaching episodes observed in the field. Corals adapted to high thermal environments, e.g. at Enewetak, Jamaica and Oman, also bleach and die when seasonally high temperature norms are exceeded by a few degrees (Jokiel & Coles 1990; Goreau *et al.* 1993; Salm 1993). Visible radiation sometimes kills shallow-living corals (Brown *et al.* 1994), but UVR, especially

UVB (290–320 nm), is the most damaging, with effects sometimes extending to 20 m depth (Gleason & Wellington 1993; Schick *et al.* 1996). However, most reported instances of UVB damage have occurred under high temperature conditions, rendering problematic the detection of UVB damage alone (Glynn 1996; Brown 1997).

Scale of disturbances

Compared with most disturbances, such as corallivore outbreaks, diseases, eutrophication and sedimentation, which are generally limited spatially, warming-induced bleaching is pervasive and global in extent. Since the first documented coral reef bleaching event in 1963, 63 severe to very severe (≥50% of corals bleached) events have been reported worldwide through 1990. From 1991 to 1998, 123 events were observed, 56 of which occurred during the 1997/98 ENSO event (Fig. 1). The only major tropical regions lacking reports of severe bleaching are the west African coast and western Australia, regions probably not so intensively studied as elsewhere. The apparent increase in bleaching incidents is related in part to recent ENSO activity, which is largely correlated with high temperature and to a lesser degree irradiance stress, but an unknown amount of this increase must also be attributed to a heightened awareness and more consistent reporting of the disturbance.

ENSO disturbances are not limited to coral reefs, but have impacts on other tropical marine taxa in temperate and subpolar ecosystems, and among terrestrial biotas. Phytoplankton and benthic plant communities that experience severe reductions in productivity, due to reduced nutrient availability in 1982/83 (Feldman *et al.* 1984), created trophic shortfalls in diverse herbivore and carnivore taxa. In the Galápagos Islands and along the South American coast, numerous herbivores (suspension-feeding invertebrates, fishes, marine iguanas) and carnivorous species (fishes, sea birds, pinnipeds) experienced extreme population declines, reproductive failures and mass mortalities (Robinson & del Pino 1985; Glynn 1990*b*). All changes were not deleterious among all species, however; some bivalves, crustaceans and fishes actually increased in abundance during and shortly following ENSO 1982/83 (Arntz & Tarazona 1990). Although beyond the scope of this chapter, it should be noted that ENSO events also cause biotic dislocations in such disparate ecosystems as temperate kelp forests, benthic and pelagic subarctic fisheries, desert floras and rain forests.

Coral mortality and persistence of coral reefs

The following examines briefly the longer term consequences of significant and widespread coral mortality vis-à-vis the persistence of coral reefs and their potential for continued reef building. Following the 1982/83 disturbance, the most obvious impact affecting eastern Pacific coral reefs was bioerosion (Glynn 1988; Scott *et al.* 1988; Guzmán & Cortés 1992; Eakin 1996; Reaka-Kudla *et al.* 1996). Bioerosion accelerated the deterioration of reefs by the direct removal of coral/algal deposits and the weakening of reef frameworks. These degradative processes have continued for nearly two decades, in many instances eliminating coral reefs locally and rendering substrates unsuitable for coral recruitment. In the Galápagos Islands, successful recruitment now occurs not on reef deposits but mainly on the summits of smooth basalt boulders, microhabitats that cannot be scaled by grazing *Eucidaris*. Former coral reef communities of high topographic complexity are now patches of coral debris of low relief covered with turf algae or crustose coralline algae, both supporting large populations of cidarid echinoids. Comparable post-ENSO habitats at many mainland locations are populated by large numbers of diadematid echinoids.

The loss of live corals harbouring obligate metazoan symbionts, such as crustaceans and molluscs, led to rapid declines in several associated species (Glynn 1985*b*). Obligate corallivores would also be expected to show decreases in abundance as their trophic resources were diminished. The sea star corallivore *Acanthaster* has steadily declined on reefs in Panamá after the 1982/83 ENSO disturbance (Fong & Glynn 1998), but it is not known if this response is related to lower coral abundance or some other factor. A gastropod corallivore (*Jenneria pustulata* (Lightfoot)) that feeds preferentially on pocilloporid corals suffered high mortality during the bleaching event, but has not recovered to pre-disturbance abundances even though its chief prey has recovered. However, a dominantly corallivorous pufferfish, *Arothron meleagris* (Bloch & Schneider), which feeds preferentially on pocilloporid corals, switched its diet to coralline algae when the former became less abundant (Guzmán & Robertson 1989). Obligate fish corallivores, such as chaetodontids, disappeared from reefs killed by *Acanthaster* outbreaks in the southern Ryukyu Islands, Japan (Sano *et al.* 1987). Reef-associated fishes in the eastern Pacific and elsewhere often do not decline in abundance following

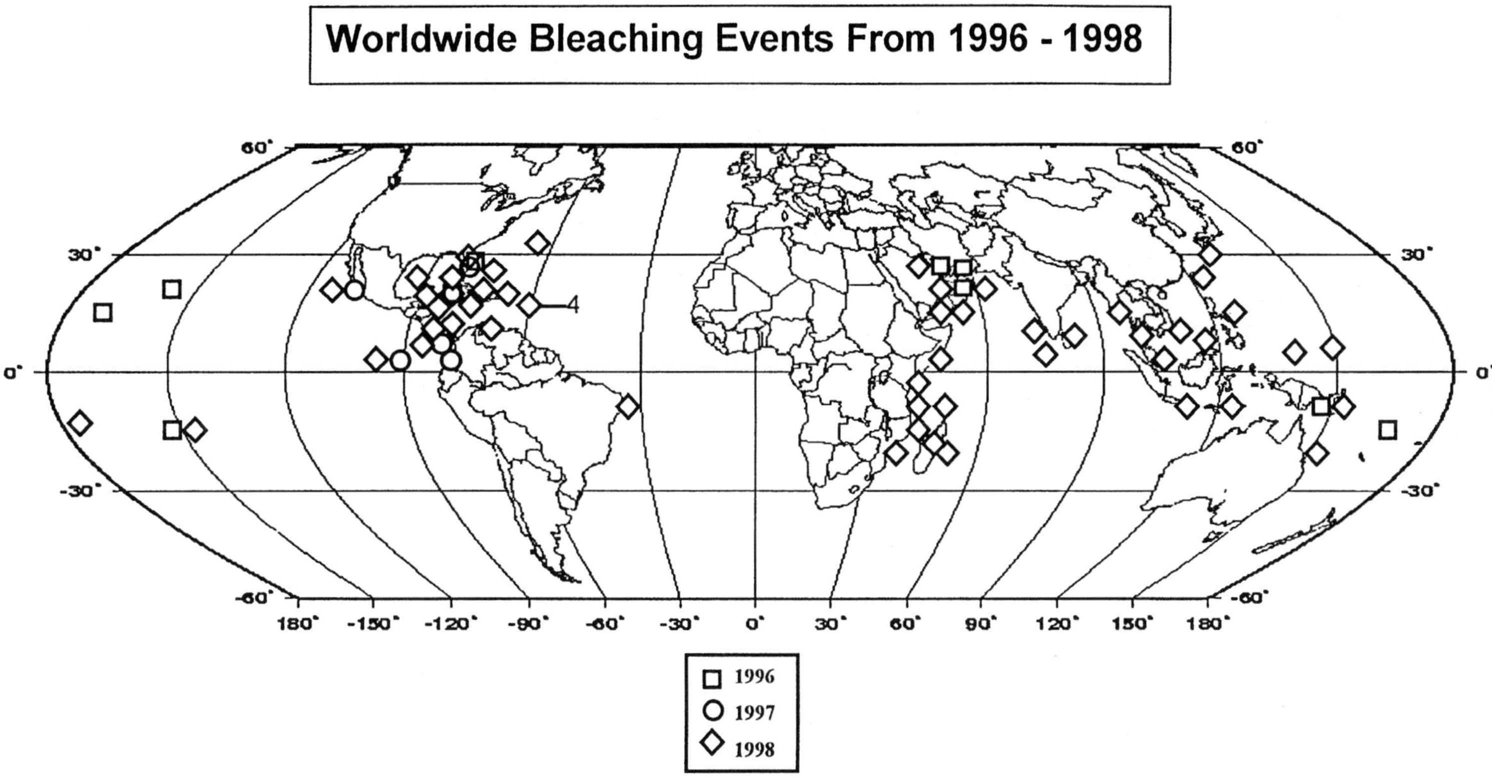

Fig. 1. Worldwide distribution of coral reef bleaching events 1996–1998. Only strong to severe events with bleaching ≥50% of total cover are noted. These records are largely from Wilkinson (1998) and the Coral-List Server (1999).

severe disturbances, but undergo redistributions, moving to rock reefs or other habitats of high relief (Jones 1991). Whether such coral reef associates would have an equal chance of preservation in the absence of reef frameworks is unknown. This raises the interesting question: when coral reef habitats disappear, do most associated species also become extinct or do they persist elsewhere in habitats of similar or diminished structural complexity? The generally long delay in the re-establishment of coral reefs following extinctions (10^6–10^7 years), and the concomitant changes in taxonomic composition, would argue for the former scenario.

About one-third of eastern Pacific zooxanthellate corals experienced severe declines in abundance and local to regional scale extinctions (Glynn 1997). Several of these species have not attained pre-1983 abundances, and many undergoing recovery have again experienced high mortalities or have disappeared following the 1997/98 ENSO. Three of four species that disappeared regionally in 1983 have not returned although these species still occur abundantly in the western Pacific. One of the eastern Pacific hydrocoral endemics (*Millepora boschmai* (de Weerdt & Glynn)) that disappeared in 1983 was found again in 1992 (Glynn & Feingold 1992), but this small population (five colonies) bleached and died during the 1997/98 ENSO event. This could possibly represent the first documented example of a modern coral extinction event. Nonetheless, these species of precarious status do not contribute importantly to reef building in the eastern Pacific. The predominant framework builders construct branching (*Pocillopora*) and massive (*Porites*, *Pavona*, *Gardineroseris*) colonies. All of these taxa virtually disappeared in the Galápagos Islands and at Cocos Island after 1983, and the minimal recovery observed subsequently has been largely nullified by the 1997/98 ENSO disturbance. It is now recognized that the major framework builders, i.e. broadcast spawners of massive colony morphology, are not recruiting rapidly enough to sustain reef growth in many areas (Kinzie 1999). Whether this is a temporary natural condition or a sympton of current global reef decline is an open question.

Whether or not coral symbioses can evolve rapidly enough to meet the apparent changes towards global greenhouse conditions is an intriguing question subject to much recent speculation. Reef-building scleractinian corals have been found to host a genetically diverse assemblage of endosymbiotic zooxanthellae (Rowan & Powers 1991; Rowan & Knowlton 1995; Baker & Rowan 1997). The possibility of short-term shuffling of symbionts, an acclimatization response whereby bleached corals are repopulated by different genetic strains of zooxanthellae more resistant to high temperature/irradiance stress, has been proposed but not yet demonstrated (Buddemeier & Fautin 1993; Ware *et al.* 1996). Evolutionary change involving zooxanthellae, corals and the establishment of a stable symbiotic association is subject to some serious constraints. For example, sexual reproduction, which increases genetic diversity and enhances the potential for evolutionary change, has not been observed in zooxanthellae, and is often of secondary importance in several coral species that rely dominantly on asexual propagation. While the generation time of zooxanthellae (days) is favourable for rapid evolution, that of corals often spans decades. Moreover, the disproportionately high fecundity of large corals and the potential for interbreeding among overlapping generations, do not promote high genetic diversity or rapid evolution. Potts & Garthwaite (1991) concluded that the scope of certain life history attributes has allowed little evolutionary differentiation in some species since the Miocene, but that their influence on other species has permitted rapid speciation in the late Quaternary.

Some additional mechanisms that might enhance genetic diversity and permit rapid evolutionary rates in scleractinian corals have been hypothesized recently. For example, some long-lived corals may become genetic mosaics through the accumulation of somatic mutations, thus increasing variation in the gene pool. Fautin (1997) has argued that selection in such corals could act immediately on favourable mutations, thereby promoting accelerated evolution. Another factor that would increase the genetic diversity of corals is hybridization, which has been observed experimentally between both closely related and distantly related congeneric species, and between species from different genera (Richmond 1997). Hybrid larvae originating from congeneric crosses have successfully settled, and some colonies have survived for two years before their accidental death (Willis *et al.* 1993). Buddemeier *et al.* (1997) maintain that certain genetic and reproductive traits of corals, together with reticulate evolution (involving frequent hybridization) and multiple symbiosis, should allow for both rapid acclimatization and adaptation to environmental change. Considering the relatively rapid projected changes in global warming (tens to hundreds of years), it remains to be seen if the necessary evolutionary adaptations required to cope with these changes can be realized.

Characteristics of extinction-prone species

Some attempts have been made to characterize the traits of species that are susceptible to extinction (e.g. Boucot 1990). Maynard Smith (1989) has underlined the importance of biological and ecological factors in conjunction with physical change to explain extinction events. The causes of extinction, even in well studied examples, are complex and the factors responsible for the demise of a species may differ totally from the events that caused its population decline. For example, among numerous contemporary extinctions in the terrestrial realm, it has not been possible to distinguish accurately between proximate intrinsic population dysfunction, such as demographic stochasticity or genetic deterioration, and extrinsic forces, such as extreme physical perturbations or increased predation, in causing the disappearance of species (Simberloff 1986). Cognizant of such complicating factors, perhaps some insight can be gained by comparing these recognized traits of extinction-prone species with those species populations that demonstrated notable declines or local to regional scale extinctions during recent ENSO bleaching events.

Among the more commonly recognized traits of extinction-prone sessile species are (a) stenotopy, (b) endemism, (c) restricted range, (d) small population size, (e) low vagility, and (f) low intrinsic population growth. Several of the eastern Pacific zooxanthellate scleractinian and hydrocoral species that experienced extreme population reductions or extinctions during the 1982/83 ENSO episode (ten of 33 species) exhibited some of these traits (Glynn 1997). The restricted habitats of these corals, i.e. their general confinement to clear oceanic water with good circulation, suggests condition (a) stenotopy, i.e. tolerances to relatively narrow environmental conditions. At least two of the three traits (b), (c) and (d) also were shared by the most severely impacted coral populations. Our meagre knowledge of dispersal and recruitment in eastern Pacific corals limits an assessment of the roles of (e) and (f) in extinction events. Little is known about the vagility or dispersal potential of corals, but preliminary studies indicate high fecundities and low recruitment success by sexual means (Glynn *et al.* 1994, 1996).

To this list may be added the zooxanthellate condition of reef-building corals because organisms harbouring endosymbiotic algae, including all zooxanthellate cnidarians and other taxa (Foraminifera, Porifera, Mollusca), are especially sensitive to stressful warming events. Also, coral species exhibiting high rates of calcification usually suffer higher mortality than slower growing corals, an observation that applies worldwide (Glynn 1993). Notable temperature drops, 8–10°C below the lower normal thermal limits of corals, cause bleaching and mortality, but not nearly at the scale of high temperature stress events (Coles & Fadlallah 1991). Finally, species living in close association with corals, such as obligate symbionts (crustaceans), micropredators (molluscs), and some larger corallivores (molluscs, sea stars and fishes) have been greatly reduced in abundance or have disappeared with the loss of their hosts and/or principal trophic resource. However, many coral reef associates are not strictly dependent upon live coral and may take up refuge and survive in non-reef settings should this biotope disappear.

Species inhabiting upwelling environments may also have an increased chance of surviving biotic crises, due to elevated nutrient supplies that enhance larval development, recruitment and growth (Vermeij 1986). Except for zooxanthellate corals, which generally prefer low nutrient environments, eastern Pacific upwelling centres have been shown to harbour numerous Neogene taxa – *Pelliseria* (a mangrove genus), molluscs, echinoids, an ectoproct family and balanoid barnacles – that became extinct in the western Atlantic. Comparisons of ENSO effects in upwelling and non-upwelling environments also support Vermeij's (1986) hypothesis that upwelling centres do not play an important role as refuges for zooxanthellate corals. The four coral species that experienced local to regional extinctions (*Acropora valida* (Dana), *Porites (Synaraea) rus* (Forskål), *Millepora platyphylla* (Ehrenberg), *Millepora boschmai* (de Weerdt & Glynn)) were known only from non-upwelling localities. Further, two species confined to upwelling environments (*Pocillopora inflata* (Glynn), *Siderastrea glynni* (Budd & Guzmán)) are either uncommon or rare, and it is not known if these had broader distributions from which they were eliminated by some severe disturbance event.

Evidence of past ENSO activity

In order to link extinction events with episodes of ocean warming, it will be necessary to identify skeletal and taphonomic signatures related to heat stress and demonstrate their association with coral reef death assemblages. Since coral reef bleaching and tissue death can result from a variety of stressors, including sudden low temperature excursions, special care must be exercised to demonstrate specific cause and effect relationships. Some markers and other sorts of evidence that may help in this analysis are examined below.

ENSO events can be identified in corals from a combination of stable oxygen ($\delta^{18}O$) isotope thermometry and alterations in skeletal growth (Carriquiry *et al.* 1988; Druffel *et al.* 1990; Wellington & Dunbar 1995; Linsley *et al.* 1999). In addition, the presence of high density fluorescent bands (Scoffin *et al.* 1989) and trace metals (Shen *et al.* 1992), such as cadmium and barium, can provide information on nutrient levels and coastal rainfall, which are strongly influenced by ENSO. At sites where salinity variation is minimal, it is possible to distinguish strong to very strong ENSOs from lesser magnitude events. Severe ENSO events do not leave an oxygen isotopic record in all corals in heavily impacted reef areas because of cessation of growth or greatly reduced skeletogenesis. However, stress bands are usually present in such corals and can serve to identify periods of suboptimal growth. By comparing $\delta^{8}O$ records and skeletal features (such as stress bands and growth discontinuities) across sites, it may be possible to correlate ENSO events regionally, thus demonstrating widespread disturbances.

Another feature of potential use are skeletal protuberances on massive corals that develop from regenerating live tissue patches that survive bleaching events or tissue death. These protuberances or lobes often occur on the sides of colonies, arising from vertically oriented tissues that survive severe bleaching, but they may also cover the summits of colonies, developing from remnant tissues in depressions and fissures. In some examples, rims may form around the periphery of colonies, surrounding the skeletons and other traces of epifauna that invaded centrally located dead surfaces (Colgan 1990). Such growth abnormalities may be diagnostic of coral bleaching, and thus serve as markers to help identify warming disturbances. Regenerating lobes with recently dead summits and extensive internal bioerosion of massive corals in Panamá and the Galápagos Islands have captured the severe 1982/83 and 1997/98 ENSO events (Figs 2 and 3).

Large scale coral mortality in Costa Rica (Carriquiry *et al.* 1988), Panamá and the Galápagos Islands (Glynn 1988), a result of the 1982/83 ENSO bleaching event, caused sudden massive increases in dead reef frameworks, which were then exposed to intense bioerosion from increases in various bioeroding taxa. A preliminary off-reef coring study in Costa Rica revealed high abundances of coral grains in surface sediments, indicative of accelerated post-ENSO bioerosion. Since different bioeroder taxa (e.g. sponges, lithophagine gastropods, echinoids and fishes) generate grains of varying size fractions, the analysis of reef sediments could provide information on the identities of the species responsible. The increase in abundances of bioeroders following mass coral mortality, e.g. echinoids (Glynn 1988; Eakin 1996), lithophagine bivalves and clionid sponges (Scott *et al.* 1988; Colgan 1990), may be detected in reef sediments. Walbran *et al.* (1989) and Gordon & Donovan (1992) offered evidence that the skeletal elements of *Acanthaster* and echinoids, respectively, were broadly representative of the abundances and distribution of these taxa on living reefs. While such studies may provide estimates of the relative abundances of bioeroders, they must be preceded by careful taphonomic analyses (Pandolfi 1992).

Estimates of the ages of damaged or dead corals and reef frame blocks can sometimes help bracket periods of former ENSO events, which can then be investigated utilizing methods of higher resolution. From the ages of old massive corals that experienced partial to total mortality in the Galápagos Islands during the 1982/83 ENSO, it was concluded that a comparable disturbance had not occurred in this area for at least 200 years and possibly for as long as 400 years (Glynn 1990*a*; Dunbar *et al.* 1994). The thickness of pocilloporid reef frame blocks in the Galápagos also suggests uninterrupted reef accumulation over periods of 110–120 years and 180–200 years with one reef demonstrating continuous growth during the last 500 years. Comparable measurements on Panamanian reefs in a non-upwelling area suggest uninterrupted framework accumulation of 135–175 years with the longest record of about 300 years before the 1983 disturbance. Finally, as storm deposits have been utilized as markers of past El Niño flooding on mainland Peru, the identification and dating of coral skeletal debris on Galápagos beaches may offer clues of the timing and intensity of ENSO-related marine disturbances.

Another approach to determine the timing of the last severe ENSO event has utilized changes in the spatial pattern of coral barriers that surround and protect massive corals from *Acanthaster* predation (Glynn 1985*b*). Before 1983, several old colonies of *Gardineroseris* were surrounded by pocilloporid corals that prevented access by *Acanthaster*. Pocilloporid coral mortality was pronounced, and resulted in the elimination of this biotic barrier. The exposed *Gardineroseris,* which suffered only partial ENSO-related mortality, was then attacked by *Acanthaster*, resulting in additional extensive

Fig. 2. *Gardineroseris planulata* (Dana) with lobes formed from tissue surviving the 1982/83 ENSO and dead summits resulting from the 1997/98 ENSO. Portions of the dead, eroded colony surface have been invaded by branching and encrusting colonies of other coral species (left foreground). The longest axis of the summit of the labelled lobe is about 30 cm. Uva Island reef, Gulf of Chiriquí, Panamá, 6 m depth, 15 May 1999.

partial mortality. Core drilling showed that the largest of these corals grew uninterruptedly for 192 years before the 1982/83 ENSO event. However any one of these various methods reveals past disturbance events, their relationship with ENSO warming will be strengthened by the application of multiple approaches and broad-scale sampling encompassing numerous reefs.

Ecological/palaeoecological implications

Whether the causes of catastrophic coral loss are due to climatically related ocean warming or other factors, numerous reef-associated species – facultative and obligate coral symbionts, parasites, corallivores, coelobites, and epibiotic species utilizing reef habitats as refugia – would appear poised to experience severe declines and local extinctions. With continuing bioerosion, loss of suitable settling surfaces and reduced recruitment (e.g. due to low coral species abundances and fragmented populations, and unfavourable environmental conditions affecting fecundity and reproductive activities), coral reef frameworks would disintegrate and eventually disappear, further depleting the availability of this habitat (Glynn & Colgan 1992). Habitat loss is now recognized as a key factor responsible for declines in biodiversity and species extinctions (Maynard Smith 1989; Wilson 1992; Carlton 1993).

Widespread coral reef bleaching has been reported only recently, since the early 1980s. If global mean surface temperatures increase by 2 to 4°C over the next 100 years, as predicted for low, best estimate and high climate sensitivities by the Intergovernmental Panel on Climate Change scenario (Houghton *et al.* 1992), then mass coral reef extinction events would probably be initiated in less than 100 years. This scenario depicts a fleeting biotic crisis that could potentially result in extinctions during brief 10^2 to 10^3 year intervals. Conversely, excluding putative bolide impact events, mass extinctions in the fossil record are coarse, occurring over periods of a few to 10–15 Ma (Sepkoski 1986; Stanley 1987; Copper 1994). Therefore, it is possible that some marine extinction events attributed to periods of refrigeration, rapid sea-level fall or other causes, could have occurred during relatively brief periods of warming in, for example, Late Devonian, middle-Late Cretaceous or Early Eocene times.

A sudden, severe and brief temperature rise

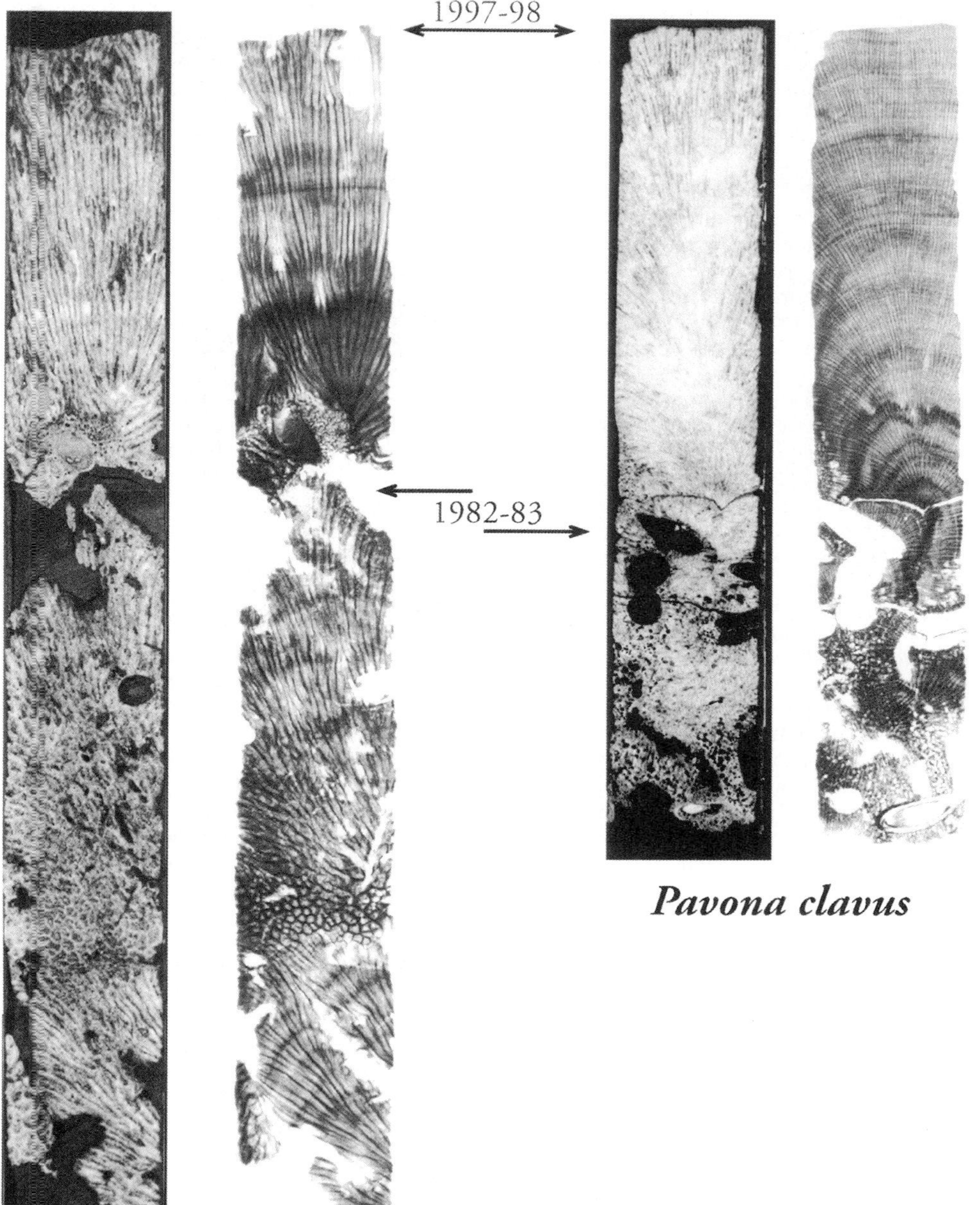

Fig. 3. Longitudinal sections of 5 cm diameter cores drilled through the lobes of *Gardineroseris planulata* and *Pavona clavus*. Cores were impregnated with blue-dyed resin S 40 (Silmar®) and then cut parallel to the linear growth axes. X-rayed sections (right) of the regenerating lobes that formed after the 1982/83 ENSO event are paired with their respective photographic images (left). Arrows denote the two recent ENSO disturbances and their corresponding skeletal growth discontinuities. Extensive bioerosion, due to clionid sponges, polychaetous annelids and lithophagine bivalves, is evident in the skeleton killed during the 1982/83 bleaching event.

was postulated by Emiliani *et al.* (1981) to explain the widespread extinctions and biological stresses at the Cretaceous/Tertiary boundary. These workers calculated that an oceanic bolide impact would have caused a global surface temperature increase exceeding 10°C and that such a stressful warming episode could have persisted for months to years. They further examined the short-term, maximum temperature tolerance limits of several taxa with Late Cretaceous affinities and calculated the percentage of genera that successfully crossed the Cretaceous/Tertiary boundary. Mindful of the many shortcomings of this approach, e.g. the taxonomic uncertainties and poor fossil record during this period, they nonetheless concluded that a heat shock was probably responsible for the observed extinctions. Survival of five neritic (to 200 m depth) benthic taxa ranged from 0 to 43% with no rudists and only 20% of reef-building coral genera surviving to the early Cenozoic. The surviving coral taxa, however, evidently did not build reefs again until the Oligocene. Early Paleocene reefs were built dominantly by coralline algae, sponges and bryozoans (Fagerstrom 1987; Wood 1999).

Since the temporal resolution of mass extinction events often spans 5–15 Ma, it is possible that several relatively minor perturbations, probably qualitatively and quantitatively dissimilar, have contributed to the longer-duration major extinctions. For example, the Late Devonian crisis, including the Kellwasser to Hangenberg events, occurred over a 14 Ma period, demonstrating a stepped series of extinction events probably related to marine regression and perhaps El Niño-like disturbances such as elevated sea temperatures, stratification and anoxic conditions (Schindler 1990; Wood 1999). The latest Permian (Changxingian) mass extinction, spanning 3–5 Ma, coincided with a period of intense volcanic activity, global warming, a major transgression, and marine anoxic conditions (Hallam & Wignall 1997; Wood 1999). Similarly, the sudden high temperature shock proposed by Emiliani *et al.* (1981) followed a period of marine regression, which resulted in a substantial reduction of shallow water environments with probably significant effects on the biota. Such multifaceted crises might be how processes operating on ecological timescales would translate into those events that are recorded in the geological record. In this light, the current short-term coral mortalities, population reductions, local and regional extinctions and reef degradation would represent one brief perturbation of possibly numerous forthcoming events that may span thousands to millions of years.

Tropical marine biotas demonstrate a high vulnerability to extinctions (Ricklefs *et al.* 1990; Jablonski 1991). In light of the causal connection between elevated sea-surface temperature/irradiance and contemporary bleaching/mortality, current ecological evidence supports an important role for warming events as ubiquitous agents impacting coral reefs. Much of the focus on the causes of reef extinctions in the palaeontological literature has been on global ocean cooling, brought on by climatic thermal deterioration (Fischer & Arthur 1977; Fagerstrom 1987; S. Stanley 1987, 1990). However, global warming-induced anoxia/dysoxia has recently been considered as a viable cause of mass extinctions in the latest Ordovician, latest Permian and latest Palaeocene (Hallam & Wignall 1997). Also, G. Stanley (1997) has noted that it has not been possible to relate major Phanerozoic reef extinctions unequivocally to icehouse periods. The disappearance of reef units at Late Triassic and latest Cretaceous periods occurred during greenhouse conditions. A recent palaeobotanical study offers evidence of a fourfold increase in atmospheric CO_2 concentration and suggests an associated 3–4°C 'greenhouse' warming across the Triassic–Jurassic boundary (McElwain *et al.* 1999). If a temperature rise of this magnitude occurred in ancient tropical marine waters, it would probably cause significant reef-building coral mortality, based on extant coral thermal tolerance limits. But it is cautioned that serial meteoric impacts may also be common to many (but not all) of these extinction events (Kauffman 1986; Bice *et al.* 1992; McRoberts & Newton 1995). Offshore core drilling near Caribbean and Pacific reefs has revealed nearly continuous sequences of reef carbonates that formed when sea level stood at 110–130 m below present sea level (Fairbanks 1989; Bard *et al.* 1996). This period of reef growth occurred near the last glacial maximum and during the Younger Dryas event, when sea temperatures were several degrees lower than today's. These findings suggest that the often-cited limiting effects of glacial cooling on coral reefs should be re-evaluated.

Knowledge of the zooxanthellate coral fauna and reef-building record of the Pliocene and Quaternary sheds little light on understanding the causes of the current global degradation of coral reefs. Both Caribbean and Pacific studies suggest that regional disturbances and not global scale effects were responsible for the changes noted in the fossil record. In the Caribbean region, where fossil reef assemblages are reasonably well known, it has been found that Holocene reef-building corals arose during a faunal turnover period of 4–1.5 Ma, roughly contemporaneous with the closure of the Central American Isthmus (3.5–3.0 Ma; Coates *et al.* 1992) and just prior to the initiation of frequent sea-level oscillations and temperature fluctuations associated with the northern hemispheric glaciation events of the Pleistocene (Budd *et al.* 1996, 1998; Jackson & Budd 1996). Approximately 80% of the 100 Early Pliocene species became extinct, and 60% of the nearly 70 extant species originated during post-Pliocene time. A high proportion of the faunal turnover evidently occurred before the onset of northern hemisphere glaciations at approximately 2.4 Ma (Budd *et al.* 1996). The Pleistocene history of Caribbean coral reefs is characterized by nearly complete stability of its species pool, coral community structure, and possibly overall rates of reef calcification (Mesolella 1967; Geister 1984). Only the genera *Stylophora* and *Pocillopora* became extinct in the early and late Pleistocene respectively (Budd *et al.* 1994). Although the Plio-Pleistocene fossil record of the Indo-Pacific region is very incomplete compared with the Caribbean, studies in Papua New Guinea also indicate stability of the coral species pool over this period (Veron & Kelley 1988; Pandolfi 1996).

Jackson (1994) and Pandolfi (1996) cite evidence of 5–6°C lower tropical sea-surface temperatures (SSTs) than today during interstadial and glacial periods in the Pleistocene (e.g. Beck *et al.* 1992; Guilderson *et al.* 1994; McCulloch & Mortimer 1994). In terms of current high thermal stress levels, such temperatures would not closely approach SST thresholds that are now causing coral bleaching and mortality events. Thus, the Holocene reef-building coral fauna appears to have evolved during a time of frequent and rapidly changing sea levels and comparatively low SSTs that characterized Pliocene time. If further studies support the prevalence of relatively low tropical SSTs during the past 1–2 Ma, then it is possible that Holocene corals are now experiencing unprecedented thermal stress conditions.

The possibility that geologically fleeting periods of global sea warming may cause biotic crises in tropical regions is provocative and worthy of further study. Whether we are presently witnessing merely background marine extinctions, mass mortality events or are entering a period of mass coral reef extinctions and degradation is a challenging question of unprecedented importance.

The information and ideas presented in this paper were enhanced substantially by my students and colleagues: S. Colley-Theodosiou, J. Cortés, L. D'Croz, C. M. Eakin, J. S. Feingold, P. Fong, R. N. Ginsburg, H. Guzmán, C. Hüerkamp, C. Jiménez, I. G. Macintyre, P. Martínez, J. L. Maté, G. Podestá, R. H. Richmond, F. Rivera, T. Smith, B. Vargas-Angel, H. R. Wanless and G. M. Wellington. For insightful comments, I thank A. C. Baker, M. W. Colgan and D. F. McNeill. Thanks are also due A. Buck, R. E. Dodge, M. Hyatt, D. F. McNeill and T. Smith for help in drilling and processing the cores. J. Hendee and G. Morisseau-Leroy kindly arranged to make bleaching records available through the NOAA Coral Health and Monitoring Program. I am grateful for the opportunity offered by E. Insalaco and P. Skelton to present an ecological perspective on modern mass coral mortalities that might help shed light on some Phanerozoic extinction events. Research support was provided by the Smithsonian Institution, US National Science Foundation (Biological Oceanography Program) grants OCE-9314798 and OCE-9711529, and National Geographic Society grants 5208–94 and 5969–97.

References

ARNTZ, W. E. & TARAZONA, J. 1990. Effects of El Niño 1982–83 on benthos, fish and fisheries off the South American Pacific coast. *In*: GLYNN, P. W. (ed.) *Global Ecological Consequences of the 1982–83 El Niño-Southern Oscillation.* Elsevier Oceanography Series, **52**, Amsterdam, 323–360.

BAKER, A. C. & ROWAN, R. 1997. Diversity of symbiotic dinoflagellates (zooxanthellae) in scleractinian corals of the Caribbean and eastern Pacific. *Proceedings of the 8th International Coral Reef Symposium*, **2**, 1301–1306.

BARD, E., HAMELIN, B., ARNOLD, M., MONTAGGIONI, L., CABLOCH, G., FAURE, G. & ROUGERIE, F. 1996. Deglacial sea-level record from Tahiti corals and the timing of global meltwater discharge. *Nature*, **382**, 241–244.

BECK, J. W., EDWARDS, R. L., ITO, E., TAYLOR, F. W., RECY, J., ROUGERIE, F., JOANNOT, P. & HENIN, C. 1992. Sea-surface temperature from coral skeletal strontium/calcium ratios. *Science*, **257**: 644–647.

BICE, D. M., NEWTON, C. R., MCCAULEY, S., REINERS, P. W. & MCROBERTS, C. 1992. Shocked quartz at the Triassic–Jurassic boundary in Italy. *Science,* **225**, 443–446.

BOUCOT, A. J. 1990. Phanerozoic extinctions: how similar are they to each other? *In*: KAUFFMAN, E. G. & WALLISER, O. H. (eds) *Extinction Events in Earth History: Proceedings of the Project 216, Global Biological Events in Earth History.* Springer, Berlin, **30**, 5–30.

BROWN, B. E. 1987. Worldwide death of corals – natural cyclical events or man-made pollution?*Marine Pollution Bulletin*, **18**, 9–13.

—— 1997. Coral bleaching: causes and consequences. *Coral Reefs*, **16** (suppl.), S129–S138.

——, DUNNE, R. P., SCOFFIN, T. P. & LE TISSIER, M. D.A. 1994. Solar damage in intertidal corals. *Marine Ecology Progress Series*, **105**, 209–218.

BUDD, A. F., STEMANN, T. A. & JOHNSON, K. G. 1994. Stratigraphic distributions of genera and species of Neogene to Recent Caribbean reef corals. *Journal of Paleontology*, **68**, 951–977.

——, JOHNSON, K. G. & STEMANN, T. A. 1996. Plio-Pleistocene turnover and extinctions in the Caribbean reef-coral fauna. *In*: JACKSON, J. B. C., BUDD, A. F. & COATES, A. G. (eds) *Evolution and Environment in Tropical America.* University of Chicago, Chicago, 168–204.

——, PETERSEN, R. A. & MCNEILL, D. F. 1998. Step-wise faunal change during evolutionary turnover; a case study from the Neogene of Curaçao, Netherlands Antilles, *Palaios*, **13**, 170–188.

BUDDEMEIER, R. W. & FAUTIN, D. G. 1993. Coral bleaching as an adptative mechanism: a testable hypothesis. *Bioscience*, **43**, 320–326.

——, R. W., FAUTIN, D. G. & WARE, J. R. 1997. Acclimation, adaptation and algal symbiosis in reef-building corals. *In:* HARTOG, J. C. DEN (ed.) *Proceedings of the 6th International Conference on Coelenterate Biology.* National Natuurhistorisch Museum, Leiden, 71–76.

CARLTON, J. T. 1993. Neoextinctions of marine invertebrates. *American Zoologist*, **33**, 499–509.

CARRIQUIRY, J. D., RISK, M. J. & SCHWARCZ, H. P. 1988. Timing and temperature record from stableisotopes of the 1982–1983 El Niño warming event in eastern Pacific corals. *Palaios*, **3**, 359–364.

CHADWICK-FURMAN, N. E. 1996. Reef coral diversity and global change. *Global Change Biology,* **2**, 559–568.

COATES, A. G., JACKSON, J. B.C., COLLINS, L. S., CRONIN, T. M., DOWSETT, H. J., BYBELL, L. M., JUNG, P. & OBANDO, J. A. 1992. Closure of the Isthmus of Panama: the near-shore marine record of Costa Rica and western Panama. *Geological Society of America Bulletin*, **104**, 814–828.

COFFROTH, M. A., LASKER, H. R. & OLIVER, J. K. 1990. Coral mortality outside of the eastern Pacificduring 1982–83: relationship to El Niño. *In*: GLYNN, P. W. (ed.) *Global Ecological Consequencesof the 1982–83 El Niño-Southern Oscillation.* Elsevier Oceanography Series **52**, Amsterdam, 141–182.

COLES, S. L. & FADLALLAH, Y. H. 1991. Reef coral survival and mortality at low temperatures in the Arabian Gulf: new species-specific lower temperature limits. *Coral Reefs*, **9**, 231–237.

COLGAN, M. W. 1990. El Niño and the history of eastern Pacific reef building. *In*: GLYNN, P. W. (ed.) *Global Ecological Consequences of the 1982–83 El Niño-Southern Oscillation.* Elsevier Oceanography Series, **52**, Amsterdam, 183–232.

COPPER, P. 1994. Ancient reef ecosystem expansion and collapse. *Coral Reefs,* **13**, 3–11.

Coral-List Server. 1999. http: //www.coral.noaa.gov/lists/coral-list.html. Information on specific sites is available at http: //www.coral.noaa.gov/glynn/

DICKINS, J. M. 1984. Evolution and climate in the Upper Palaeozoic. *In*: BRENCHLEY, P. J. (ed.) *Fossils and Climate.* Wiley, Chichester, 317–327.

DRUFFEL, E. R.M., DUNBAR, R. B., WELLINGTON, G. M. & MINNIS, S. A. 1990. Reef-building corals and identification of ENSO warming episodes. *In:* GLYNN, P. W. (ed.) *Global Ecological Consequences of the 1982–83 El Niño-Southern Oscillation.* Elsevier Oceanography Series, **52**, Amsterdam, 233–253.

DUNBAR, R. B., WELLINGTON, G. M., COLGAN, M. W. & GLYNN, P. W. 1994. Eastern Pacific sea surface temperature since 1600 A. D.: the $\delta^{18}O$ record of climate variability in Galápagos corals. *Paleoceanography*, **9**, 291–315.

EAKIN, C. M. 1996. Where have all the carbonates gone? A model comparison of calcium carbonate budgets before and after the 1982–1983 El Niño at Uva Island in the eastern Pacific.*Coral Reefs*, **15**, 109–119.

——, SMITH, D. B., GLYNN, P. W., D'CROZ, L. & GIL, J. 1989. Extreme tidal exposures, cool upwelling and coral mortality in the eastern Pacific (Panamá). *Association of Marine Laboratories of the Caribbean,* **22**, 29.

EMILIANI, C., KRAUS, E. B. & SHOEMAKER, E. M. 1981. Sudden death at the end of the Mesozoic. *Earth and Planetary Science Letters,* **55**, 317–334.

FAGERSTROM, J. A. 1987. *The Evolution of Reef Communities*. Wiley, New York.

FAIRBANKS, R. G. 1989. A 17,000-year glacio-eustatic sea level record: influence of glacial melting rates on the Younger Dryas event and deep-ocean circulation. *Nature*, **342**, 637–642.

FAUTIN, D. G. 1997. Cnidarian reproduction: assumptions and their implications. *In:* HARTOG, J. C. den (ed.) *Proceedings of the 6th International Conference on Coelenterate Biology*. National Natuurhistorisch Museum, Leiden, 151–162.

FELDMAN, G., CLARK, D. & HALPERN, D. 1984. Satellite color observations of the phytoplankton distribution in the eastern equatorial Pacific during the 1982–1983 El Niño. *Science*, **226**, 1069–1071.

FISCHER, A. G. & ARTHUR, M. A. 1977. Secular variations in the pelagic realm. *In:* COOK, H. E. & ENOS, P. (eds) *Deep-water Carbonate Environments.* Society of Economic Paleontologists and Mineralogists, **25**, 19–50.

FITT, W. K., & WARNER, M. E. 1995. Bleaching patterns of four species of Caribbean reef corals. *Biological Bulletin*, **189,** 298–307.

FONG, P. & GLYNN, P. W. 1998. A dynamic size-structured population model: does disturbance control size structure of a population of the massive coral *Gardineroseris planulata* in the eastern Pacific? *Marine Biology*, **130**, 663–674.

GEISTER, J. 1984. Récifs Pleistocenes de la Mer des Caraibes: aspects geologiques et paleoecologiques. *In*: GEISTER, J. & HERB, R. (eds) *Geologie et paleoecologie des récifs, 3ème cycle romand en sciences de la terre*. Institute de Geologie, Université de Berne, Switzerland, 3.1–3.34.

GILI, E., MASSE, J.-P. & SKELTON, P. W. 1995. Rudists as gregarious sediment-dwellers, not reef-builders, on Cretaceous carbonate platforms. *Palaeogeography, Palaeoclimatology, Palaeoecology,* **118**, 245–67.

GLEASON, D. F. & WELLINGTON, G. M. 1993. Ultraviolet radiation and coral bleaching. *Nature,London*, **365,** 836–838.

GLEESON, M. W. & STRONG, A. E. 1995. Applying

MCSST to coral-reef bleaching. *Advances in Space Research*, **16**, 151–154.

GLYNN, P. W. 1984. Widespread coral mortality and the 1982–83 El Niño warming event. *Environmental Conservation*, **11**, 133–146.

—— 1985*a*. Corallivore population sizes and feeding effects following El Niño (1982–83)associated coral mortality in Panamá. *Proceedings of the Fifth International Coral Reef Congress, Tahiti*, **4**, 183–188.

—— 1985*b*. El Niño-associated disturbance to coral reefs and post disturbance mortality by *Acanthaster planci*. *Marine Ecology Progress Series*, **26**, 295–300.

—— 1988. El Niño warming, coral mortality and reef framework destruction by echinoid bioerosion in the eastern Pacific. *Galaxea*, **7**, 129–160.

—— 1990*a*. Coral mortality and disturbances to coral reefs in the tropical eastern Pacific. *In:* GLYNN, P. W. (ed.) *Global Ecological Consequences of the 1982–83 El Niño-Southern Oscillation.* Elsevier Oceanography Series, **52**, Amsterdam, 55–126.

—— (ed.) 1990*b*. *Global ecological consequences of the 1982–83 El Niño-Southern Oscillation.* Elsevier Oceanography Series, **52**, Amsterdam.

—— 1993. Coral reef bleaching: ecological perspectives. *Coral Reefs*, **12**, 1–17.

—— 1996. Coral reef bleaching: facts, hypotheses and implications. *Global Change Biology*, **2**, 495–509.

—— 1997. Eastern Pacific reef coral biogeography and faunal flux: Durham's dilemma revisited. *Proceedings of the Eighth International Coral Reef Symposium, Panamá*, **1**, 371–378.

—— & COLGAN, M. W. 1992. Sporadic disturbances in fluctuating coral reef environments: El Niño and coral reef development in the eastern Pacific. *American Zoologist*, **32**, 707–718.

—— & D'CROZ, L. 1990. Experimental evidence for high temperature stress as the cause of El Niño-coincident coral mortality. *Coral Reefs*, **8**, 181–191.

—— & FEINGOLD, J. S. 1992. Hydrocoral species not extinct. *Science*, **257**, 1845.

——, IMAI, R., SAKAI, K., NAKANO, Y. & YAMAZATO K. 1993. Experimental responses of Okinawan (Ryukyu Islands, Japan) reef corals to high sea temperature and UV radiation. *Proceedings of the Seventh International Coral Reef Symposium, Guam*, **1**, 27–37.

——, COLLEY, S. B., EAKIN, C. M., SMITH, D. B., CORTÉS, J., GASSMAN, N. J., GUZMÁN, H. M., DEL ROSARIO, J. B. & FEINGOLD, J. S. 1994. Reef coral reproduction in the eastern Pacific: Costa Rica, Panamá, and Galápagos Islands (Ecuador). II. Poritidae. *Marine Biology*, **118**, 191–208.

——, ——, GASSMAN, N. J., BLACK, K., CORTÉS, J. & MATÉ, J. L. 1996. Reef coral reproduction in the eastern Pacific: Costa Rica, Panamá, and Galápagos Islands (Ecuador). III. Agariciidae (*Pavona gigantea* and *Gardineroseris planulata*). *Marine Biology*, **125**, 579–601.

GORDON, C. M. & DONOVAN, S. K. 1992. Disarticulated echinoid ossicles in paleoecology and taphonomy: the last interglacial Falmouth Formation of Jamaica. *Palaios*, **7**, 157–166.

GOREAU, T. J. & HAYES R. L. 1994. Coral bleaching and ocean 'hot spots'. *Ambio*, **23**, 176–180.

——, ——, CLARK, J. W., BASTA, D. J. & ROBERTSON, C. R. 1993. Elevated sea surface temperatures correlate with Caribbean coral reef bleaching. *In*: GEYER, R. A. (ed.) *A Global Warming Forum, Scientific, Economic and Legal Overview*. CRC, Boca Raton, 225–255.

GRIGG, R. W. & BIRKELAND, C (eds) 1997. *Status of Coral Reefs in the Pacific*. University of Hawaii Sea Grant College Program.

GUILDERSON, T. P., FAIRBANKS, R. G. & RUBENSTONE, J. L. 1994. Tropical temperature variations since 20,000 years ago: modulating interhemispheric climate change. *Science*, **263**, 663–665.

GUZMÁN, H. M. & CORTÉS, J. 1992. Cocos Island (Pacific of Costa Rica) coral reefs after the 1982–83 El Niño disturbance. *Revista de Biología Tropical*, **40**, 309–324.

—— & ROBERTSON, D. R. 1989. Population and feeding responses of the corallivorous pufferfish *Arothron meleagris* to coral mortality in the eastern Pacific. *Marine Ecology Progress Series*, **55**, 121–131.

——, ——, RICHMOND, R. H. & GLYNN, P. W. 1990. Coral mortality associated withdinoflagellate blooms in the eastern Pacific (Costa Rica and Panamá). *Marine Ecology Progress Series*, **60**, 299–303.

HALLAM, A. & WIGNALL, P. B. 1997. *Mass Extinctions and their Aftermath.* Oxford University, Oxford .

HARMELIN-VIVIEN, M. L. & LABOUTE, P. 1986. Catastrophic impact of hurricanes on atoll outer reef-slopes in the Tuamotu (French Polynesia). *Coral Reefs*, **5**, 55–62.

HOEGH-GULDBERG, O. & SALVAT, B. 1995. Periodic mass-bleaching and elevated sea temperatures: bleaching of outer reef slope communities in Moorea, French Polynesia. *Marine Ecology Progress Series*, **121**, 181–190.

HOUGHTON, J. T., CALLANDER, B. A. & VARNEY, S. K. (eds) 1992. *Climate Change 1992: the supplementary report to the IPCC Scientific Assessment.* Cambridge University, Cambridge.

JABLONSKI, D. 1991. Extinctions: a paleontological perspective. *Science*, **253**, 754–757.

JACKSON, J. B.C. 1994. Constancy and change of life in the sea. *Philosophical Transactions of the Royal Society of London* B, **344**, 55–60.

—— & BUDD, A. F. 1996. Evolution and environment: introduction and overview. *In*: JACKSON, J. B. C., BUDD, A. F. & COATES, A. G. (eds) *Evolution and Environment in Tropical America.* University of Chicago, Chicago, 1–20.

——, JUNG, P. & FORTUNATO, H. 1996. Paciphilia revisited: transisthmian evolution ofthe *Strombina* group (Gastropoda: Columbellidae). *In*: JACKSON, J. B.C., BUDD, A. F. & A. G. COATES (eds) *Evolution and Environment in Tropical America.* University of Chicago, Chicago, 234–270.

JOHANNES, R. E. 1975. Pollution and degradation of coral reef communities. *In*: FERGUSON WOOD, E.

J. & JOHANNES, R. E. (eds) *Tropical Marine Pollution*. Elsevier Oceanography Series, **12**, Amsterdam, 13–51.

JOKIEL, P. L. 1980. Solar ultraviolet radiation and coral reef epifauna. *Science*, **207**, 1069–1071.

—— & Coles, S. L. 1990. Response of Hawaiian and other Indo-Pacific reef corals to elevated sea temperatures. *Coral Reefs*, **8**, 155–162.

JONES, G. P. 1991. Postrecruitment processes in the ecology of coral reef fish populations: a multifactorial perspective. *In*: SALE, P. F. (ed.) *The Ecology of Fishes on Coral Reefs*. Academic Press, San Diego, 294–328.

KAUFFMAN, E. G. 1986. High-resolution event stratigraphy: regional and global Cretaceous bio-events. *In:* WALLISER, O. (ed.) *Lecture Notes in Earth Sciences, Global Bio-Events: A Critical Approach*, **8**, Springer-Verlag, New York, 279–335.

—— & JOHNSON, C. C. 1988. The morphological and ecological evolution of Middle and Upper Cretaceous reef-building rudistids. *Palaios*, **3**, 194–216.

—— & —— 1997. Ecological evolution of Jurassic–Cretaceous Caribbean reefs. *Proceedings of the Eighth International Coral Reef Symposium, Panamá*, **2**, 1669–1676.

KINZIE, III, R. A. 1999. Sex, symbiosis and coral reef communities. *American Zoologist*, **39**, 80- 91.

KLEYPAS, J. A., BUDDEMEIER R. W., ARCHER, D., GATTUSO, J.-P., LANGDON, C. & OPDYKE, B. N. 1999*a*. Geochemical consequences of increased atmospheric carbon dioxide on coral reefs. *Science*, **284**, 118–120.

——, MCMANUS, J. W. & MEÑEZ, L. A.B. 1999*b*. Environmental limits to coral reef development: where do we draw the line? *American Zoologist*, **39**, 146–159.

LABOUTE, P. 1985. Evaluation of damage done by the cyclones of 1982–1983 to the outer slopes of the Tikehau and Takapoto Atolls (Tuamotu Archipelago). *Proceedings of the Fifth International Coral Reef Congress, Tahiti*, **3**, 323–329.

LESSER, M. P., STOCHAJ, W. R., TAPLEY, D. W. & SCHICK, J. M. 1990. Bleaching in coral reef anthozoans: effects of irradiance, ultraviolet radiation, and temperature on the activities of protective enzymes against active oxygen. *Coral Reefs*, **8**, 225–232.

LINSLEY, B. K., MESSIER, R. G. & DUNBAR, R. B. 1999. Assessing between-colony oxygen isotope variability in the coral *Porites lobata* at Clipperton Atoll. *Coral Reefs*, **18**, 13–27.

MCCULLOCH, M. T. & MORTIMER, G. 1994. High fidelity Sr/Ca record of sea surface temperatures: 1982–83 El Niño and MIS-5e. *Abstracts of the Eighth International Conference on Geochronology, Cosmochronology, and Isotope Geology*. US Geological Survey, Circular **1107**, 210.

MCELWAIN, J. C., BEERLING, D. J. & WOODWARD, F. I. 1999. Fossil plants and global warming at the Triassic–Jurassic boundary. *Science*, **285**, 1386–1390.

MCROBERTS, C. A. & NEWTON, C. R. 1995. Selective extinction among end-Triassic European bivalves. *Geology*, **23**, 102–104.

MAYNARD SMITH, J. 1989. The causes of extinction. *Philosophical Transactions of the Royal Society of London*, **B325**, 241–252.

MESOLELLA, K. J. 1967. Zonation of uplifted Pleistocene coral reefs on Barbados, West Indies. *Science*, **156**, 638–640.

MOORE, H. B. 1972. Aspects of stress in the tropical marine environment. *Advances in Marine Biology*, **10**, 217–269.

PANDOLFI, J. M. 1992. A palaeobiological examination of the geological evidence for recurring outbreaks of the crown-of-thorns starfish, *Acanthaster planci* (L.). *Coral Reefs*, **11**, 87–93.

—— 1996. Limited membership in Pleistocene reef coral assemblages from the Huon Peninsula, Papua New Guinea: constancy during global change. *Paleobiology*, **22**, 152–176.

PODESTÁ, G. P. & GLYNN, P. W. 1997. Sea surface temperature variability in Panamá and Galápagos: extreme temperatures causing coral bleaching. *Journal of Geophysical Research*, **102**, 15749–15759.

POTTS, D. C. & GARTHWAITE, R. L. 1991. Evolution of reef-building corals during periods of rapid global change. *In:* DUDLEY, E. C. (ed.) *The Unity of Evolutionary Biology. Proceedings of the 4th International Congress of Systematic and Evolutionary Biology*, **1**, Dioscorider Press, Portland, Oregon, 170–178.

PRAHL, H. VON 1986. Crecimiento del coral *Pocillopora damicornis* durante y después del fenómeno El Niño 1982–1983 en la Isla de Gorgona, Colombia. *Boletín ERFEN*, **18**, 11–13.

PROTHERO, D. R. 1994. The Late Eocene-Oligocene extinctions. *Annual Review of Earth and Planetary Sciences*, **22**, 145–165.

REAKA-KUDLA, M. L., FEINGOLD, J. S. & GLYNN, P. W. 1996. Experimental studies of rapid bioerosion of coral reefs in the Galápagos Islands. *Coral Reefs*, **15**, 101–107.

RICHMOND, R. H. 1997. Reproduction and recruitment in corals: critical links in the persistence of reefs. *In:* BIRKELAND, C. (ed.) *Life and Death of Coral Reefs*. Chapman & Hall, New York, 175- 197.

RICKLEFS, R. E., BUFFETAUT, E., HALLAM, A., HSU, K., JABLONSKI, D., KAUFFMAN, E. G., LEGENDRE, S., MARTIN, P., MCLAREN, D. J., MYERS, N. & TRAVERSE, A. 1990. Biotic systems and diversity – report of working group 4, Interlaken workshop for past global changes. *Palaeogeography, Palaeoclimatology, Palaeoecology (Global and Planetary Change Section)*, **82**, 159–168.

ROBINSON, G. 1985. The influence of the 1982–83 El Niño on Galápagos marine life. *In*: ROBINSON, G. & DEL PINO, E. M. (eds) *El Niño in the Galápagos Islands: the 1982–1983 Event*. Charles Darwin Foundation for the Galápagos Islands, ISALPRO, Quito, Ecuador, 153–190.

—— & DEL PINO, E. M. (eds) 1985. *El Niño en Las Islas Galápagos: el evento de 1982–1983*. Fundación Charles Darwin para Las Islas Galápagos, Quito, Ecuador.

ROWAN, R. & KNOWLTON, N. 1995. Intraspecific diversity and ecological zonation in coral-algal symbiosis. *Proceedings of the National Academy of Science USA*, **92**, 2850–2853.

——, ——, BAKER, A. & JARA, J. 1997. Landscape ecology of algal symbionts creates variation in episodes of coral bleaching. *Nature*, **388**, 265–269.

—— & POWERS, D. A. 1991. A molecular genetic classification of zooxanthellae and the evolution of animal-algal symbioses. *Science*, **251**, 1348–1351.

SALM, R. V. 1993. Coral reefs of the Sultanate of Oman. *Atoll Research Bulletin*, **380**, 1–85.

SANO, M., SHIMIZU, M. & NOSE, Y. 1987. Long-term effects of destruction of hermatypic corals by *Acanthaster planci* infestation on reef fish communities at Iriomote Island, Japan. *Marine Ecology Progress Series*, **37**, 191–199.

SCHICK, J. M., LESSER, M. P. & JOKIEL, P. L. 1996. Effects of ultraviolet radiation on corals and other coral reef organisms. *Global Change Biology*, **2**, 527–545.

SCHINDLER, E. 1990. Die Kellwasser-Krise (hohe Frasne-Stufe, Ober-Devon). *Göttinger Arbeiten zur Geologie und Paläontologie*, **46**, 1–115.

SCOFFIN, T. P., TUDHOPE, A. W. & BROWN, B. E. 1989. Fluorescent and skeletal density banding in *Porites lutea* from Papua New Guinea and Indonesia. *Coral Reefs*, **7**, 169–178.

SCOTT, P. J.B., RISK, M. J. & CARRIQUIRY, J. D. 1988. El Niño, bioerosion and the survival of east Pacific reefs. *Proceedings of the Sixth International Coral Reef Symposium, Townsville,* **2**, 517–520.

SEPKOSKI, J. J., Jr. 1986. Phanerozoic overview of mass extinction. *In*: RAUP, D. M. & JABLONSKI, D. (eds) *Patterns and Processes in the History of Life.* Dahlem Konferenzen, Springer-Verlag, Berlin, 277–295.

SHEN, G. T., COLE, J. E., LEA, D. W., LINN, L. J., MCCONNAUGHEY, T. A. & FAIRBANKS, R. G. 1992. Surface ocean variability at Galápagos from 1936–1982: calibration of geochemical tracers in corals. *Paleoceanography*, **7**, 563–588.

SIMBERLOFF, D. 1986. The proximate causes of extinction. *In*: RAUP, D. M. & JABLONSKI, D. (eds) *Patterns and Processes in the History of Life.* Springer, Berlin, 259–276.

SMITH, S. V. & BUDDEMEIER, R. W. 1992. Global change and coral reef ecosystems. *Annual Review of Ecology and Systematics*, **23**, 89–118.

STANLEY, G. D. JR. 1992. Tropical reef ecosystems and their evolution. *In*: *Encyclopedia of Earth System Science*. Academic Press, New York, **4**, 375–388.

—— 1997. Evolution of reefs of the Mesozoic. *Proceedings of the Eighth International Coral Reef Symposium, Panamá*, **2**, 1657–1662.

STANLEY, S. M. 1984. Marine mass extinctions: a dominant role for temperature. *In*: NITECKI, M. H. (ed.) *Extinctions*. University of Chicago, Chicago, 69–117.

—— 1986. Anatomy of a regional mass extinction: Plio-Pleistocene decimation of the western Atlantic bivalve fauna. *Palaios*, **1**, 17–36.

—— 1987. *Extinction*. Scientific American Library, New York.

—— 1990. Delayed recovery and the spacing of major extinctions. *Paleobiology*, **16**, 401–414.

—— & CAMPBELL, L. D. 1981. Neogene mass extinction of western Atlantic molluscs. *Nature*, **293**, 457–459.

—— & HARDIE, L. A. 1998. Secular oscillations in the carbonate mineralogy of reef-building and sediment-producing organisms driven by tectonically forced shifts in seawater chemistry. *Palaeogeography, Palaeoclimatology, Palaeoecology,* **144**, 3–19.

SZMANT, A. M. & GASSMAN, N. J. 1990. The effects of prolonged 'bleaching' on the tissue biomass and reproduction of the reef coral *Montastrea annularis*. *Coral Reefs,* **8**, 217–224.

VERMEIJ, G. J. 1986. Survival during biotic crises: the properties and evolutionary significance of refuges. *In*: ELLIOTT, D. K. (ed.) *Dynamics of Extinction.* Wiley, New York, 231–246.

VERON, J. E.N. & KELLEY, R. 1988. Species stability in reef corals of Papua New Guinea and the Indo-Pacific. Memoir of the Association of Australasian Palaeontologists, Memoir, **6**, Brisbane, Australia, 1–69.

WALBRAN, P. D., HENDERSON, R. A., JULL, A. J.T. & HEAD, M. J. 1989. Evidence from sediments oflong-term *Acanthaster planci* predation on corals of the Great Barrier Reef. *Science*, **245**, 847–850.

WARE, J. R., FAUTIN, D. G. & BUDDEMEIER, R. W. 1996. Patterns of coral bleaching: modeling the adaptive bleaching hypothesis. *Ecological Modeling*, **84**, 199–214.

WATERHOUSE, J. B. 1973. The Permian–Triassic boundary in New Zealand and New Caledonia and its relationship to world climatic changes and extinction of Permian life. Canadian Society of Petroleum Geologists, Memoir, **2**, 445–464.

WELLINGTON, G. M. & DUNBAR, R. B. 1995. Stable isotopic signature of El Niño-Southern Oscillation events in eastern tropical Pacific reef corals. *Coral Reefs*, **14**, 5–25.

WILKINSON, C. R. (ed.) 1998. *Status of Coral Reefs of the World: 1998*. Australian Institute of Marine Science, Queensland, 1–184.

WILLIAMS, E. H. & BUNKLEY-WILLIAMS, L. 1990. The world-wide coral bleaching cycle and related sources of coral mortality. *Atoll Research Bulletin*, **335**, 1–71.

WILLIS, B. L., BABCOCK, R. C., HARRISON, P. L. & WALLACE, C. C. 1993. Experimental evidence of hybridization in reef corals involved in mass spawning events. *Proceedings of the 7th International Coral Reef Symposium, Guam*, **1**, 504 (abstract).

WILSON, E. O. 1992. *The Diversity of Life*. Norton, New York.

WOOD, R. 1999. *Reef Evolution*. Oxford University, Oxford.

Isolated carbonate platforms of Belize, Central America: sedimentary facies, late Quaternary history and controlling factors

EBERHARD GISCHLER[1] & ANTHONY J. LOMANDO[2]

[1]*Geologisch-Paläontologisches Institut der Universität, Senckenberganlage 32–34, 60054 Frankfurt/Main, Germany*

(e-mail: gischler@em.uni-frankfurt.de)

[2]*Chevron Overseas Petroleum Inc., 6001 Bollinger Canyon Road, San Ramon, CA 94583, USA*

(e-mail: alom@chevron.com)

Abstract: The closely spaced, isolated carbonate platforms of Glovers Reef, Lighthouse Reef and Turneffe Islands (Belize) differ significantly with regard to geomorphology and distribution of sedimentary facies, especially in platform interiors. Glovers Reef has a deep (18 m) lagoon with 860 more or less randomly distributed patch reefs. The interior of Lighthouse Reef is characterized by a linear trend of hundreds of coalescing patch reefs that separate a deeper (8 m) eastern and a shallower (3 m) western lagoon. Within both Glovers and Lighthouse Reefs, non-skeletal (peloidal) wackestone and packstone are found in shallow (<5 m) water depths. Deeper lagoon parts are characterized by mollusc-foram wackestones. Facies belts are circular in Glovers Reef, as opposed to linear in Lighthouse Reef. The Turneffe Islands platform has up to 8 m deep interior lagoons that are surrounded by large land/mangrove areas. These lagoons have restricted circulation, are devoid of coral patch reefs, and are dominated by organic-rich wackestone.

These differences are largely a consequence of variations in antecedent topography and in exposure to waves and currents from platform to platform. Differences in elevation and relief of the Pleistocene basement are most likely controlled by differential subsidence and latitudinal variation in karstification. Differences in exposure to waves and currents are created as the Turneffe Islands platform is in a leeward position, protected from the open Caribbean Sea by Lighthouse Reef to the east. Variation in Pleistocene elevation caused the platforms to flood successively at different rates during Holocene sea-level rise. However, all reefs investigated kept pace with the rising Holocene sea level. Sediment is now largely bypassing the margins and filling in platform interior lagoons.

These examples indicate how local attributes of antecedent topography and exposure to waves and currents can be at least as important as globally operating factors such as sea-level fluctuations. This observation should be kept in mind when interpreting palaeo-relationships in the geologic record.

The isolated carbonate platforms of Glovers Reef, Lighthouse Reef and the Turneffe Islands are part of the 600 km long reef system offshore from Belize and the eastern Yucatan Peninsula (Fig. 1). This reef system consists of the Belize Barrier Reef, the fringing reefs that can be followed from Reef Point on Ambergris Cay to Cancun, along with a number of isolated carbonate platforms. From north to south these are: Arrowsmith Bank, a drowned platform; the reef-fringed Tertiary-Pleistocene island of Cozumel; and the reef-fringed carbonate platforms Banco Chinchorro, the Turneffe Islands, Lighthouse Reef and Glovers Reef.

A series of NNE-trending tilted fault-blocks characterizes the passive margin of Belize and the eastern Yucatan, forming the basement of the reef system (Dillon & Vedder 1973). Neotectonic activity on the fault-blocks is presumably caused by the nearby lithospheric plate boundary between the North American and Caribbean plates. Deep boreholes on the Turneffe Islands and Glovers Reef penetrated 1030 and 560 m respectively, of Tertiary and younger reefal and shallow-water carbonates, before reaching Late Cretaceous to Eocene age siliciclastics (Fig. 2).

Even though Glovers Reef, Lighthouse Reef and the Turneffe Islands have a common geological framework, including the same Holocene sea-level rise, they are considerably different with regard to geomorphology and sedimentary facies. In this paper, we consider the controlling factors of Holocene reef and platform development, including antecedent topography, exposure to waves and currents, and sea level.

From: INSALACO, E., SKELTON, P. W. & PALMER, T. J. (eds) 2000. *Carbonate Platform Systems: components and interactions*. Geological Society, London, Special Publications, **178,** 135–146. 0305–8719/00/$15.00

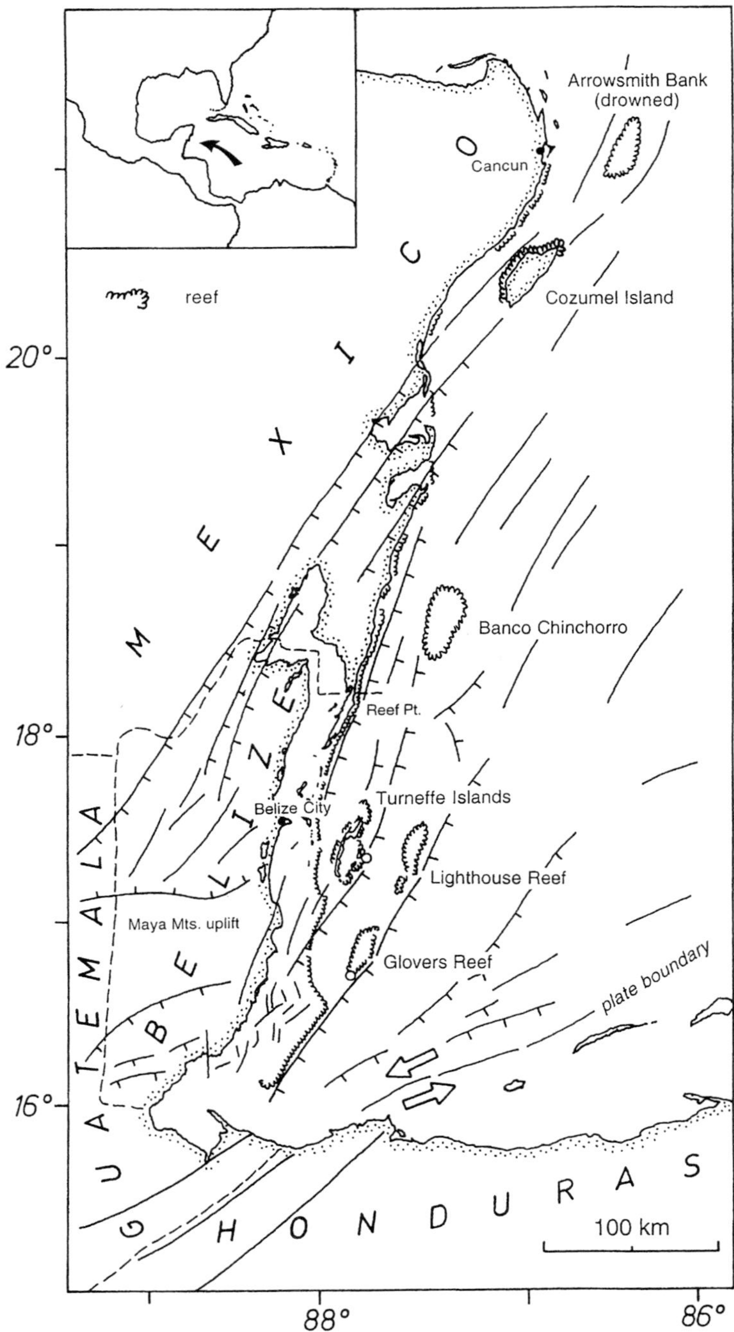

Fig. 1. Geological–tectonic setting of isolated carbonate platforms of Belize (after Dillon & Vedder 1973; Purdy 1974*b*; Case & Holcombe 1980). Location of deep boreholes on Glovers Reef and the Turneffe Islands are marked by open circles (after Deal 1983).

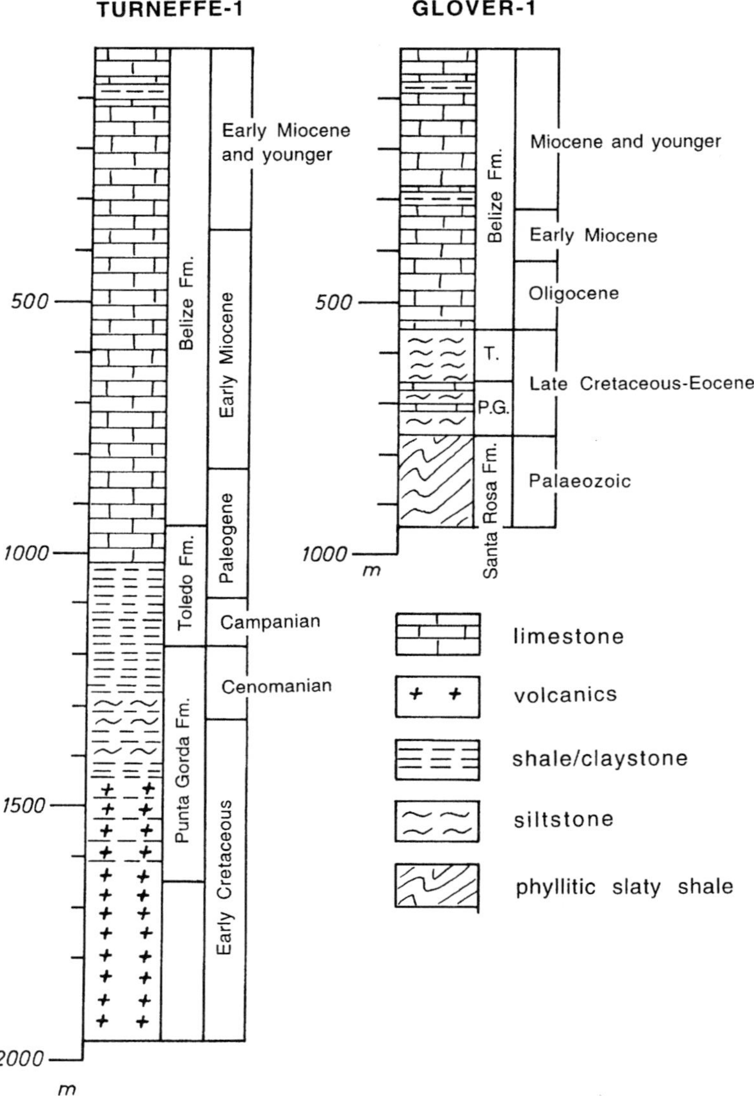

Fig. 2. Description of deep boreholes on Turneffe Islands and Glovers Reef (compiled from unpublished data of Royal Dutch Shell, The Hague).

Geomorphology

The carbonate platforms offshore from Belize are truly isolated in that they are surrounded by deep water (Fig. 3). Surface-breaking reef margins enclose lagoons that identify the platforms as atolls (Stoddart 1962). James *et al.* (1976) and James & Ginsburg (1979) described marginal reefs of one of the platforms in detail. Major differences among the three platforms are found in platform interiors, as described below.

Glovers Reef covers 260 km^2 with only 0.2% of land in the form of small, marginal sand–shingle cays. The lagoon is up to 18 m deep, has open circulation, and there are over 850 patch reefs of coral (Wallace & Schafersman 1977) some of which form linear trends. Lighthouse Reef has an area of 200 km^2 with 2.9% of land: there are three marginal, small sand–shingle cays and two large mangrove–sand cays in the platform interior. The well circulated lagoon is divided into two halves by a linear

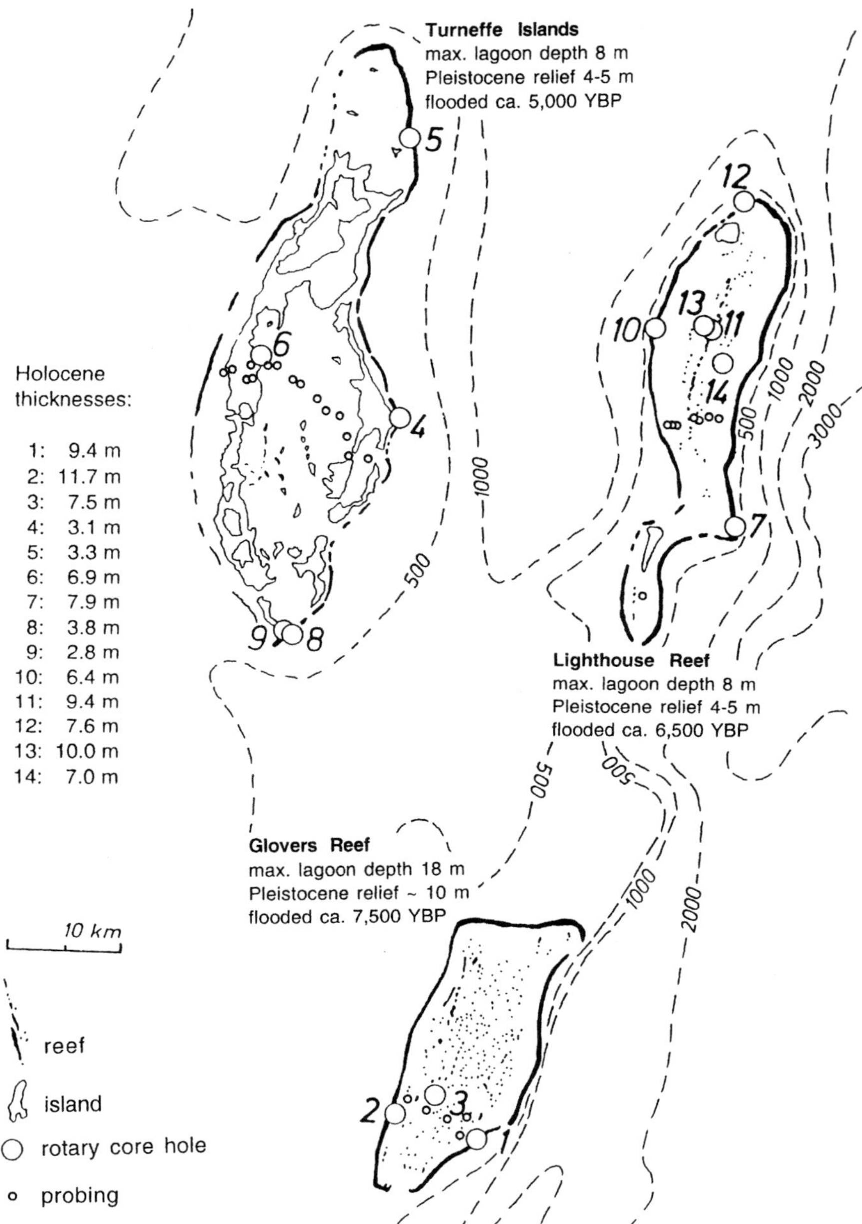

Fig. 3. Maps of isolated carbonate platforms offshore from Belize. Water depth between Turneffe Islands and Belize Barrier Reef is 250 m; that between Glovers Reef and the barrier reef amounts to over 400 m. Localities of rotary core holes are indicated by large circles (after Gischler & Hudson 1998; Gischler & Lomando 1999). Holes 1, 2, 4, 5, 7, 8, 10 and 12 were drilled into marginal reefs, holes 3, 11, 13 and 14 into lagoon patch reefs. Hole 6 was drilled into the lagoon floor, hole 9 on an island. Localities of probing with a 4 m long aluminium rod are indicated by small circles. The results of probing are included in the construction of cross-sections shown in Fig. 6.

trend of coalescing coral patch reefs. The eastern lagoon reaches 8 m in depth whereas the western lagoon is only 3 m deep. The Turneffe Islands platform covers 525 km^2 with 22% of land. The land area comprises numerous small sand cays on the windward margin. Large mangrove cays and mangrove–sand cays enclose two large restricted lagoons that have rare coral patch reefs. Lagoon depths reach a maximum of 8 m. The northernmost part of the platform has open circulation with abundant coral patch reefs.

Surface sediments

Sediments of the platform margins are skeletal (coral–red alga–*Halimeda*) grainstones (Fig. 4). In contrast, major sediment differences occur in the platform interiors: Glovers Reef has a circular facies pattern with mixed peloidal–skeletal wackestones and packstones in shallow lagoon parts and skeletal (mollusc–foram–*Halimeda*) wackestones in deeper (>5 m) areas. Similar facies are found in Lighthouse Reef, but their distribution is linear. Mixed peloidal–skeletal compositions are found in the western lagoon, whereas mollusc–foram compositions occur in the eastern lagoon. Larger areas are represented by grainstone textures compared to Glovers Reef. Restricted lagoons in the Turneffe Islands platform are covered by *Halimeda* wackestones, rich in organic matter. Total organic carbon values average 5.6% with maximum values of 15% (Gischler & Lomando 1999).

Nine surface sediment samples from Glovers Reef were radiometrically dated (Table 1). Sediment from the deep forereef is modern. Sediment samples from the deep interior lagoon are not older than 300 years, but sediment from the marginal reef and backreef areas are significantly older with ages approximating 600–700 and 1000 years BP, respectively. These ages are comparable to radiometric ages of beachrock in Belize (Table 1), the material of which is largely derived from marginal reef areas (Gischler & Lomando 1997). These limited data are nonetheless consistent in indicating not only the similarity of measured ages from similar reef environments but also the clear differences between sediment ages from different environments. This is the case even though the dated sediment represents a mixture of ages from younger and older constituent grains.

Subsurface data

The Holocene thickness of marginal reefs is highest on Glovers Reef and lowest on the

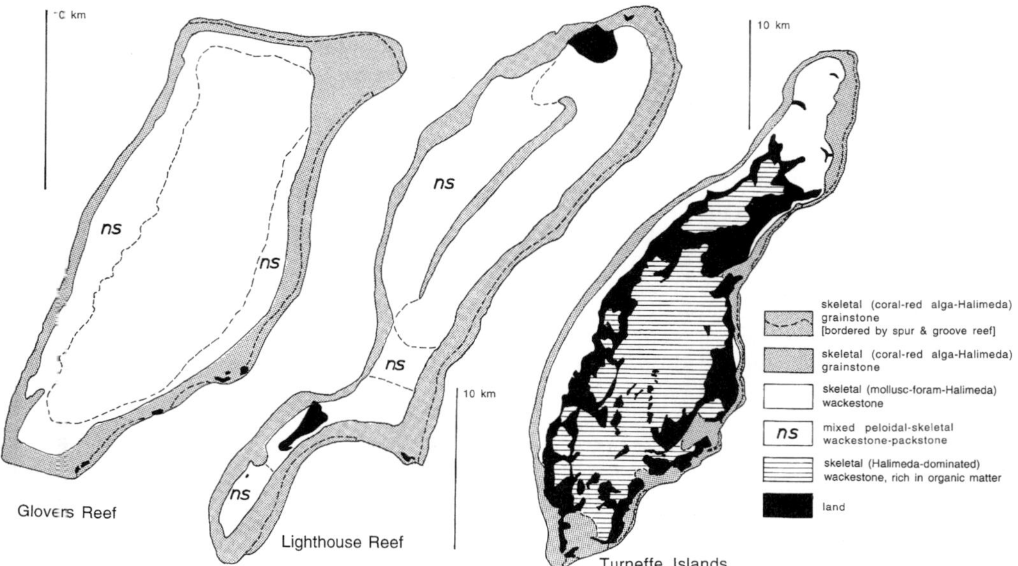

Fig. 4. Sedimentary facies of isolated carbonate platforms (after Gischler 1994; Gischler & Lomando 1999). Maps are based on quantitative textural and compositional analysis of 345 surface sediment samples. Note scale differences for each of the platforms.

Table 1. *Radiometric ages of sediment samples from Glovers Reef*

Sample environment	Depth (m)	Conventional age (years BP)	2-sigma range
Sediment sample			
G 150, windward reef	0.5	920 ± 70	AD 1315–1515
G 101, leeward reef	1	850 ± 60	AD 1405–1565
G 124, windward back reef	2	1380 ± 40	AD 960–1085
G 135, leeward back reef	6	1400 ± 50	AD 905–1085
G 17, windward fore reef	27	260 ± 60	'modern'
G 1, leeward fore reef	40	220 ± 60	'modern'
G 116, deep lagoon	11	420 ± 50	AD 1805–1950
G lag 1, deep lagoon	17	470 ± 40	AD 1720–1950
G lag 2, deep lagoon	17	510 ± 40	AD 1695–1950
Beachrock from cays on windward platform margins			
Glovers Reef, Long Cay	sea level	1.550 ± 60	AD 710–990
Glovers Reef, Middle Cay	sea level	1.230 ± 60	AD 1045–1290
Glovers Reef, Southwest Cay	sea level	1.320 ± 60	AD 980–1220
Lighthouse Reef, Halfmoon Cay	sea level	1.910 ± 70	AD 345–650

Dates of beachrock from Glovers and Lighthouse Reefs from Gischler & Lomando (1997). Conventional ages are adjusted for reservoir correction. The 2-sigma range has 95% probability. (Dating was performed by BETA ANALYTIC INC., Miami)

Turneffe Islands (Figs 3, 5, 6). Radiometric dating of corals and peats from rotary drill cores shows that Glovers Reef was the first platform to be flooded by the rising Holocene sea, about 7500 years BP. The Turneffe Islands platform was flooded last, a little less than 6000 years BP (Table 2; Fig. 7). Pleistocene relief is highest on Glovers Reef (*c*. 10 m) compared to Lighthouse Reef and Turneffe Islands (*c*. 4–5 m) (Fig. 6). Probing in lagoon interiors has shown that Holocene sediment thicknesses rarely exceed 4 m (Gischler & Hudson 1998).

The Holocene reef facies is represented either by coral boundstone, grain to rudstone, or unconsolidated sand and rubble. Holocene reefs are generally situated on Pleistocene reef limestone, with the exception of the patch reef of hole 13 that developed on a topographic low (Fig. 6). Radiometric dates from diagenetically unaltered corals of the Pleistocene in Belize suggest that the foundations of Holocene reefs were deposited during the last interglacial highstand of sea level (oxygen isotope stage 5e) (Gischler *et al.* 2000).

Discussion: controlling factors of late Quaternary development

Antecedent topography

The importance of antecedent topography for Holocene reef development and sedimentation on the Belize shelf was demonstrated by Purdy (1974*a*, *b*). He showed that Holocene reefs are situated on Pleistocene highs whereas unconsolidated sediment accumulated in topographic lows of the Pleistocene. Pleistocene relief was created during sea-level lowstands by karstification which was directed in its expression by underlying structure. Halley *et al.* (1977) and Shinn *et al.* (1982) showed that Holocene reefs in the central Belize shelf area are located on Pleistocene reefs. This possibility was also considered earlier by Purdy (1974b). Choi & Holmes (1982) and Choi & Ginsburg (1982) interpreted Holocene reefs in the southern shelf to superpose clastics, based on seismic data. The importance of underlying structure for the development of Quaternary reefs offshore from Belize was again stressed by Lara (1993) for the southern shelf and Lomando *et al.* (1995) for the isolated carbonate platforms.

Results of drilling show that Pleistocene elevation is different among the three platform margins (Figs 3, 5, 6). The decrease of Pleistocene elevation from Lighthouse Reef to Glovers Reef, which are situated on the same fault-block trend, might be caused by a southward tilt. Dill (1977) showed that giant Pleistocene stalactites in the 125 m deep sinkhole in the eastern lagoon of Lighthouse Reef (south of our hole 14), are tilted 10–15° towards the north. Differential subsidence along and among the offshore fault-blocks is also indicated by the results of radiometric dating of uppermost-Pleistocene sections. Pleistocene deposits on the northern barrier reef and the platforms differ by as much as 10 m in elevation, but have the same age (Gischler *et al.* 2000).

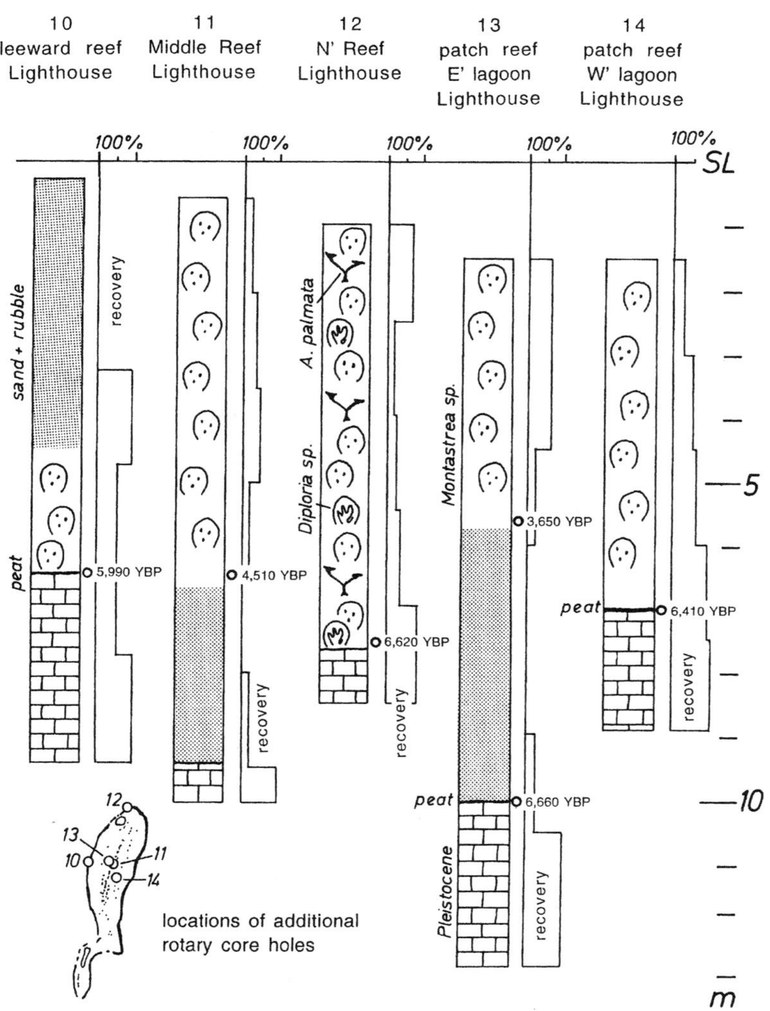

Fig. 5. Description of rotary cores on Lighthouse Reef (core hole numbers 10–14) that were drilled in addition to those described in Gischler & Hudson (1998) (core hole numbers 1–9). For complete data on radiometric dates see Table 2.

The differences in Pleistocene relief between the northern two platforms and Glovers Reef might be a consequence of an increase in the intensity of Pleistocene karstification. Data on precipitation on the isolated platforms proper are not available, but the precipitation rates on the relatively flat relief of the mainland adjacent to the two northern platforms amount to 1500 mm per year at the most (Purdy 1974*b*). In contrast, the annual rainfall on the mainland adjacent to Glovers Reef exceeds 4500 mm per year (Purdy 1974*b*) reflecting the 1160 m elevation of the Maya Mountains.

Following the model of Purdy (1974*a*,*b*), we interpret linear morphological features on the platforms as an expression of underlying structure. These are the NNE-trending lagoonal patch reefs in Lighthouse and Glovers Reef that parallel the fault-block system (Fig. 1). Northwest-trending features such as the island chain in the main lagoon of the Turneffe Islands (Fig. 3, west of borehole 4) may also be an expression of structural control in the form of underlying wrench faults (Lomando *et al.* 1995).

Sea level

The response of all the platform reefs to rising sea level was to keep up (*sensu* Neumann & Macintyre 1985; Davies & Montaggioni 1985).

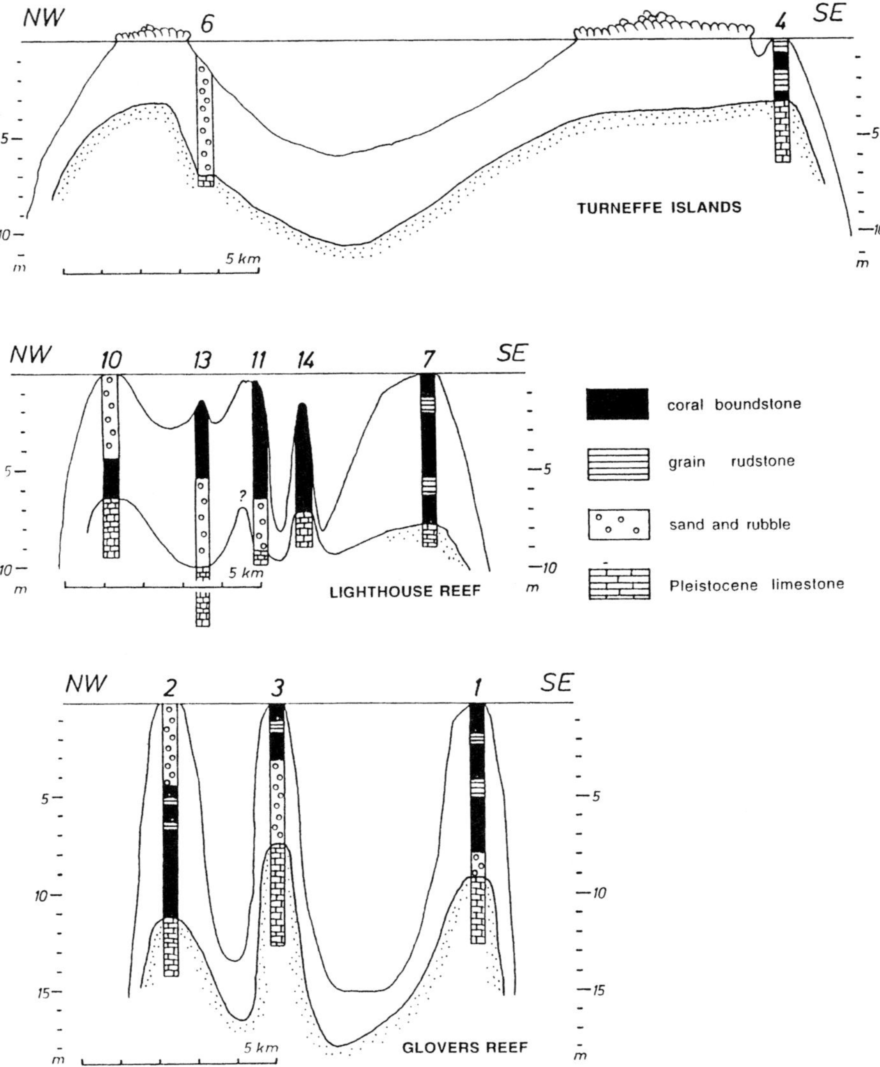

Fig. 6. Cross-sections through isolated carbonate platforms based on rotary core drilling, probing with 4 m long aluminium rod in unconsolidated sediment in platform interiors, and lagoon bathymetries. Locations of probing are shown on Fig. 3. Holocene sediment thicknesses obtained from probing are projected on sections that are drawn from borehole to borehole. Question mark indicates that Pleistocene elevation under Holocene patch reef is probably higher than observed in the hole, as we drilled near the reef margin.

All radiometric dates plot close to the Belize sea-level curve (Macintyre *et al.* 1995) or within a 3 m envelope around it. This is remarkable in that the three platforms were flooded more or less successively (Fig. 7) and had to keep pace with different rates of sea-level rise.

The relatively old ages of marginal reef sediment (Table 1) in combination with radiometric dates from core material (Fig. 7) suggest that the margin of Glovers Reef 'caught up' with sea level early because of lack of accommodation space. In general, the margins of the Belize isolated platforms can be interpreted as areas that are largely bypassed by sediment filling in interior lagoons (Gischler & Lomando 1999).

Exposure to waves and currents

The three platforms are situated within the trade wind belt with winds from the northeast and east

Table 2. *Radiometric dates from rotary cores through isolated platforms offshore from Belize*

Sample (hole no., depth below sea level)	Conventional age ± std dev. (years BP)	2-sigma range
Turneffe Islands		
8, 3.8 m	5,850 ± 50	4825–4575 BC
Lighthouse Reef		
10, 6.5 m	5,990 ± 60	5030–4765 BC
11, 6.5 m	4,510 ± 60	2890–2560 BC
12, 7.5 m	6,620 ± 60	5275–5015 BC
13, 5.5 m	3,650 ± 70	1765–1420 BC
13, 10.0 m	6,660 ± 50	5605–5450 BC
14, 7.0 m	6,410 ± 50	5435–5255 BC

These data are additional to those published in Gischler & Hudson (1998)

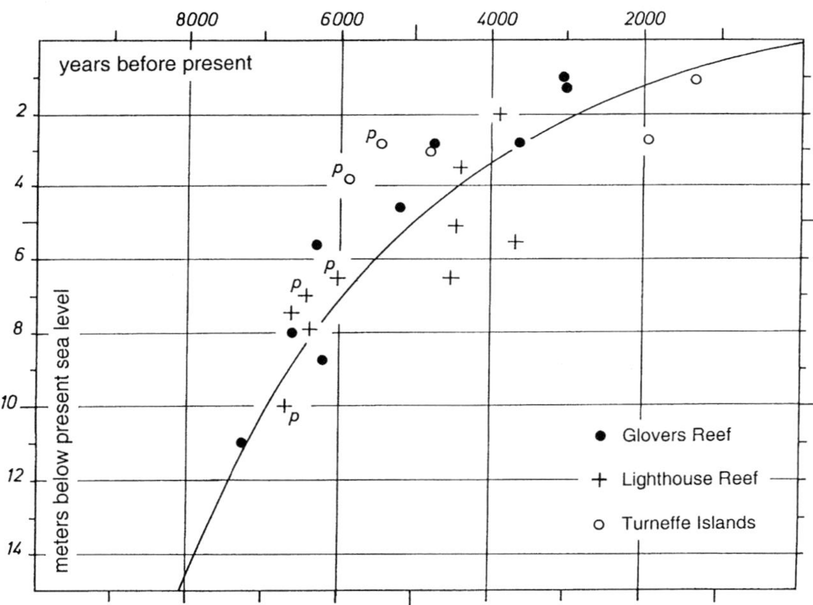

Fig. 7. Radiometric dates from core material (coral and peat) plotted on minimum western Atlantic and Belize sea-level curve of Macintyre *et al.* (1995). Data from Gischler & Hudson (1998) and Table 2. Peat samples are marked by 'p'.

predominating. Mean wave approach is 75° (Burke 1982). Glovers Reef, Lighthouse Reef and the northernmost part of Turneffe Islands are open to the Caribbean Sea (Fig. 8). The part of Turneffe Islands that has extensive mangrove areas receives only modified and impeded wave forces as it is protected by Lighthouse Reef to the east. This configuration suggests that extensive mangrove growth is only possible in rather protected areas. The same is true for the Belize shelf. The central barrier reef that is protected by the isolated platforms to the east (zone 3) has the highest abundance of mangrove islands (Fig. 8).

An alternative interpretation for the extensive Turneffe Islands mangrove development relates this mangrove development to the fact that this platform was flooded last during the lowest rate of Holocene sea-level rise (Fig. 7). If this were the case one would also have to presume a lower Pleistocene elevation and an earlier date of flooding of the open, northernmost part of the platform. The subsurface data,

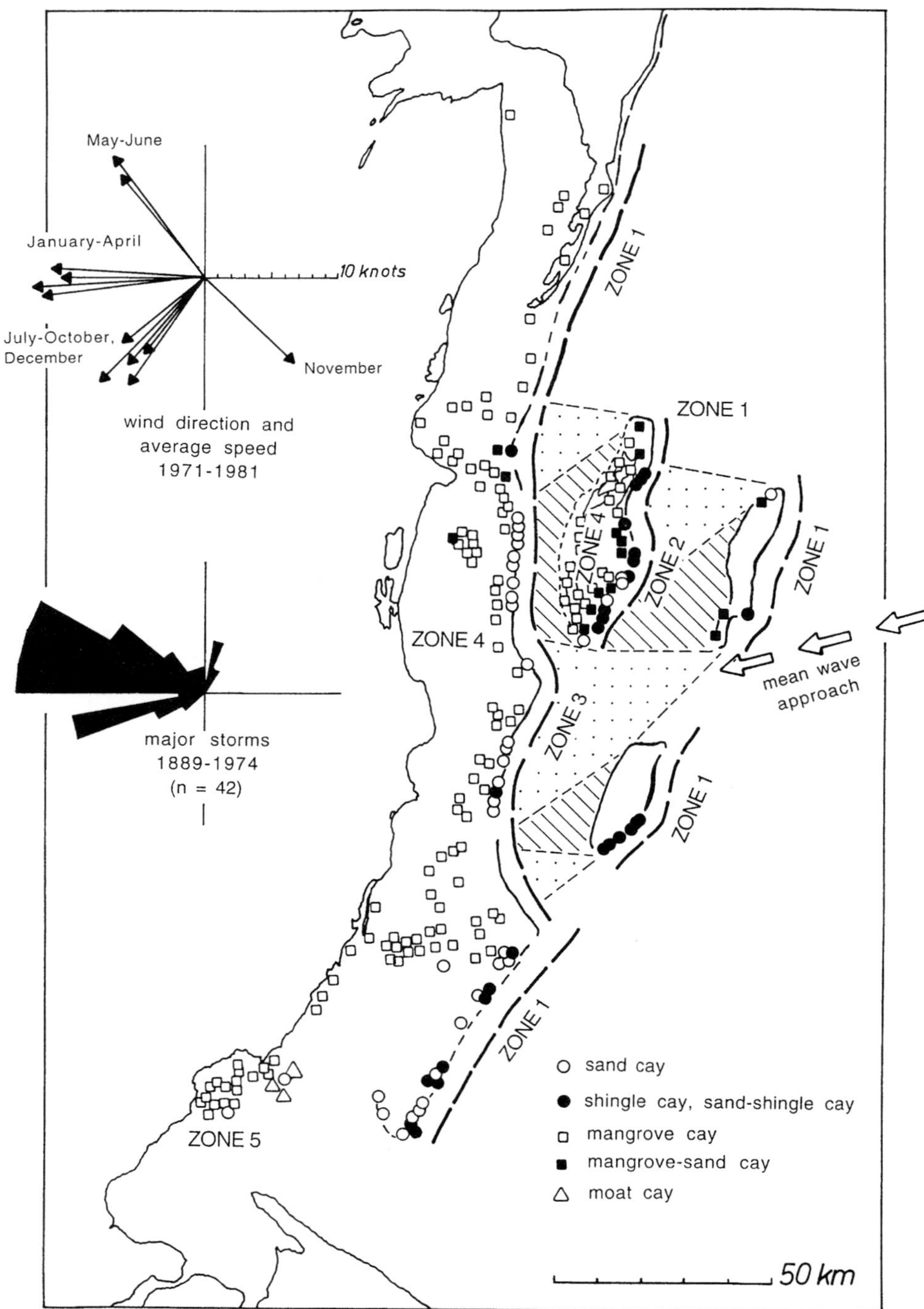

Fig. 8. Wave-sheltered positions of Turneffe Islands due to position of Lighthouse Reef and sheltered areas of central barrier reef by isolated platforms (after Burke 1982; Gischler & Hudson 1998). Stippled areas: modified wave force; hatched areas: maximum impedance of wave force. Wave energy zones and resulting island types after Stoddart (1962, 1965). Zone 1, maximum and submaximum wave energy; zone 2, high wave energy; zone 3, medium wave energy; zone 4, low wave energy; zone 5, moderate wave energy.

especially relating to borehole 5, argue against this interpretation.

Conclusions

Locally operating factors such as antecedent topography and exposure to waves and currents can be used to explain the variation in geomorphology and facies among the closely spaced, isolated carbonate platforms of offshore Belize. Holocene sea-level rise was of importance in creating the accommodation space for reef and platform development. However, this factor by itself does not explain the described differences of geomorphology and facies from platform to platform. The implication of this modern example suggests the potential influence of local controlling factors in ancient examples that might not always be easy to recognize.

We are deeply indebted to our 'chief of drilling operations' H. Hudson (Key Largo) for his invaluable work. We thank E. Shinn (St Petersburg) who let us use his wireline drill equipment, and the Comparative Sedimentology Laboratory of the University of Miami, especially R. Ginsburg, G. Eberli and D. McNeill for giving us access to all kinds of equipment and letting us use facilities. We thank Captain M. Jackson (Dangriga, Belize) for navigation and technical assistance. G. Meyer (Frankfurt/Main) helped us collect and catalogue hundreds of sediment samples and assisted during drilling. E. Hudson (Miami), A. Buck (Miami), B. Goodwin (Key Largo) and H. Legarre (La Habra) are also thanked for their help during drilling operations. G. Ingram (The Hague) helped the senior author to gain access to the deep Shell wells on Glovers Reef and Turneffe Islands. Investigations were supported by the German Research Foundation (DFG) and Chevron Overseas Petroleum Inc. The helpful comments of reviewers L. Montaggioni and E. Purdy are gratefully acknowledged.

References

BURKE, R. B. 1982. Reconnaissance study of the geomorphology and benthic communities of the outer barrier reef platform, Belize. *In*: RÜTZLER, K. & MACINTYRE, I. G. (eds) *The Atlantic Barrier Reef Ecosystem at Carrie Bow Cay, Belize*. Smithsonian Contributions to the Marine Sciences, **12**, 509–526.

CASE, J. E. & HOLCOMBE, T. L. 1980. *Geologic-Tectonic Map of the Caribbean Region*. US Geological Survey Misc. Invest. map I-1100, scale 1: 2,500,000.

CHOI, D. R. & GINSBURG, R. N. 1982. Siliciclastic foundations of Quaternary reefs in the southernmost Belize lagoon, British Honduras. *Geological Society of America, Bulletin*, **93**, 116–126.

—— & HOLMES, C. W. 1982. Foundations of Quaternary reefs in south-central Belize lagoon, British Honduras. *AAPG Bulletin*, **66**, 2663–2681.

DAVIES, P. J. & MONTAGGIONI, L. 1985. Reef growth and sea-level change: the environmental signature. *Proceedings of the 5th International Coral Reef Symposium, Tahiti*, **3**, 477–515.

DEAL, C. S. 1983. Oil and gas developments in South America, Central America, Caribbean area, and Mexico in 1982. *AAPG Bulletin*, **67**, 1849–1883.

DILL, R. F. 1977. The blue holes: geologically significant submerged sinkholes and caves off British Honduras and Andros, Bahama Islands. *Proceedings of the 2nd International Coral Reef Symposium, Miami*, **2**, 237–242.

DILLON, W. P. & VEDDER, J. G. 1973. Structure and development of the continental margin of British Honduras. *Geological Society of America, Bulletin*, **84**, 2713–2732.

GISCHLER, E. 1994. Sedimentation on three Caribbean atolls: Glovers Reef, Lighthouse Reef and Turneffe Islands, Belize. *Facies*, **31**, 243–254.

—— & HUDSON, J. H. 1998. Holocene development of three isolated carbonate platforms, Belize, Central America. *Marine Geology*, **144**, 333–347.

—— & LOMANDO, A. J. 1997. Holocene cemented beach deposits in Belize. *Sedimentary Geology*, **110**, 277–297.

—— & —— 1999. Recent sedimentary facies of isolated carbonate platforms, Belize-Yucatan system, Central America. *Journal of Sedimentary Research*, **69**, 747–763.

——, ——, HUDSON, J. H. & HOLMES, C. W. 2000. Last interglacial reef growth beneath Belize barrier and isolated platform reefs. *Geology*, **28**, 387–390.

HALLEY, R. B., SHINN, E. A., HUDSON, J. H. & LIDZ, B. 1977. Recent and relict topography of Boo Bee patch reef, Belize. *Proceedings of the 3rd International Coral Reef Symposium, Miami*, **2**, 29–35.

JAMES, N. P. & GINSBURG, R. N. 1979. *The Seaward Margin of Belize Barrier and Atoll Reefs*. International Association of Sedimentologists, Special Publication, **3**.

——, ——, MARSZALEK, D. S. & CHOQUETTE, P. W. 1976. Facies and fabric specificity of early subsea cements in shallow Belize (British Honduras) reefs. *Journal of Sedimentary Petrology*, **46**, 523–544.

LARA, M. E. 1993. Divergent wrench faulting in the Belize southern lagoon: implication for Tertiary Caribbean plate movements and Quaternary reef distribution. *AAPG Bulletin*, **77**, 1041–1063.

LOMANDO, A. J., SUISINOV, K. & SHILIN, A. 1995. Reservoir architecture characteristics and depositional models for Tengiz Field, Kazakhstan. *Caspi Shelf International Science Seminar, Almaty, Proceedings*, 51–73.

MACINTYRE, I. G., LITTLER, M. M. & LITTLER, D. S. 1995. Holocene history of Tobacco Range, Belize, Central America. *Atoll Research Bulletin*, **430**, 1–18.

NEUMANN, A. C. & MACINTYRE, I. G. 1985. Reef response to sea level rise: keep-up, catch-up or give-up. *Proceedings of the 5th International Coral Reef Symposium, Tahiti*, **3**, 105–110.

PURDY, E. G. 1974*a*. Reef configurations: cause and effect. *In*: LAPORTE, L. F. (ed.) *Reefs in Time and*

Space. Society of Economic Paleontologists and Mineralogists, Special Publication, **18**, 9–76.

—— 1974*b*. Karst determined facies patterns in British Honduras: Holocene carbonate sedimentation model. *AAPG Bulletin*, **58**, 825–855.

SHINN, E. A., HUDSON, J. H., HALLEY, R. B., LIDZ, B., ROBBIN, I. G. & MACINTYRE, I. G. 1982. Geology and sediment accumulation rates at Carrie Bow Cay, Belize. *In*: RÜTZLER, K. & MACINTYRE, I. G. (eds) *The Atlantic Barrier Reef Ecosystem at Carrie Bow Cay, Belize*. Smithsonian Contributions to the Marine Science, **12**, 63–75.

STODDART, D. R. 1962. Three Caribbean atolls: Turneffe Islands, Lighthouse Reef, and Glover's Reef, British Honduras. *Atoll Research Bulletin*, **87**, 1–147.

—— 1965. British Honduras cays and the low wooded island problem. *Institute of British Geographers, Transactions and Papers*, **36**, 131–147.

WALLACE, R. J. & SCHAFERMAN, S. D. 1977. Patch-reef ecology and sedimentology of Glovers Reef Atoll. *In*: FROST, S. F., WEISS, M. P. & SAUNDERS, J. B. (eds) *Reefs and Related Carbonates*. AAPG Studies in Geology, **4**, 37–53.

Reef episodes, anoxia and sea-level changes in the Frasnian of the southern Timan (NE Russian platform)

M. R. HOUSE [1], V. V. MENNER[2], R. T. BECKER[3], G. KLAPPER[4] N. S. OVNATANOVA[5] & V. KUZ'MIN[5]

[1] *School of Ocean and Earth Science, Southampton Oceanography Centre, European Way, Southampton, SO14 3ZH, UK (e-mail: dys@mail.soc.soton.ac.uk)*

[2] *Institute of Geology and Exploitation of Combustible Fuel, Fersman 50, Moscow 117312, Russian Federation*

[3] *Museum für Naturkunde der Humboldt-Universität, Paläontologisches Institute, Invalidenstrasse 43, 10115 Berlin, Germany*

[4] *Department of Geoscience, University of Iowa, Iowa City, IA 52242, USA*

[5] *All-Russia Research Geological Oil Prospecting Institute, Shosse Entuziastov 36, Moscow 105819, Russian Federation*

Abstract: The development of the Frasnian (Upper Devonian) reef complexes of the southern Timan and Pechora region of northern European Russia is described. Barrier reef complexes progressively prograded eastwards during the Frasnian but the carbonate complexes were interrupted many times by regressive events. Using new conodont and ammonoid biostratigraphical dating, the timing of reef building episodes has been established which enables international correlation with other similar Devonian areas. Basinal anoxic and hypoxic deposits associated with the reef complexes of the Domanik facies provide the major hydrocarbon source rocks of the region and the palaeoenvironmental interpretation of these is discussed. Initial transgressions appear to have been associated with the global Taghanic Onlap of the late Givetian. The new level for the base of the Frasnian and Upper Devonian lies in the Timan Formation, after the deposition of which marine conditions mostly prevailed in the area examined until the late Frasnian when a sharp regression occurred with no evidence of the typical Kellwasser facies of Western Europe and other areas. Transgressive pulses initiated ammonoid biofacies in the Regional Sargaev Stage and the widespread Timan Event was marked by the spread of *Timanites* faunas. A significant deepening event which initiated the Domanik facies correlates approximately with the Middlesex black shale of New York and the main development of the Domanik facies with the Rhinestreet black shale of New York. There are faunal and floral peculiarities of the area, shown by endemic genera and rather different ranges of cosmopolitan species than elsewhere, which complicates precise international correlation. Nevertheless, several of the main sea-level deepening pulses of the Frasnian, documented in North America, Western Europe, North Africa and Australia, are recognizable and these are thought to represent eustatic events.

With current scientific interest in establishing past sea-level, climate and palaeogeographical changes, carbonate and reef complexes provide a particularly valuable source of factual information. Early cementation of reef carbonates often results in successions near their original thickness and this enables the relations of platform, reef margin, marginal slope and basin deposits to be established and their changes through time documented. This study of reef developments and their interpretation in the Late Devonian of northern European Russia has concentrated on the data known in surface outcrops and in the subsurface from boreholes resulting from the search for hydrocarbons. The area investigated covers the southern part of the Timan Range and Pechora River Basin (Fig. 1) and has concentrated on the deposits of Frasnian age where the most celebrated reef, anoxic and hypoxic facies are developed in the Southern Timan. It is the anoxic sediments that provide the primary source rocks for the hydrocarbons of the region which contribute so greatly to national wealth. From Mediaeval time the Timan–Pechora area has been known for the natural occurrence of crude oil, and recent investigation has been partly related to the development of the area as an oil-bearing province. Summaries of this aspect include reviews in English by Ulmishek (1982). This has led to much work on microfossils in order to interpret borehole sequences. The new results

From: INSALACO, E., SKELTON, P. W. & PALMER, T. J. (eds) 2000. *Carbonate Platform Systems: components and interactions*. Geological Society, London, Special Publications, **178**, 147–176. 0305–8719/00/$15.00

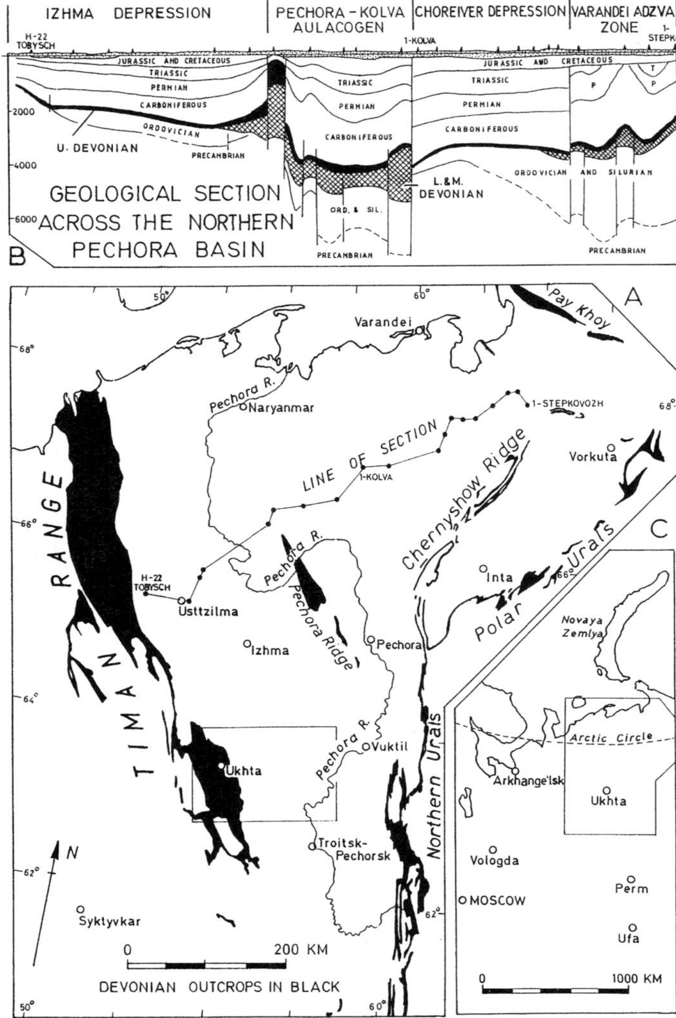

Fig. 1. (A) Map of the northeastern part of European Russia showing the Timan Range and Pechora River Basin and in black the outcrop of Devonian rocks. **(B)** A geological cross-section along the line marked on map A showing the structure of the basin. It is the southern part of the Basin which is discussed in this paper. **(C)** Map showing the location of the Timan–Pechora region.

from conodont and ammonoid studies enable the sea-level changes and periods of anoxia and hypoxia to be reviewed in relation to documentation of these events in other areas of the world. This allows a new understanding of processes that controlled the spatial and stratigraphical distribution of reef complexes and organic-rich basinal facies.

The main aims of the project here reported were several. First, to correlate sections in different facies of the Frasnian (Upper Devonian) using lithological, palaeoecological and sedimentological criteria. Second, to establish the sequence of conodont and ammonoid complexes in order to provide a biostratigraphical framework for comparison with globally known facies sequences of similar age. Third, to refine knowledge of the history of the basin and in particular to establish the limits of sedimentary rhythms and their sequence-stratigraphic interpretation. Fourth, to determine the periods and setting of the main anoxic/hypoxic events. Fifth, to correlate the Frasnian succession of the Southern Timan, and the stages of development and the main geological and bioevents of the Timan–Pechora area with sections and events in the Frasnian stage of the Ardennes–Rhenish area of Europe and other areas of the world, especially New York, North Africa and Western Australia, where similar documentation and interpretation has already been attempted (House & Kirchgasser 1993; Kirchgasser *et al.* 1997; Becker *et al.* 1993, 1997; Becker & House 1995, 1997).

Historical review

The Devonian rocks of the Timan–Pechora area (Fig. 1) have been famous since the publication of a survey by Keyserling (1844, 1846) which followed recommendations by Sir Roderick Murchison after his geological tour of Russia, at the invitation of the Czar, which immediately preceded the foundation of the Devonian System in 1839 (Rudwick 1985). Keyserling described the Devonian sequence of the Ukhta region, established the rock succession and described many fossils. Other reviews were by Tschernyshev (1884, 1890, 1915). Palaeontological work was followed in the last century by notable studies of brachiopods by Tschernyshev (1884) and later by studies of spiriferid brachiopods by Nalivkin (1936) and more general reports by A. I. Lyashenko (1956, 1959, 1969, 1973, 1985) and Nefiodova (1955). Bivalves were monographed by Zamyatin (1911). Important studies on the dacryoconarids led to a major monograph by G. P. Lyashenko (1959). There are studies of the rich ostracod faunas by Fokin (1975, 1977) and others. Conodont studies will be referred to later. Spore studies include work by Raskatova (1969), Medyanik (1981), Kushnareva & Raskatova (1980), Medyanik & Yatskevitch (1981) and there is a recent review by Avchimovitch *et al.* (1993). Studies on the reef facies include that of Maximova (1970) and Kushnareva & Matviyevskaya (1966, 1973).

Current stratigraphical terminology remains much as proposed by Tikhonovitch (1941), Kushnareva (1959), Kushnareva *et al.* (1979) and the Stratigraphic Dictionary (1975). There is a recent revision by Menner *et al.* (1992).

Reviews of the structure, especially in relation to oil production include works by Tikhonovich (1930, 1941), Likharev (1931) and Tszyu & Kossovy (1973) and there is a review in English by Ulmishek (1982).

Structural setting

The Timan–Pechora province occupies the northeastern margin of the East European Platform. The western part is represented by the Timan range (Fig. 2) eastward from which is the extensive Pechora Plate. The Pechora Plate is divided into several tectonic zones including the Izhma Depression or basin, with the Omra-Luza Zone to the south. The Kolwa Aulacogen includes the Pechora–Kozhwa Swell and the Kolwa Swell with the Denisov Trough or depression between them. To the northeast are the Choreiver Depression or basin and the Varandei-Adzva Zone. To the east of the Pechora Plate the Pre-Uralian and Pre-Pakhoian foredeeps border on the western slopes of the Urals and Paykhoy fold zone. The accumulation of Palaeozoic deposits in the Timan Range and the Pechora Plate developed in connection with the adjacent Urals basin.

In western regions of the Timan–Pechora province terrigenous red-coloured deposits accumulated predominantly from the beginning of the Ordovician to the Early Carboniferous. On the bulk of the Pechora Plate shelf areas were developed; this setting extended far to the east and especially thick developments occur in the basins of the Pechora–Kolva Aulacogen. There similar developments can be seen in sections of the forefolds of the Urals (Eletz Facies). Yet further to the east the shelf deposits were replaced by continental slope sediments (Lemva Facies) of the Urals and Pay Khoy.

Setting of Devonian sedimentation

The Devonian rocks of the southern Timan–Pechora area rest upon a nearly peneplanated surface of Precambrian rocks which comprise metamorphic complexes and metasediments of the Riphaean. In certain parts earlier Lower Palaeozoic rocks occur (Fig. 1) and the Middle Devonian may be represented by spore-bearing, mostly fresh-water deltaic deposits. The Late Devonian development in the southern Timan–Pechora area is initiated by spreads of basalt lava and volcanic ash of Dzh'er age associated with tensional rift fracturing. Subsequently a great variety of facies types is found. Following the Dzh'er are terrigenous and calcareous sandstones of Timan and Ust'yarega time with the earliest marine transgression commencing with neritic arenites and argillites of the Timan Regional Stage. The initial transgression may be part of the widespread Taghanic Onlap (Johnson 1970; House 1975, 1983; see later discussion).

Within the Pechora Plate the Upper Devonian shows the greatest facies variety. The early stages show rift movements with active magmatism. Later large shallow water areas, with barrier and isolated reef and bank carbonate

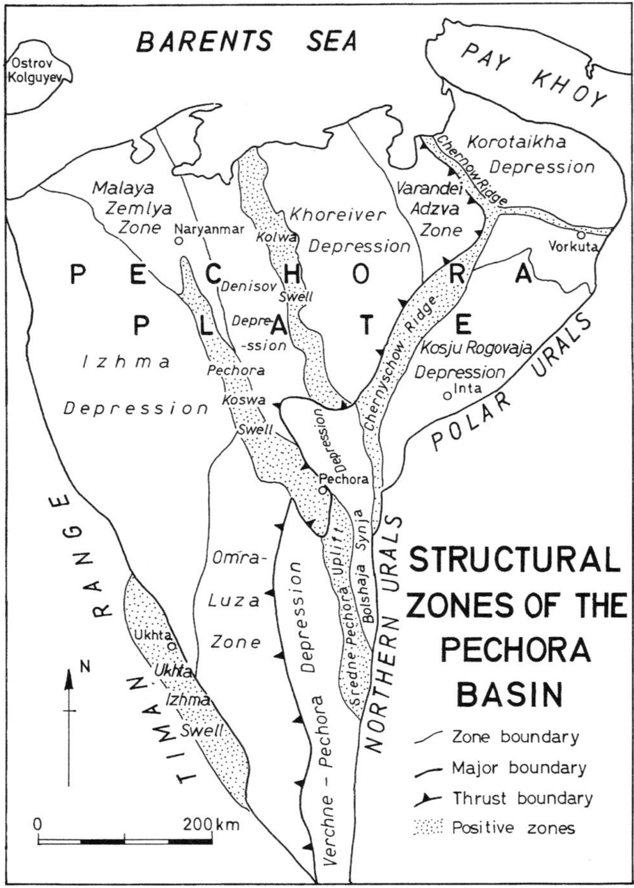

Fig. 2. Diagram showing the main geological structural zones in the Timan and Pechora Basin.

massifs developed over the shelf zones. Large starved basin areas occurred with accumulations of the bituminous deep-water facies of the Domanik facies. The isopachytes on the distribution of the Domanik facies of the Upper Devonian reflect the oil- and gas-bearing capacity of the entire Ordovician to Lower Carboniferous complex.

Devonian deposits of the Timan–Pechora region occur up to depths of 12 km and there are outcrops only within the Timan, Chernyschow and Chernow Ridges (Fig. 2). Two factors are responsible for this. In the Late Palaeozoic and Early Mesozoic, in connection with the closure of the Urals basin to the east, foredeeps were formed in which Permian and Triassic terrigenous molasse of 5–9 km thickness accumulated. Secondly, the peculiarities of sedimentation in the Pechora Plate, with replacement of marine sediments by terrestrial deposits, were associated with inversion of ancient troughs and by thrust dislocations along deep fault zones.

In the search for oil and gas in the deeper areas of the Devonian many boreholes have been drilled and much seismic work has been accomplished. From these data a comprehensive regional model for the history of sedimentation in the Timan–Pechora basin has been developed. At the present time work is focused on the refining of local key sections and the establishment of more precise regional and local correlations using especially the conodont and ammonoid scales; the work reported here is part of this and is concerned especially with the southern area. This work follows recommendations of the Subcommission on Devonian Stratigraphy (SDS) of the International Union of Geological Sciences (IUGS) and the Devonian Commission of Russia (DCR).

Following the recommendations of SDS and

DCR, the primary work for this project has been made in the Ukhta anticline and adjacent areas where different facies of the Devonian are developed. In this area shallow-water shelf terrigenous and carbonate deposits with reefs of different types occur and there are basins associated with anoxic or hypoxic facies; these last form the primary petroleum source rocks of the area. The rocks are rich in the fossils required for palaeoenvironmental interpretation and biostratigraphic correlation and using these characteristics the distinctive facies of the area can be related to sea-level and climate changes in other areas of the world. Organic remains include rich faunas of brachiopods, ostracods, radiolarians, goniatites, dacryoconarids, conodonts and spores. The exposures at outcrop are complemented by data on sections in numerous boreholes: this enables reliable correlation to be made of the different facies developments, delineation of time blocks of sedimentary sequences, and establishment of a framework for the elucidation of sea-level changes in the basins.

Because the carbonate facies provides the best framework for establishing deepening events, this will be used as the framework for discussing the Frasnian development of the reef facies. But faunas of particular use in global correlation occur in reef margin or basinal facies and it is convenient to comment on them particularly.

General stratigraphy of the Frasnian succession

The setting of Frasnian deposition and reef accumulation is illustrated in Figs 3 and 4. The stratigraphical terms used in the Timan–Pechora area are shown in Figs 5 and 7. Frasnian time commences within the Timan regional stage and is taken to terminate with Savinobor and Izhma Formations. In practice these levels are currently mostly recognized by the complexes of brachiopods, ostracods and spores, rather than by the goniatites and conodonts which, however, contribute so much to the global time correlation which is the focus of this contribution.

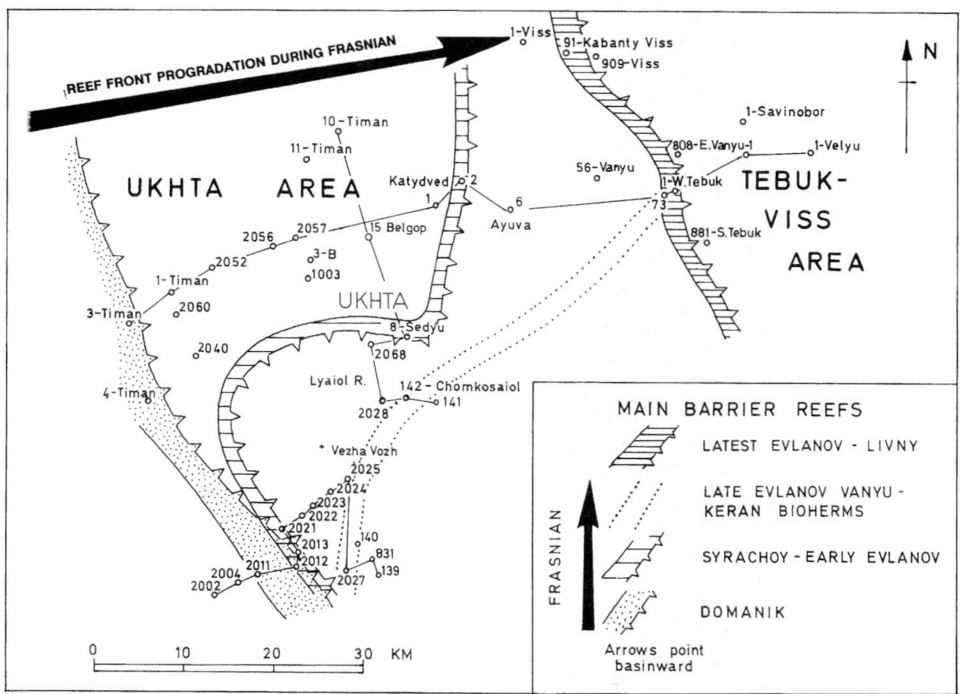

Fig. 3. Map showing the position of boreholes in the Ukhta and Tebuk–Viss area of the southern Timan–Pechora Basin and the position of boreholes referred to in the text. The position of the Frasnian barrier reef at three periods is marked, illustrating the eastward progradation of the reef complex during the Frasnian; in each case basinal deposits are to the east. Note that the barrier reefs may at times have been discontinuous. Geological cross-sections along the western lines indicated are shown in Fig. 4. The west to east line of sections forms the basis for the diagram illustrating the development of the reef complex shown in Fig. 6.

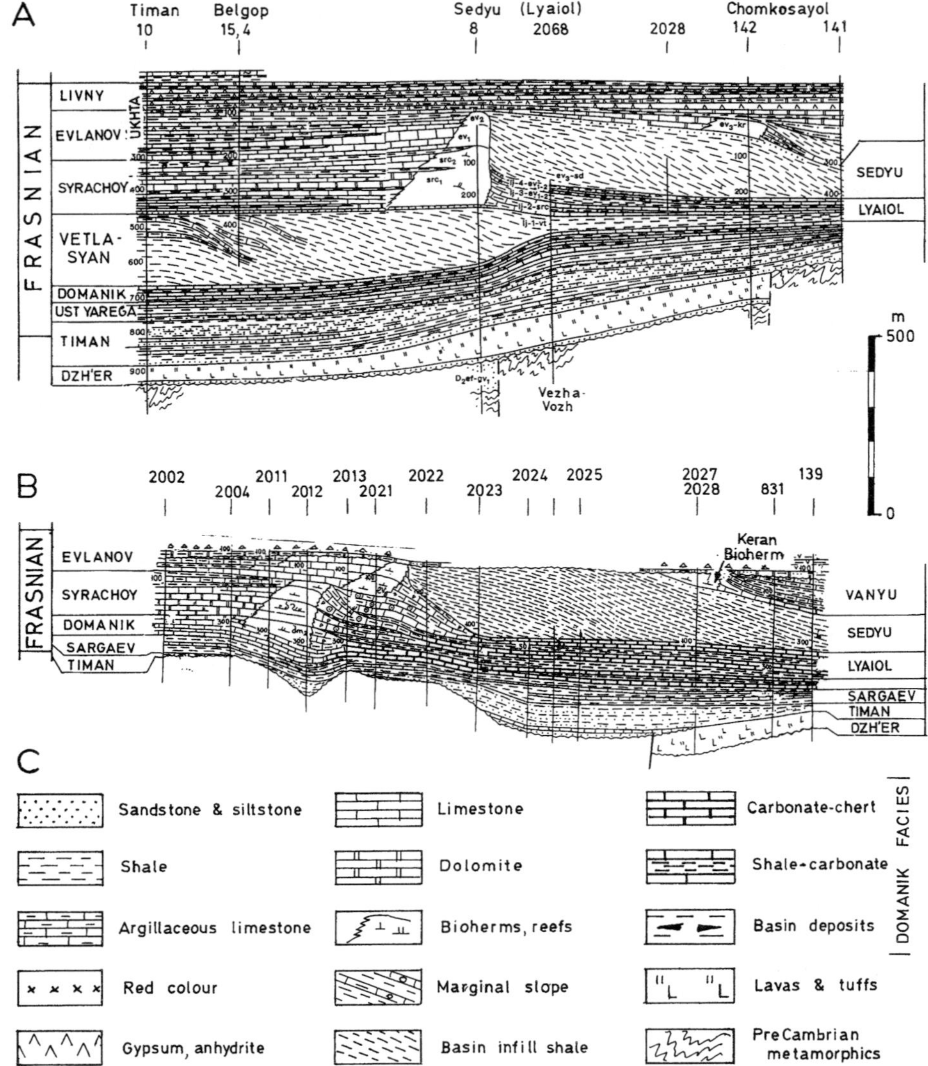

Fig. 4. Two geological cross-sections showing the structure of the Frasnian bioherms, biostromes and infill deposits. **(A)** Section from 10-Timan borehole to 141-Chomkosayol borehole. **(B)** Section from boreholes 2002 to 139: the position of boreholes is shown in Fig. 3. **(C)** Key to the ornament used in the sections (modified from Menner *et al.* 1992).

The development of a formal system of local subdivisions, formations and members in the Frasnian Stage of the southern Timan, as for the entire Timan–Pechora area, is still at an early stage, but most units can be correlated with the regional stages (Fig. 5). Formerly the base of the Frasnian was taken at the surface of regional unconformity at the base of the Dz'her Formation but following the acceptance by IUGS of a higher level to define the stage boundary, the level now lies within the Timan Formation according to Ovnatanova *et al.* (1999*a*,*b*). But as conodont facies of the Timan Formation are entirely within the *Polygnathus* biofacies, the exact position of the boundary cannot be determined.

The Dzh'er Formation in sections of the Ukhta anticline is composed of a basal layer of quartz sandstones with an overlying series of basalt lavas and tuffs reaching thicknesses of

150 m. Eastward, and outside of the Ukhta palaeorift, the formation is represented by alternating sandstones, siltstones and mudstones with an admixture of tuffaceous material. The most ancient terrigenous deposits (the Yaran and lower Dzh'er horizons) are found only in small troughs and representing local fresh-water palaeolacustrine environments. These correspond to shallow-water fluviatile regimes of the late Givetian through the Yaran, Dzh'er and lowest Timan horizons. At about the junction of the early and late Dzh'er Formation, there is evidence of active movement of tectonic blocks and local erosional washout phenomena. In rifts of the Timan and Pechora-Kolvin tectonic zones, intertongues of lava flows and tuff accumulations occur. Sandstones, siltstones and clays of the late Dzh'er Formation are distributed over a large area and they rest unconformably on Middle Devonian rocks of different ages.

Timan regional stage

This is typically represented, after the regional break in sedimentation in the middle of the late Givetian, by the establishment of a new sedimentary regime. This was initiated by a transgressive period which reached its acme in dark-coloured clays with *Buchiola* and *Pterochaenia*, indicators of deeper neritic facies, in the lower part (65–100 m) of the Timan Formation. Open marine deposits of shallow-water type are indicated by associations of the brachiopods *Schizophoria*, *Uchtella* and *Uchtospirifer* and by the abundant occurrence of ostracods and conodonts. In other areas the Timan Formation is represented by marine clays with thin beds of sandstones with occasional tuffs to a thickness of 80–170 m. The lower Timan Formation is referred to the local brachiopod zone of *Uchtospirifer nalivkini*. An ostracod complex including *Ornatella multiplex* continues from the earlier Dzh'er units.

The upper part (40–50 m) of the Timan Formation is much more widely distributed in the south Timan area than is the lower part, but it contains more red-coloured beds; the facies is of red and green-grey clays, siltstones, sandstones and rare limestones. By the very latest Timan time, when the red clays were replaced by green clays, marine conditions flooded over all the formerly elevated islands in the shelf zone of the Timan–Pechora area. The upper Timan Formation is referred to the brachiopod zone of *Uchto. timanicus*. The Upper Timan deposits represent a deepening which extends from the upper Timan into the Sargaevo regional stage. Yudina & Moskalenko (1994, 1998) have noted similar brachiopods to those of the lower division but including *Devonoproductus*, *Pseudatrypa*, *Spinatrypa* and *Komispirifer*, most elements of which also occur higher in the upper Timan Formation. Brachiopod coquina beds represent lag deposits or tempestites. These two Timan sequences cover an interval from the late Givetian to an early part of the Frasnian stage.

Sargaevo regional stage

In the Ust'yarega Formation there is evidence that the basin deepened; rich associations of brachiopods and ostracods (especially entomozoids) occur and goniatites first appear. The Ust'yarega is composed of grey clays with levels of siltstones at the base and limestone lenses in the upper part which have the abundant brachiopods *Hypothyridina calva* and *Mucrospirifer novosibiricus*. For ostracods this is the *Cavellina chvorostanensis*/*Richterina scabrosa* fauna. Sedimentation on the greater part of the Pechora Plate and in the southern Timan predominantly comprised clays of shallow-water type. Only in the extreme northeast and east were carbonates formed and locally, in depressions, thin deposits of dark sediments of domanikoid type occur. A goniatite fauna with *Hoeninghausia* appears in the early part of the middle member (or upper member of Kuz'min *et al.* 1997), which is followed by levels rich in *Timanites*. In terms of biofacies, these and the terminal Ust'yarega *Komioceras* Beds are similar to the deeper-water, open marine, pelagic cephalopod limestones of the subsequent Domanik Formation. The deposits of the Ust'yarega Formation represent a transgression with a progressive relative rise in sea level but there are somewhat neglected calcarenite beds in the highest part which represent again a shallower bioclastic interval (Fig. 8). There has been dispute in the past as to where the top of the Ust'yarega should be drawn and the *Komioceras* beds have been taken by some as the base of the Domanik. Here these are included in the Ust'yarega Formation and the top of the formation is recognized in the Chut' River section at the transition from calcarenites (Fig. 8, Beds F to G) to the overlying green shales (Bed H).

Domanik regional stage

In the Domanik Formation there is evidence for an extremely rapid and significant rise in sea level which caused major changes in the sedimentation of large areas within the shelf. Generally a facies of shallow water with terrigenous input was replaced by very different basinal conditions which have discrete facies zones and sharp lateral variations. In the middle Timan and the western sides of the southern Timan, this division is represented by shallow-water shelf sediments in which clays alternate with biogenic limestones. Eastwards, in the back-reef zone, there are a series of limestone/dolomites up to 40 m thick which contain corals, stromatoporoids and brachiopods of the *Cyrtospirifer disjunctus*/*Anathyris helmerseni* Zone. The back-reef sequences change eastward into the barrier reef zone. The lower part is composed of alternations of dark grey stromatoporoid limestones with thin siliceous/carbonate/clay layers with coquinas of dacryoconarids. The upper part is made up of light grey limestones, sometimes with dark zones and striped cavernous dolomites which preserve biohermal forms in some places. Often in the bioherms several phases of reef formation can be recognized. The thickness of the reef complex known as the Timan reef in the Ukhta region reaches 150 m. A more easterly starved basin became the site of hypoxic to anoxic conditions and of the distinctive domanikoid facies.

The establishment of a barrier of reefs occurred in Domanik time (Fig. 3) in the west of the area only, and it may well be discontinuous along its length. Also, the reefoid facies developed especially in the late part of Domanik time (Figs 4B, 5). The reef massifs had rather steep frontal slopes and were replaced seaward by the starved basin deposits of the Domanik facies; in typical sections this facies is represented by dark grey siliceous/carbonate rocks enriched in organic carbon and containing layers and nodules of limestones with goniatites, nautiloids,

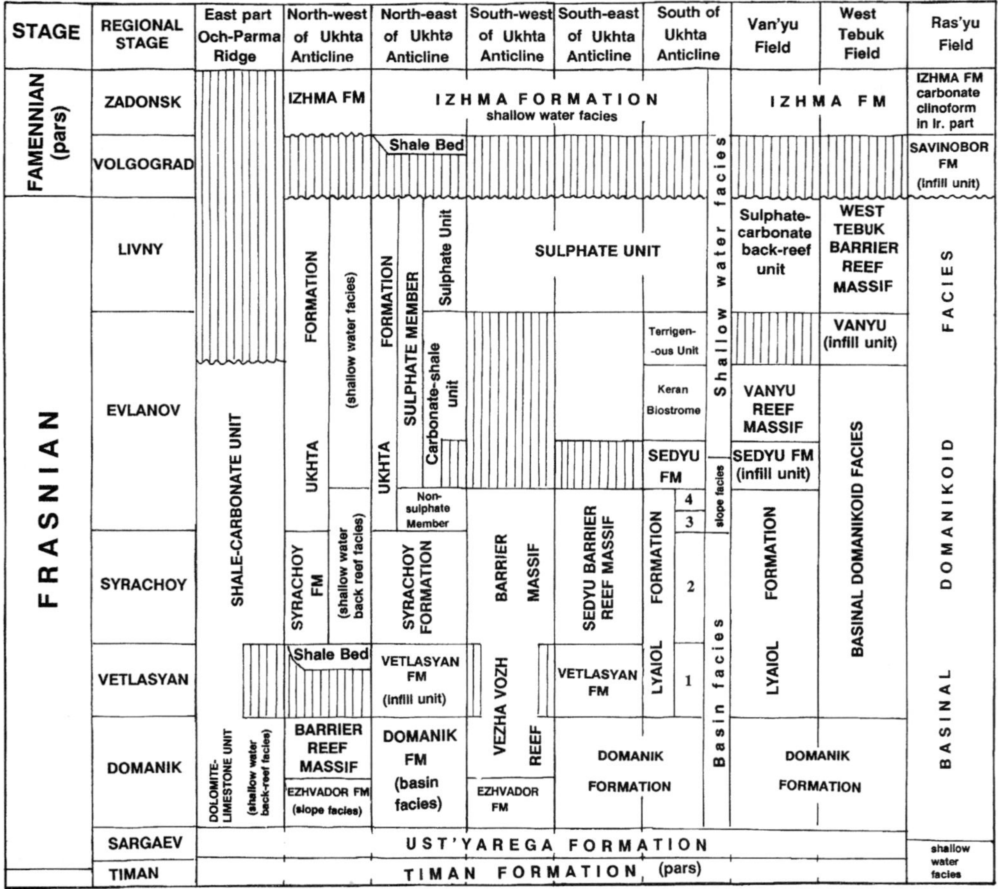

Fig. 5. Diagram illustrating the stratigraphic terminology and biostratigraphic divisions of the late Givetian, Frasnian and early Famennian deposits of the southern Timan–Pechora Basin.

tentaculitids, ostracods and entomozoids with several levels of clays. The total thickness of the Domanik facies is about 50–70 m decreasing to the south and east to 20–30 m. In the stratotype exposures in the Ukhta area the formation is divided into several parts. The base is taken at a green shale above a calcarenite unit (Fig. 8). The formation is divided into three divisions.

The marked development of anoxic and hypoxic sediments, which characterize the domanikoid facies, is especially well developed in the lower part of the Domanik Formation and is characterized by an abundance of laminae high in organic matter, often siliceous and with common Radiolaria, and by an almost complete absence of benthos; however, there are some specialized bivalves and brachiopods including rare rhynchonellids. The sea at that time was around its highest level. Water circulation in the lower part of the water mass in the basin was probably virtually eliminated and stagnant conditions resulted. During deposition of the middle and upper parts of the Domanik Formation, when sea level may have risen even more, circulation appears to have resumed weakly and a scarce benthos occurred. Similarities are striking with the lithofacies and biofacies of the famous Kellwasser limestones of Western Europe and North Africa.

The sea-level rises probably corresponded to periods in the middle and late Domanik when the growth of bank and reefs became especially active. Thus the carbonate ramp developed in Domanik time formed a sequence in which thin carbonates and clays near the boundary between the Ust'yarega and Domanik Formations in the western Ukhta region formed the transgressive series which continued through the basinal deposition of the Domanik Formation, although there appear to have been pulses of particular deepening.

Vetlasyan regional stage

The reef growth of Domanik time terminated suddenly and this is thought to be due to regression. Vetlasyan time led to the accumulation of a thick series of clays but these seem to have formed under a sediment starvation regime and eastern basinal areas did not receive much sediment. The Vetlasyian, in Timan sections of the shallow water zone, is represented by a thin (1–10 m) clay unit with siltstones and spores of the *Cymbosporites vetlasjanicus* Zone. Over the Timan barrier reef zone these clays are often wholly pinched out. In front of the Timan barrier reef, clay deposits of the Vetlasyan reach thicknesses of 150–200 m. Here they define the Vetlasyan Formation (suite). The dark grey clays at the base contain spores of the *bellus* Zone. The middle and upper part of the Vetlaysian is made up of greenish-grey clays with layers of siltstones and sometimes of limestones thought to have formed in shallower water. Within the starved basin areas, the thickness of Vetlaysian deposits thins to 30 m and even 15 m.

Syrachoy regional stage

The Syrachoy division, in shallow-water shelf facies, starts with a number of sandstones which transgressively overlap the thick series of the Vetlaysian and Domanik Formations below. Upwards, limestone layers, marls and clays alternate. Eastwards, towards the edge of the shallow-water shelf, the number and thickness of carbonate beds increases and is illustrated in the quarries of Podgorny, Syrachoy and Belgop, and in the exposures along the Ukhta River and in numerous boreholes. The thickest carbonate beds occur in the middle part of the Syrachoy Formation and the upper boundary is drawn at a thin dark red unit.

In facies of the back-reef zone the Syrachoy division is distinguished by an abundance and variety of benthos including stromatoporoids, tabulates, rugosans, brachiopods, bivalves, ostracods, crinoids and algae. The brachiopods belong to the *Nervostrophia latissima*/*Adolfia siratschoica* Zone. The total thickness of the Syrachoy in the Ukhta region is 60–120 m. Eastwards and southeastwards the shallow-water shelf facies is replaced by dolomites making up the lower part of the Sedyu and the middle part of the Vezhavozh barrier massifs with thicknesses of 100–150 m. In the basinal facies the unit is represented by Member 2 of the Lyaiol Formation which is about 20–30 m thick.

Problems remain on the determination of the upper boundary of the Vetlasyan, Syrachoy and lower Evlanovo sequences.

Evlanovo regional stage

The Evlanovo and Livny regional stages in sections of the western shallow-water facies correspond to the Ukhta Formation (Figs 4,5) which has a thickness of 150–240 m. In sections through the lower part of the Evlanovo (Ukhta Formation, Fig. 4) some red-coloured beds occur in the back-reef zone. The corresponding fore-reef and more basinal equivalents (Member 4 of the Lyaiol Formation) show an abundance of brachiopods, including coquinas of rhynchonellids, which suggests shallower water than in Member 2.

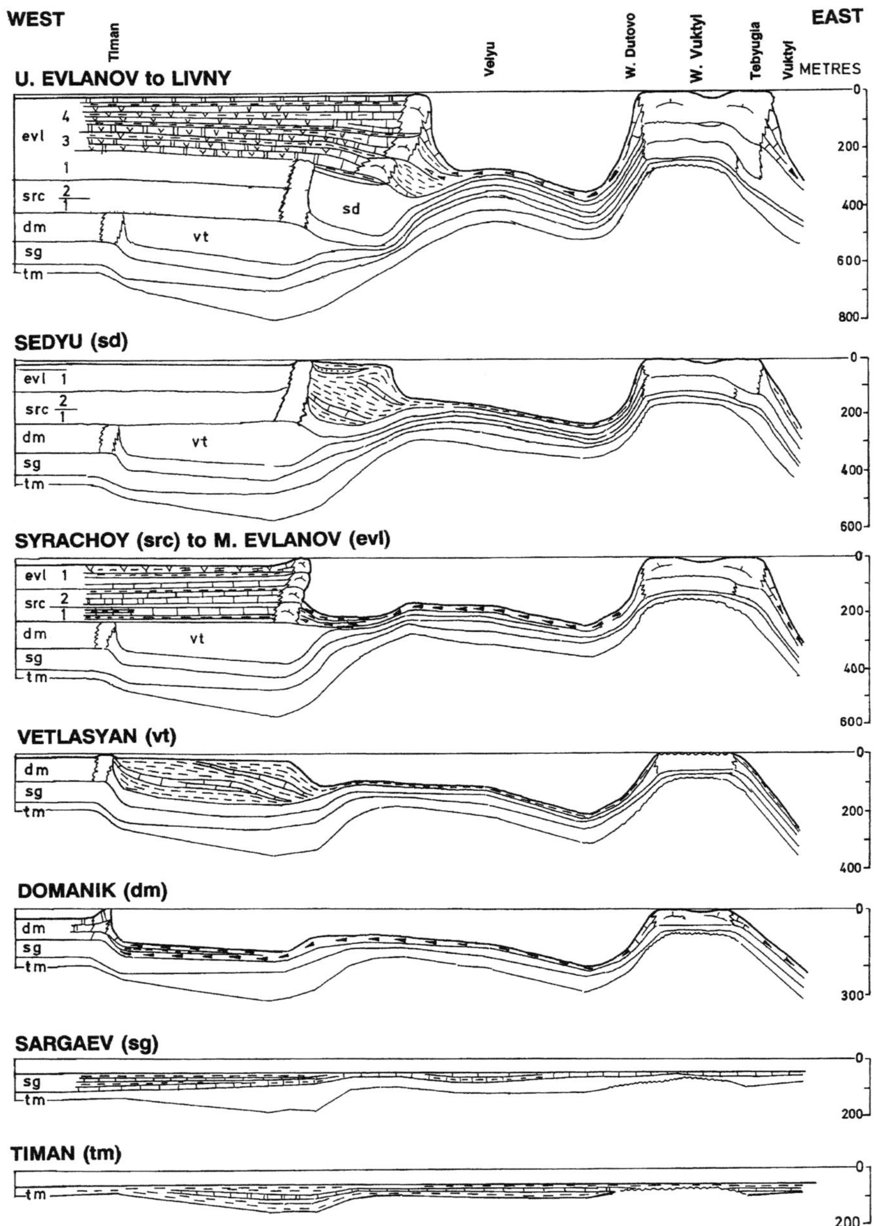

Fig. 6. Diagram illustrating the development of the Frasnian reef complexes of the southern Timan–Pechora Basin at the time of the regional stages given on the left. Sedimentological key as in Fig. 4. Line of the boreholes on which the interpretation is based is shown in Fig. 3.

In the lower part of the Ukhta Formation there are up to 100 m of limestones, marls and clays showing rhythmic alternations. The percentage of clay is higher than in the Syrachoy Formation and there are fewer organic remains and a sparser diversity. This appears to have resulted from different salinities. The upper part of the Sedyu barrier reef is the stratigraphic analogue of the lower part of the Uhkta Formation and Evlanovo regional stage. It is separated from the Syrachoy reef below by a non-sequence and is composed predominantly of algal limestones with a fenestral texture, oncolite limestones and layers of branching stromatoporoids. In the southwest area of the Ukhta anticline, the Syrachoy–Lower Evlanovo barrier reef massif built upon the Timan barrier massif of Domanik age (Figs 2, 4B).

In basinal areas to the east, beyond the fore-reef margin, the lower part of the Ukhta Formation is represented by the clay-carbonate Members 3 and 4 of the Lyaiol Formation (about 20 m thick). The facies, and the abundance of brachiopods and coquinas of *Cariorhynchus*, testify to shallower depths of formation than for Lyaiol Member 2. Ammonoid faunas of Member 4 are of low diversity.

The shale beds in the middle part of typical sections of the shallow-water shaly deposits of the Ukhta Formation probably correspond to the Sedyu (up to 150 m) and Vanyu (up to 100 m) infill shale units developed southward and eastward from the Syrachoy–early Ukhta barrier reef zone; the Keran (up to 40 m) and the Vanyu (up to 60 m) are clinoform carbonate units and not barrier reefs. Equivalents of these units in basinal facies of the Domanik facies have not yet been properly elucidated.

Livny regional stage

This corresponds to the upper part of the evaporite unit of the Ukhta Formation and to the final period of reef development in the Ukhta region which has been named the West Tebuk Reef Massif (Fig. 5) or Zapadno-Tebuk or Pozdne-Ukhta reefs. These are thought to be an extension of the previously established Keran and Vanyu biostromes. The corresponding back-reef facies are carbonate and evaporitic sulphate rocks with carbonates in basinal sections. Lithological peculiarities of the back-reef sulphate series were earlier thought to suggest that there was a fall in sea level after the formation of the Keran biostrome. However, the new precise age determination of the Zapadno-Tebuk reef massif suggests a correlation with a deepening event at approximately the level of the latest Frasnian. This interpretation is supported by rich deeper-water conodont faunas (Kuz'min *et al.* 1998) from carbonates at the slope foot of the Tebuk Massif which overlie argillites with rare sandstone interbeds.

Frasnian/Famennian boundary

The boundary between the Frasnian and Famennian in the Ukhta region corresponds with a non-sequence between the Ukhta and Izhma Formations. In a basal clay layer (about 2.0 m thick) of the Izhma Formation spores belonging to the local *imperpetuus* Zone occur which indicate the Volgograd regional stage and the lower part of the Famennian. Eastwards, in the Tebuk area, in boreholes 909, 881, 888 and others (Klapper *et al.* 1996; Kuz'min *et al.* 1998), the boundary between the stages coincides with the lower boundary of the Savinobor Formation. The Savinobor Formation, which is up to 180 m thick, contains brachiopods, miospores and ostracods of Famennian type and *triangularis* Zone conodonts.

The argillites of the lower part of the Savinobor Formation give indications of deeper-water facies in the early part and shallower-water facies later. In areas of shelf and those above earlier reefs a stratigraphical break corresponds to the early part, and in some areas, the whole of the Savinobor Formation.

Comments on the Frasnian reef complexes

The position of successive barrier reef fronts in the carbonate complex of the southern Timan during the Frasnian is illustrated in Fig. 3 which shows how the reefs prograded in stages eastwards through the Frasnian over earlier reef and basin infill deposits. The development of the carbonate reef complex of the southern Timan is also illustrated here by two cross-sections through the region (Fig. 4A,B) along lines shown on Fig. 3. Terminology used in this discussion is illustrated in Fig. 4. The sequential history of reef development is shown in Fig. 6. An interpretative analysis of the reef development in sequence stratigraphical terms is discussed in a later section and illustrated in Fig. 10.

The beginning of the formation of reefs and organogenic carbonate banks was associated with a rise of sea level in the Domanik. At this time the shelf was divided into shallow-water zones along which barrier reefs were formed along the seaward edge and large deeper-water starved depressions. The reefs were terminated by regression and a discontinuation of the reefs and banks.

In the southern Timan, in the Tebuk area as well as in other regions of the Timan–Pechora province, three phases of large, lengthy barrier massifs of Frasnian age were developed – those of the Domanik, the Syrachoy–early Evlanovo (Syrachoy–early Ukhta) and latest Evlanovo–Livny (late Ukhta) – and a zone of small biostromes and carbonate clinoforms of the Late Evlanovo (Keran and Vanyu biostromes) (Figs 4, 5).

In areas of clay accumulation terraces, subsequent barrier reefs formed on their seaward margins, and thus later reefs are shifted east or southeast in relation to earlier reefs. In areas near tectonic faults, some biohermal massifs developed through several phases without lateral shift.

The thickness of the reefs and banks was conditioned by the added influence of sea-level rise, tectonic subsidence of the zone, and the period of formation of the massif. Therefore the thickness of massifs varies from 50 to 500 m. The barrier reefs of the Domanik are up to 150 m in thickness, those of the Syrachoy and the early Evlanovo up to 270 m, and those of the latest Evlanovo–Livny up to 210 m.

Early Frasnian reef massifs are made up of organogenic bedded limestones, developing into bioherms with biostromal parts abundant in stromatoporoids and blue-green and red algae. The upper parts of massifs are represented by layered algal limestones with fenestral textures. Frontal areas of the massifs are often strongly dolomitized. At the early and sometimes middle stages of the Domanik, reefs are black-coloured bedded limestones approaching the basinal deposits.

The palaeoecology of the reefs changed through time. The reef massifs of the Domanik are dominated by colonies of stromatoporoids. Those of the Syracho–Evlanovo are dominantly of branching stromatoporoids and algae, whilst those of the Livny are primarily composed of algal and fenestral limestones.

Anoxic facies and their interpretation

The major period of formation of anoxic/hypoxic sediments during the Frasnian of the area was during the Domanik Regional Stage. A major rise in sea level led to the formation of an extensive starved basin east of the Domanik reef front (Figs 3, 4 and 6). The characteristics of this interval are the development of laminites in the form of bituminous, brown/grey siliceous shales, interbedded siliceous limestones and chert lenses. More oxic limestone levels occur, with limestone concretions at certain levels. The fauna is rich in Radiolaria, probably the source of the silica. Conodonts were mostly only extracted from the limestones. Goniatites occur especially in calcareous beds which are atypical and which may contain a restricted benthos including specialized bivalves and rare rhynchonellids pointing to a limited oxygen level at the sediment–water interface; endobenthos is generally lacking.

The major development of this facies is in the Domanik Formation but there are other limited levels of similar type, notably in Member 2 of the Lyaiol Formation, when a similar starved basin occurred east of the Syrachoy reef and when considerable depths for the basin of up to at least 200 m are interpreted (Fig. 6). Other occurrences at similar settings but with more limited horizons of this type occur in Evlanovo time.

Any interpretation of the high primary kerogen content and source rock quality of the bituminous shales and limestones in the Domanik facies has to consider several factors: (1) the primary source of the nutrient-rich waters initiating the system; (2) the nature of the primary organic input resulting from productivity in upper waters; (3) the sedimentation rate; and (4) the stagnant sea-floor conditions which resulted in the lack of normal aerobic biodegradation processes. Some models for such situations are given by Schwarzkopf (1993).

Domanik-type shales accumulated in an offshore region of the deeper shelf within a pelagic setting. The environment was characterized by starved sedimentation which prevented dilution of primary organic material by sediment. Laminites and lack of endobenthos or even of all epibenthos indicate a hostile, anoxic to dysoxic sea floor which lacked significant currents and turbulence to disturb such an environment. The situation would have been well below maximum storm wave base. Bioturbation, aerobic biodegradation and recycling of organic matter by benthic consumers was kept to a minimum. The organic input must have been so high that it considerably exceeded the resorption potential of anaerobic bacteria on the sea floor and in the sediment (Pedersen & Calvert 1992). Evidence for high organic productivity is directly given by blooms of palynomorphs, Radiolaria, and by the mass occurrences of cephalopods, conodonts and bivalves in carbonate beds. Such carbonate facies are very similar to the many highly eutrophic black limestones of Western Europe and North Africa, especially the late Frasnian Kellwasser beds (Becker & House 1994*b*) or black limestone of the European Kačák Event (House 1996) and other Devonian events (House 1985; Walliser 1996).

A high terrestrial nutrient input could theoretically explain the higher productivity level of offshore regions of the shallow intracratonic Timan–Pechora Basin since most plate-tectonic palaeoclimatic reconstructions (Parrish 1982) would not identify the area as one of constant open marine or coastal upwelling. As will be shown later, the spreading of hypoxic facies occurred during times of eustatic sea-level rises which would have increased the areas of starved, quiet-water sedimentation, but this would not necessarily increase primary organic productivity in the whole basin. A role for global climatic change leading to vertical and lateral movements of water masses and recurrent incursion of nutrient-rich water masses has to be considered in addition (Becker 1992; Becker & House 1994*b*). Black shale episodes probably reflect eutrophication events as in other Frasnian epicratonic seas, such as in eastern North America, northern European Russia and areas of the western Gondwanaland margins. Attention should be drawn to the anomalous Devonian palaeogeography with most continents confined to one hemisphere, which makes upwelling comparisons with the present day very uncertain.

Conodont biostratigraphy

The correlation of Timan Frasnian calcareous deposits by means of conodonts commenced with the early studies of Khalymbadzha (1981), Ovnatanova (1976) and Kushnareva *et al.* (1978). In subsequent papers (e.g. Ovnatanova & Kononova 1984; Ovnatanova & Kuz'min 1991; Yatskov & Kuz'min 1992), both standard and regional Timan zones have been used. Following preliminary new regional successions published by Kuz'min (1997) and Kuz'min & Yatskov (1997), the Frasnian conodont sequence in the Timan–Pechora Basin has been extensively reviewed by Ovnatanova *et al.* (1999*a*,*b*). They propose a succession of conodont assemblages numbered 0 to XI and present detailed correlations with the earlier Frasnian standard zonation (Ziegler 1971) and the revised standard zonation (Ziegler & Sandberg 1990). Consequently, these correlation results are not repeated here. In a paper on subsurface conodont sequences in the southern Timan and Khoreyver Basin, Klapper *et al.* (1996, p. 133–137) demonstrated correlations with the Frasnian zonation first proposed for the Montagne Noire (MN) sequence (Klapper 1989), extending from Zone 4 to Zone 13. Correlations of Timan–Pechora conodont faunas with the MN zonation are straightforward once discrepancies in taxonomy are resolved. A major step toward this end was accomplished in June 1998 when G. K. and R. T. B. studied most of the southern Timan collections together with N. S. O. and A. V. K. in Moscow. The description of the Timan conodont sequence that follows is presented, as far as possible, in the framework of the MN zonation. Shallow-water units such as the Timan, Vetlasyan, Syrachoy and Ukhta formations have low-diversity conodont faunas dominated by species of *Polygnathus* and *Icriodus* that do not allow precise correlation and are zonally undiagnostic.

Figure 7 shows the conodont alignments of southern Timan lithostratigraphic units using MN zones and the correlation with the new regional assemblage zones. Old conodont assignments are also given to illustrate the significant progress in conodont research and in order to allow an understanding of earlier literature. However, it should be emphasized that the correlation between standard zones as used regionally in the Timan–Pechora Basin and MN zones differs significantly from the correlation of the two zonal schemes based on German type sections (Klapper & Becker 1999). This discrepancy can be taken as a warning not to equate uncritically regionally used standard zones with the MN zones. To cite just one example, the base of the *Ancyrognathus triangularis* Zone correlates with the base of MN Zone 10 at the Martenberg type section, but in the Timan the much older Middle Domanik Formation, now correlated with MN Zone 6, has been assigned in the past to the *A. triangularis* Zone.

Topmost Givetian to MN Zone 2

Ovnatanova *et al.* (1999*a*) distinguish a fauna with *Polygnathus xylus*, *P. webbi* and *P. alatus* in the upper part of the Lower Timan Formation as TP-0 assemblage from the more diverse TP-I assemblage with *P. angustidiscus* and other species of *Polygnathus* in the Upper Timan and lowermost Ust'-Yarega formations. All recorded species of the genera *Polygnathus, Icriodus* and *Mehlina* from these two assemblages are zonally undiagnostic. Nor can the species of *Polygnathus* listed in TP-0 and TP-I identify the precise position of the Givetian–Frasnian boundary with confidence. Khalymbadzha (1981) mentioned *Ancyrodella binodosa* from the Timan Formation, which may include the early form of *A. rotundiloba.* The recovery of *Ancyrodella* has not been repeated subsequently, however, and the original specimens need to be restudied in the light of current taxonomy.

The oldest zonally diagnostic conodont fauna in the Timan–Pechora Province is at the Kozhim River Outcrop 3801 (Polar Urals) where *Ancyrodella rotundiloba* s.s. at the base of the section in Bed 1 (sample 213) indicates a correlation with MN Frasnian Zone 2. The species ranges higher into a massive limestone with abundant *Hoeninghausia* (near the base of Bed 12, sample 72).

MN Zone 3

Higher in the lower part of Bed 12 (sample 73) of Kozhim River Outcrop 3801, there is a fauna

stage	regional stages	formations		members	old conodont assignments	MN zones	regional assembl.	regional ammonoid zonation	intern. zones	miospore zonation
FRASNIAN	LIVNIAN	Ukhta	West Tebuk Reef		Uppermost *gigas*	13	XI	[no fauna]	I-L	*G. subsuta*
	EVLANOVIAN	Ukhta	Vanyu Reef / Sedyu		Upper *gigas*	13 ?	X	[no fauna]	I-K	*A. speciosa*
	EVLANOVIAN		Lyaiol	4	Upper *gigas*	12	IX	*Manticoceras lyaiolense*	I-J	*A. speciosa*
	EVLANOVIAN		Lyaiol	3	Upper *gigas*	11	VIII	*Manticoceras lyaiolense*	I-I	*M. radiatus*
	SYRACHOIAN	Syrachoy	Lyaiol	2	Lower *gigas*	11	VIII	*Virg. ljaschenkoae*	I-I	*M. radiatus*
	VETLASYANIAN	Vetlasyan	Lyaiol	1	Lower *gigas*	10	VII	(*Carinoceras* sp.)	I-H	*C. vetlasjanicus*
	DOMANIKIAN	Domanik		Upper	*Ag. triangularis*	8-9	VI	(*Lobotornoceras strangulatum*)	I-H	*S. bellus*
	DOMANIKIAN	Domanik		Upper	*Ag. triangularis*	7	VI	*Nordiceras timanicum*	I-G	*S. bellus*
	DOMANIKIAN	Domanik		Middle	*Ag. triangularis*	6	V	*Nordiceras timanicum*	I-F	*S. bellus*
	DOMANIKIAN	Domanik		Lower	*Po. timanicus*	5	IV	(*Pont. auritum*) *Pont. domanicense*	I-E	*G. semilucensa*
	DOMANIKIAN	Domanik		Lower	*Po. timanicus*	5	IV	*Komioceras stuckenbergi*	I-D	*P. donensis*
	SARGAEVIAN	Ust´yarega		Upper	*Ad. rotundiloba*	4	III	*Komioceras stuckenbergi* / *Tim. keyserlingi*	I-C	*C. optivus*
	SARGAEVIAN	Ust´yarega		Middle	*Ad. rotundiloba*	3	II	*Hoen. nalivkini*	I-B	*C. optivus*
	SARGAEVIAN	Ust´yarega		Lower		1-2	I	[no fauna]	I-B	*S. krestovnikovii*
	TIMANIAN	Timan		Upper		1-2 ?	I	[no fauna]	I-A	
GIV.	TIMANIAN	Timan		Lower	*Ad. binodosa*		O			

Fig. 7. Table showing the relation of the Timan–Pechora regional stage and formation terminology to the older conodont zones (Ziegler 1971), the Montagne Noire zones (Klapper 1989) and the new division using conodont complexes (Ovnatanova *et al.* 1999*a*,*b*) compared with the regional ammonoid zonation discussed herein and the international ammonoid scale of Becker *et al.* (1993) and House & Kirchgasser (1993) compared also with the miospore zonation of the area (Avchimovitch *et al.* 1993; Ku'zmin *et al.* 1998).

with abundant *Ancyrodella recta* and *Mesotaxis asymmetrica*, correlating with MN Zone 3. Still higher in Bed 12 (sample 75), *Ancyrodella recta* occurs together with *A. africana* in a fauna that also includes *Mesotaxis ovalis* and *Polygnathus dengleri*. An overlap of these two *Ancyrodella* species has not been observed previously. At present in the sections graphed for the Frasnian Composite Standard, *A. recta* does not range above a position high in MN Zone 3 and *A. africana* does not range below the base of MN Zone 4 (Klapper 1997, p. 119, 123). *Ancyrodella africana* does occur lower than the entry of *Palmatolepis transitans*, the defining species for the base of Zone 4, in sections such as Outcrop 3801 and also some sections in western Canada. But in the Canadian sections that have been graphically correlated these low occurrences of *A. africana* plot above the projected base of Zone 4, indicating that *P. transitans* at those sections enters at a higher position than its lowest base in the Composite Standard. Nevertheless, the occurrence of *A. africana* with *A. recta* in Kozhim River Outcrop 3801 may indicate a downward range extension of the species into a high part of Zone 3.

In the southern Timan, both *Ancyrodella recta* and *A. africana* enter in a brachiopod marker limestone at the top of the Lower Ust'yarega Formation (Outcrop C = 14) and similar faunas of the TP-II assemblage that also have *A. rugosa* (Ovnatanova 1999*a*) continue into the middle member of the formation.

MN Zone 4

The highest zonally diagnostic fauna in Outcrop 3801 at Kozhim River is in the highest limestone with goniatites (Bed 15, sample 235) where *Palmatolepis transitans* occurs with *Ancyrodella alata*. Faunas with *Ancyrodella africana* that also (as in Outcrop 3801) are below the entry of *P. transitans* but occur together with *Polygnathellus* n. sp. (Klapper 1997, p. 126) should be correlated with Zone 4, because the latter is restricted to that zone in Western Australia and western New York. Such faunas are known in the Polar Urals from Outcrop 5302 (Bed 20, samples 98,

112) at Syv'yu River, a tributary to Kozhim River, below the lowest *P. transitans* in Bed 21 (sample 138), which also correlates with Zone 4. Here, as in some Canadian sections, the local entry of *P. transitans* is higher than the base of Zone 4.

In the southern Timan, a sequence of Zone 4 faunas occurs in Chut' River Outcrop 7 (Kuz'min 1998, table 1). The *Timanites* Limestones of the upper member of the Ust'yarega Formation exposed in the stream at the base of the section yielded *Ancyrodella rugosa*, *A. africana* and a stratigraphically important homeomorph of *Ancyrodella binodosa* (sample K9101); *Palmatolepis transitans* was recorded slightly higher in the same bed (Kuz'min 1998, fig. 1, sample K9102). There are no confirmed occurrences of *A. rotundiloba* s.s. in any of the Outcrop 7 samples. According to the revision of the faunas, the overlying *Komioceras* Beds at the top of the Ust'yarega Formation have a different faunal association. Sample K911A has *Mesotaxis bogoslovskyi*, *Playfordia primitiva* and *Ancyrodella africana* but lacks *Palmatolepis transitans*. The next higher K911B has *P. transitans* together with *Playfordia primitiva*, *A. africana* and *A. pramosica*. The latter species is restricted to Zone 4. Samples D911 and D911A have *P. transitans* as the only species of *Palmatolepis* present. The TP-III assemblage of the upper Ust'yarega Formation (Ovnatanova *et al.* 1999*a*,*b*), characterized by *P. transitans* and *Mesotaxis bogoslovskyi*, correlates with Zone 4, but at least regionally the second species is characteristic for a higher level within the zone.

MN Zone 5

The conodont sequence at Chut' River Outcrop 7 (Fig. 8) from samples D911B to D9214 in the lower member of the Domanik Formation (Kuz'min 1998, table 1) correlates with MN Zone 5. Significant entries include *Palmatolepis punctata c.* 1.5 m above the base of the Domanik Formation (sample D911B), *P. maximovae* just above in D912, *Polygnathus vjalovi c.* 4 m above the formation base (in D917), and *Mesotaxis johnsoni* in D919 from the top of the main cliff right at the river. Distinctive new faunal elements enter also in the first (*Palmatolepis gutta* and *Polygnathus timanicus* in D9210) and in the third *Ponticeras* Bed (*Ancyrodella gigas* form 1 and *A. curvata* early form in D9213). *Ancyrognathus ancyrognathoideus* enters in a new sample taken by R. T. B. from the third *Ponticeras* Bed; its lowest occurrence in the Frasnian Composite Standard is the upper part of MN Zone 5 (Klapper 1997, p. 123). The Outcrop 7 sequence is the main basis for the TP-IV assemblage of Ovnatanova *et al.* (1999*a*,*b*) and *Palmatolepis gutta*, *Mesotaxis johnsoni* and *Polygnathus vjalovi* are restricted to this interval in the Timan sequence. It may be possible to use the staged entry of taxa within the zone for regional correlation of sections.

MN Zone 6

At the top of the Chut' River Outcrop 7 sequence, in the middle member of the Domanik Formation, in samples D9216 and D9217, respectively, the entries of *Palmatolepis bohemica* and *P. spinata* indicate a correlation with MN Zone 6. The basal part of the Middle Domanik in Outcrop 15b also yielded *Ancyrognathus ancyrognathoideus* and *Polygnathus uchtensis*. In the higher and main part of the middle member there are occurrences of *Ozarkodina trepta* (Outcrops 7k, 21k), *Ancyrognathus primus* (Outcrop 3 = 15c) and of *Palmatolepis domanicensis* s.s. in Outcrop 504b. The latter species characterize the middle Domanik Formation TP-V assemblage of Ovnatanova *et al.* (1999*a*,*b*). *Palmatolepis domanicensis* is distinguishable from the form incorrectly identified as that species in Zones 10–lower 11 by Klapper (1989) and Klapper & Foster (1993), which should instead be referred to *P. plana*, as discussed by Klapper & Becker (1999).

MN Zone 7

The basal part of the topmost carbonate unit of the upper member of the Domanik Formation in Outcrop 21 on the Domanik River and on the Ukhta River near Schudajag village yielded *Ozarkodina nonaginta*, the index species of MN Zone 7 and the characteristic species of the TP-VI assemblage of Ovnatanova *et al.* (1999*a*,*b*).

MN Zones 8–10

Higher in the topmost carbonate unit of the upper member of the Domanik Formation, the conodont fauna of MN Zone 8 is represented at Outcrop 5040 (= 13) by *Palmatolepis* aff. *P. proversa*, *P. mucronata*, *P. orbicularis*, *Ozarkodina nonaginta* and *Ancyrognathus amplicavus*. MN Zone 9 is represented in the uppermost 1–2 m of the carbonate unit at Outcrop 1904 by a fauna with *P. proversa*. Slightly higher in the same sequence at Outcrop 1904 is a fauna with *Palmatolepis luscarensis*, *P. proversa*, *P. orbicularis* and *P. amplificata*, an association correlative with MN Zone 10. A similar fauna with *P. luscarensis* also occurs at the top of the Upper

Domanik at Outcrop 5040. MN Zones 8–10 are combined into the TP-VII assemblage represented by a succession at Outcrop 13 (Ovnatanova *et al.*, 1999*b*, fig. 3). The TP-VII assemblage continues into the overlying Member 1 and probably into the basal part of Member 2 of the Lyaiol Formation (Outcrops 1354, 1905, borehole 2068).

MN Zone 11

Faunas with *Palmatolepis semichatovae*, *P. mucronata*, *P. amplificata*, *P. plana*, *P. ljaschenkoae*, *P. muelleri*, *P. timanensis*, *Polygnathus lodinensis* and *Ancyrognathus triangularis* s.s. are well developed in Member 2 of the Lyaiol Formation at Vezha-Vozh River Outcrop 8 and *P. semichatovae* ranges into the lower beds of Member 3 at Outcrop 9 (Bed 3) on the same river. Similar conodont assemblages come from sections along the Lyaiol River (e.g. Outcrops 1906, 1357, 1358). In the upper part of Outcrop 8 and the lower part of Outcrop 9, *Palmatolepis* n. sp. aff. *P. winchelli* Klapper & Lane (see Klapper & Becker 1999) joins the fauna. Its range in the Frasnian Composite Standard (Klapper 1997, p. 126) is from the upper part of Zone 11 into Zone 12, whereas *P. semichatovae* is restricted to Zone 11. This association in Member 2 and lower Member 3 of the Lyaiol Formation equates with the TP-VIII assemblage of Ovnatanova *et al.* (1999*b*, figs 4, 5).

MN Zone 12

In Vezha-Vozh Outcrop 9, the highest MN Zone 11 fauna with *Palmatolepis semichatovae* (Bed 3) is directly overlain by a Zone 12 fauna with *P. winchelli* and *P. foliacea* in Bed 4 low in Member 3 of the Lyaiol Formation. The next higher bed at Outcrop 9 (Bed 5) additionally has *P. kireevae* and *P. muelleri*. Low in Member 4 of the Lyaiol Formation at Vezha-Vozh Outcrop 10, the faunas include abundant *P. foliacea*, *P. orlovi*, *P.* n. sp. aff. *P. winchelli*, *P. winchelli*, *P. muelleri* and *Polygnathus samueli* (rare occurrence). Zone 12 faunas low in Member 4 at Outcrop 1359 on the Lyaiol River have *P. foliacea*, *P. winchelli*, *P.* aff. *P. winchelli*, *P. muelleri* and *P. gyrata*. *Ancyrognathus amana* occurs in the middle part of Member 4 exposed in Outcrop 1360. Faunas in the upper part of Member 4 at Outcrop 1908 on the Lyaiol River are similar to those lower in this member with *P. foliacea*, *P. winchelli*, *P. orlovi*, *P. muelleri* and *P.* aff. *P. winchelli*. The last three species do not range above Zone 12 in the Frasnian Composite Standard, and *P. foliacea* ranges from low in Zone 12 into the lowermost part of Zone 13 (Klapper *et al.* 1996, table 2). The TP-IX assemblage of Ovnatanova *et al.* (1999*a*,*b*, figs 4, 5) correlates with MN Zone 12.

MN Zone 13

Ovnatanova *et al.* (1999*a*,*b*) distinguish a TP-X assemblage that is mainly characterized by *Palmatolepis juntianensis*, which is restricted to Zone 13 in the Frasnian Composite Standard. The lowest *P.* cf. *P. juntianensis* are said to occur in the topmost Member 4 of the Lyaiol Formation but further taxonomic study is needed because there are similar, related forms in Zone 12 in some North American sequences. Klapper *et al.* (1996, p. 137 and fig. 3) also correlated the top of Member 4 with Zone 13 but this was not based on the occurrence of a Zone 13 marker, but rather on the graphing of Core 2023. Currently, the base of Zone 13 cannot be drawn with certainty in the southern Timan sequence.

Limestone beds of the Sedyu Formation along Sedyu River contain only species of *Polygnathus,* which may be useful for regional correlation but are untested as widespread zonal markers. The same applies to conodont faunas dominated by *Polygnathus dentimarginatus* from the lower part of the upper member of the Ukhta Formation. Correlation with Zone 13 is only based on stratigraphic superposition well above the Lyaiol Formation. In more basinal, *Palmatolepis* biofacies of the southern Timan subsurface (Tebuk–Viss region), Ovnatanova *et al.* (1999*b*; compare faunas from Bagan well 3 in Klapper *et al.* 1996 as well as borehole faunas in Kuz'min *et al.* 1998) distinguish a TP-XI assemblage with *Palmatolepis linguiformis*, *P. boogaardi* and *P. bogartensis* (= *P. rotunda* auct.), an association that correlates with the uppermost part of Zone 13.

Goniatite biostratigraphy

The southern Timan is one of the classic areas from where rich Upper Devonian ammonoid faunas have been described over the last 150 years. The first descriptions were given by Keyserling (1844, 1846), and Frasnian faunas were later monographed by Holzapfel (1899). Rather early in the research history it was realized that the Timan is characterized by a high number of endemic genera and species but some of these were later found in other parts of Russia (Urals, Eastern Siberia). The index form, *Timanites*, has now been found in Western Australia and northwest Canada. Following the discovery of several new species by G. P. Lyashenko (1957*a*,*b*),

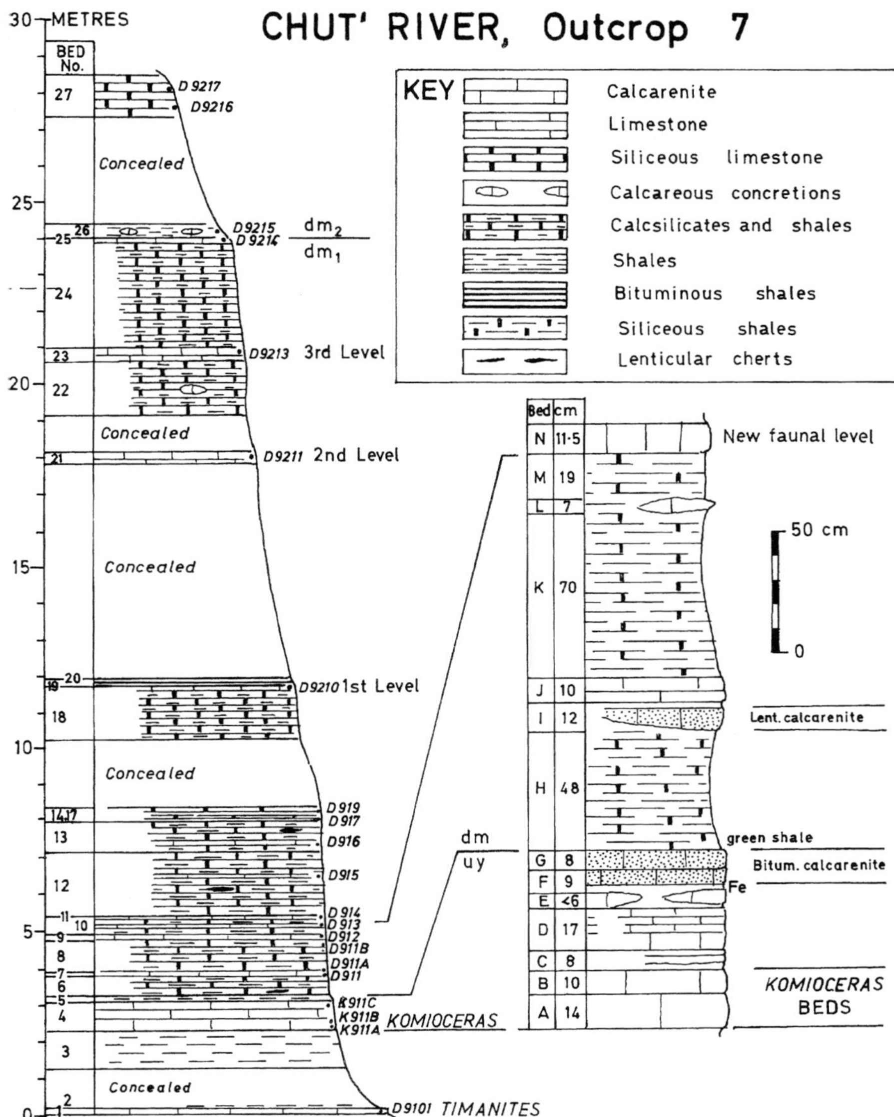

Fig. 8. Diagrams illustrating the section of the uppermost Ust'yarega Formation and early Domanik Formation along the Chut' River (Outcrop 7 of Yudina & Moskalenko 1994). The base of the Domanik Formation is drawn at the line marked dm/uy.

Bogoslovskiy (1969, 1971) provided a revision of Timan faunas and the ammonoid-bearing sections. In recent years there have been attempts to establish a regional Frasnian ammonoid zonation (Yatskov 1994; Kuz'min & Yatskov 1997) which will be superceded by a detailed revision of all southern Timan occurrences and by first records of many additional taxa (Becker *et al.* in press). Goniatites proved to be very useful for the dating of lithological successions and the new regional zonation, despite its endemic characteristics, has been correlated with the international Frasnian zonation established by Becker *et al.* (1993) in Western Australia and House & Kirchgasser (1993) in New York State. This account is given in summary form since a detailed review will appear elsewhere (Becker *et al.* in press). A range chart of typical forms is given in Fig. 9. Results can be summarized as follows.

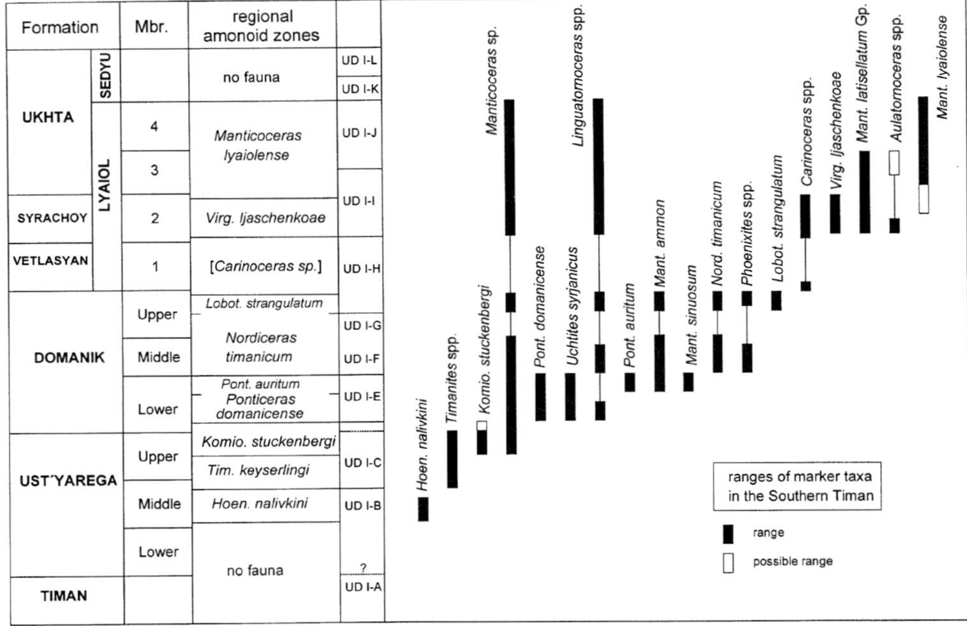

Fig. 9. Diagram illustrating the sequence of ammonoid faunas in the southern Timan area and their correlation with the generic marker zonation of Becker *et al.* (1993). Abbreviations: *Hoen.*, *Hoeninghausia*; *Komio.*, *Komioceras*; *Pont.*, *Ponticeras*; *Mant.*, *Manticoceras*; *Lobot.*, *Lobotornoceras*; *Nord.*, *Nordiceras*; *Virg.*, *Virginoceras*.

The ammonoid record begins with the appearance of oxyconic *Hoeninghausia nalivkini* in the middle Member of the Ust'yarega Formation (G. P. Lyashenko 1957a,b). Much richer contemporaneous assemblages of the regional *H. nalivkini* Zone are known from the northern Timan, particularly from the Usa River Basin (Chernyshov Ridge), as well as from the Polar Urals region within the Timan–Pechora Province (e.g., Kozhim and Syvyu River sections). By comparison with other areas, the *nalivkini* Zone with its Koenenitidae faunas correlates with Upper Devonian UD I-B, probably with a level in its upper part.

The globally widespread index genus of UD I-C, *Timanites*, enters higher in the Middle Ust'yarega Formation and becomes abundant in two *Timanites* Beds in the lower part of the upper member, for example in outcrops along the Chut' River (Bogoslovskiy 1969). Because of homonymy, the species name of the zonal marker had to be changed from *T. acutus* to *T. keyserlingi* (Miller 1938).

Timanites continues into a regional upper subdivision of UD I-C which is characterized by the entry of *Komioceras stuckenbergi*, associated with the tornoceratid *Domanikoceras timidum*, the oldest *Manticoceras* and other species. The distinction of the *K. stuckenbergi* Zone in limestones at the top of the Ust'yarega Formation was first recognized by Yatskov & Kuz'min (1992). *Komioceras* has been reported by Kushnareva *et al.* (1978) to continue into the lowest metre of the overlying Domanik Formation but this has not been substantiated since. However, there is the possibility that an upper part of the zone correlates with UD I-D, defined in North America by the entry of *Sandbergeroceras*, a genus unknown from the Russian Platform.

The main part of the lower member of the Domanik Formation is characterized by rich *Ponticeras* faunas which have been collected from a new level low in the member and from a sequence of three *Ponticeras* Beds along Chut' River (Fig. 8; Bogoslovskiy 1969; Yudina & Moskalenko 1994). Associated are the oldest *Linguatornoceras* and an oxyconic ponticeratid descendant, *Uchtites*. The regional *P. domanicense* Zone can be correlated with UD I-E although species previously included in *Probeloceras* (e.g. Bogoslovskiy 1969) do not belong to this marker genus of UD I-E. The third *Ponticeras* Bed has *Pont. auritum*, frequent *M.*

ammon and *M. sinuosum*; the latter enters in New York in the Cashaqua Shale of the higher part of UD I-E. Based on the common occurrence of manticoceratids, it seems possible to place the upper part of the Lower Domanik Formation in an upper subzone of the *domanicense* Zone.

The change from the lower to middle member of the Domanik Formation is characterized by the regional extinction of the *Ponticeras–Uchtites* faunas. The rediscovery in the Middle Domanik of the multilobed *Nordiceras timanicum*, representing probably a parallel lineage to early Beloceratidae, allows the introduction of a new *N. timanicum* Zone which is also characterized by frequent occurrences of *M. ammon* and of *Phoenixites* species. Assemblages have a very regional composition and correlation with UD I-F is based on conodont faunas only. In the higher part of the Upper Domanik Formation, faunas are still similar but contain the formerly poorly understood *Lobotornoceras strangulatum*. This easily recognizable, extremely compressed tornoceratid can be used to recognize an upper subzone of the *timanicum* Zone, correlating with the lower to main part of UD I-H.

From the very top of the Domanik Formation (Kuz'min & Yatskov 1997) and from a sandstone unit at the boundary of the Vetlasyan and Syrachoy Formations came *Carinoceras* which characterizes a short interval that is generally poor in ammonoids. Completely different rich assemblages with *C. menneri*, *Virginoceras ljaschenkoae*, various manticoceratids including *M. carinatum*, *Linguatornoceras clausum* and *Aulatornoceras* s.s. characterize Member 2 of the basinal Lyaiol Formation. The last three forms are elsewhere not known below UD I-I and such correlation is supported by diverse conodont faunas. The best faunas of the regional *V. ljaschenkoae* Zone come from outcrops along the Vezhavozh and Lyaiol Rivers.

Near the top of Member 2 of the Lyaiol Formation, oxyconic Gephuroceratidae disappear, for example in Outcrop 9 at Vezhavozh River, and the *Manticoceras* faunas become dominated partly by large-sized relatives of *M. cordatum* and *M. intumescens*. The regional extinction of oxyconic goniatites, the decline of the *M. latisellatum* Group with wide flank saddles, and the spread of the involute and strongly compressed *M. lyaiolense* allow the recognition of a new *M. lyaiolense* Zone which ranges to the top of the Lyaiol Formation. Correlation with the top of UD I-I and I-J relies on conodont data. There are no latest Frasnian ammonoids recorded from the Timan so far; typical forms occurring elsewhere in the Lower to Upper Kellwasser interval, for example *Crickites*, are lacking. The Frasnian/Famennian boundary cannot be fixed with the help of ammonoids.

Sequence stratigraphic analysis

In Figs 10 and 11 an analysis is given of the Frasnian developments of the southern areas of the Timan and Pechora Basin Frasnian succession in sequence stratigraphic terms with an interpretation of sea-level changes. The lithological and faunal basis for environmental interpretations has been given in earlier sections.

The major initiating transgression surface (TS) below the Dzh'er Formation establishes the setting for the development of the carbonate ramp and associated structures of the Timan reef developments. The age is inferred to correspond with the Taghanic Onlap of the late Givetian. This apparently followed a preceding regressive, lowstand systems tract event. Sea level during Dzh'er times was low, however, and it is the late Dzh'er which best shows the transgressive systems tract (TST).

Sargaevo regional stage

The flooding of the southern Timan–Pechora area by marine rocks represents a more successful transgressive phase which may extend into the Sargaevo regional stage. These two sequences cover the interval from the late Givetian into the earliest part of the Frasnian stage. The Ust'yarega shows deeper marine conditions and a TST (Figs 10, 11) is interpreted here. Terminating this are the calcarenites of the late Ust'yarega Formation above the *Komioceras* Beds. These are interpreted as indicating a brief shallowing event (thin lowstand system wedge, LSW).

Domanik regional stage

A significant deepening initiated the Domanik Formation and the domanikoid anoxic facies begining in the early Domanik in basinal areas east of the reef front. The development of first foundation carbonates and then reefs in late Domanik Formation (Figs 4B, 6) shows that there was progressive deepening with which the reef growth kept pace although, in the area considered, the reefs do not reach the thicknesses they do later. Basinward, where anoxic and hypoxic domanikoid facies are developed, starved basin conditions prevailed (condensed section, CS) in a basin which geometrical considerations from the borehole data suggest must have been 100 to 200 m in depth.

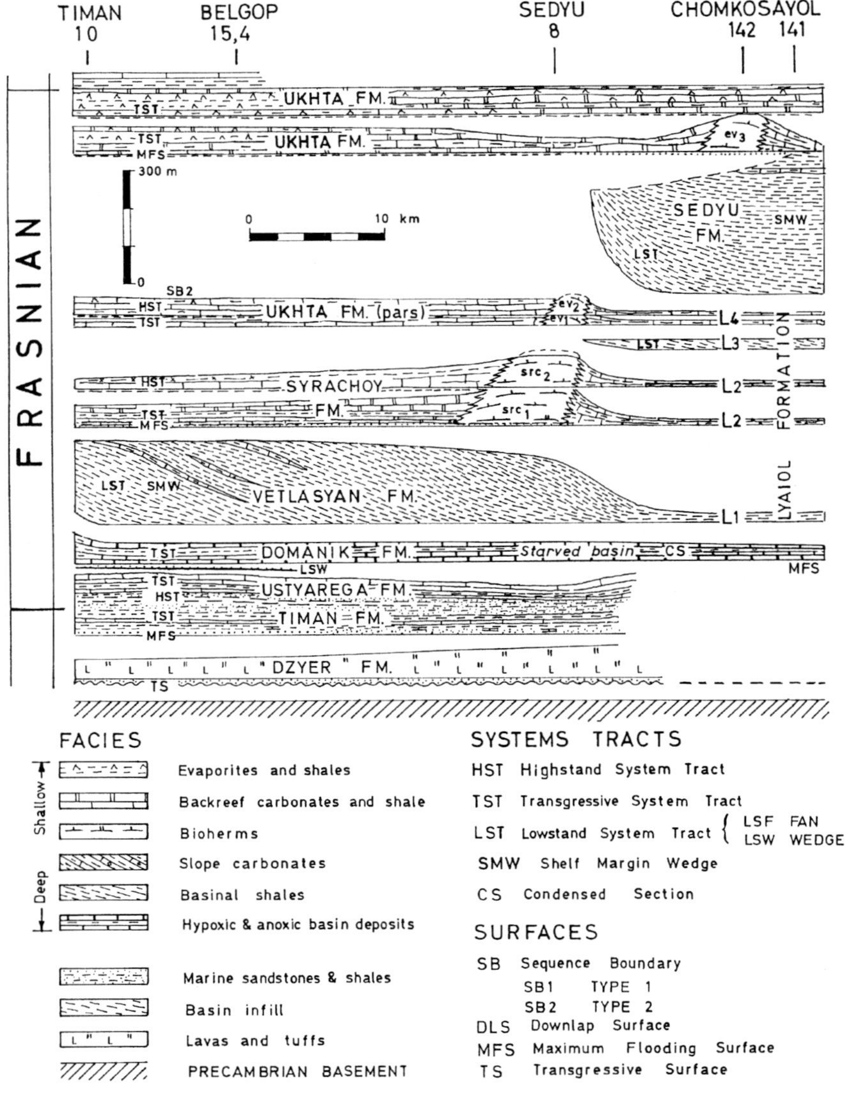

Fig. 10. Sequence stratigraphic interpretation of the reef development in the Frasnian of the southern Timan–Pechora Basin based largely on the cross-sections illustrated in Fig. 4.

The nature of the discontinuation which terminated Domanik reef growth is less easily interpreted. A regressive period is possible which eliminated reef growth and carried littoral areas well to the west. Nevertheless, the succeeding Vetlasyan infill deposits (Figs 4A, 6) correspond to a base level coincident with the top of the former reefs and the Vetlasyan Formation reaches around 200 m in thickness, appropriate to fill the basins seaward of the former Domanik reefs forming a shelf margin sediment wedge (SMW). However, the line of reef front (Fig. 3) prograded significantly.

Vetlasyan regional stage

During a highstand of sea level the accumulation of the Vetlasyan Formation continued. An accumulation terrace, or shelf wedge, was formed which progressively migrated to the east and southeast and basinward. The upper part of the Vetlasyan Formation accumulated in

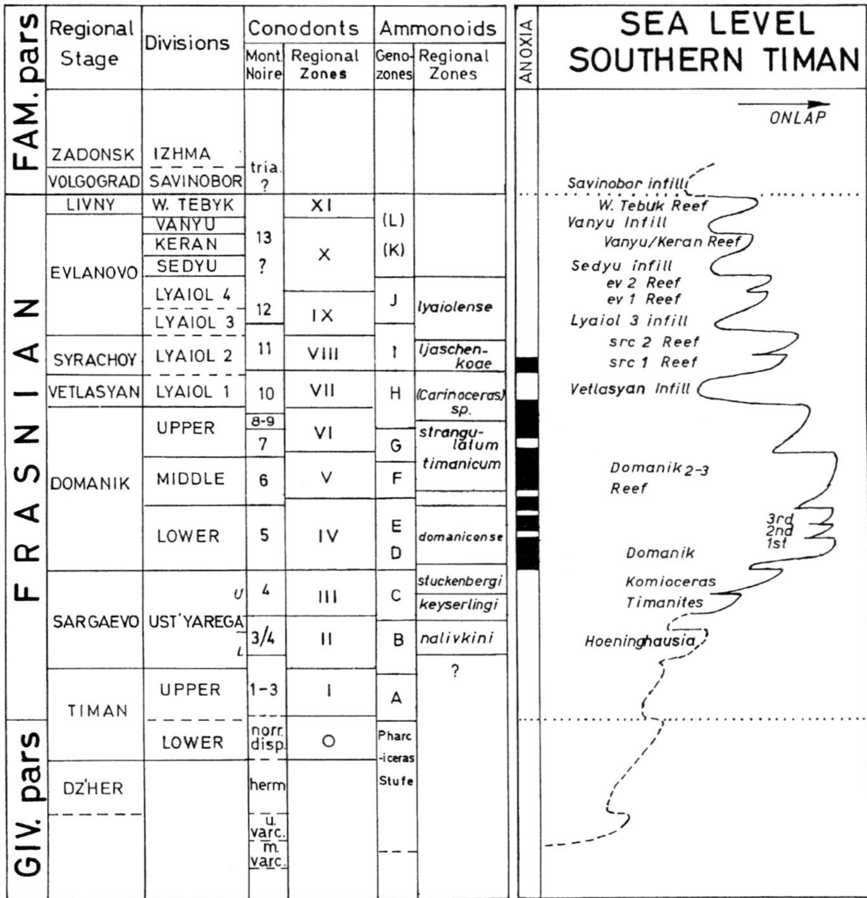

Fig. 11. Interpretation of sea-level changes in the southern Timan–Pechora Basin through the late Middle Devonian (Givetian) and early Upper Devonian (Frasnian and early Famennian).

shallow-water conditions and thinly bedded, very condensed clay sediments were deposited in the deeper-water areas where the environment was hypoxic or perhaps at times anoxic; there are analogies with the domanikoid facies which are represented by Member 1 of the Lyaiol Formation to the east.

Syrachoy regional stage

The Syrachoy rocks may therefore have been deposited on a nearly horizontal surface of earlier sediments. Reef growth was apparently initiated on the seaward margin of the Vetlasyan sea-level progradation delta top so it may be that the lateral equivalents of early Syrachoy deposition in the basin (Member 2 of Lyaiol Formation) inherited some earlier basin topography. Rapid growth of the Syrachoy reefs clearly indicates the initiation of a further TST. By the end of Syrachoy time, this reef mass stood almost 200 m high, and the seaward front appears to have been almost vertical (Figs 4A, 6) in some areas. Borehole data reveal an interruption (between src1 and src2, Figs 4A, 6) indicating a short-term sea-level fall (parasequence boundary).

The terrigenous carbonates of the basal part of the Syrachoy Formation were widely distributed over the former shallow-water shelf and, following a transgression, the basal parts of the new reef systems were initiated along the edge of the former Vetlasyan accumulation terrace (Figs 4, 6). The upper, most calcareous part of the Syrachoy Formation and the lower sulphate-free part of the Ukhta Formation (the lower part of the Evlanovo Formation) correspond to a phase of high sea level, the maximum of which appears to fall in the middle Syrachoy when the thickest limestone beds of Syrachoy Quarry (Yudina & Moskalenko 1994, p. 9) were formed, representing the thickest part of the barrier reef at that time. Laterally to the east, domanikoid facies with goniatite and rhynchonellid limestones characterizes the relatively condensed Member 2 of the Lyaiol Formation.

At the end of Syrachoy time a similar sea-level fall occurred giving a great break in the reef development and marking the base of Evlanovo time. The regressive interval is marked by intercalating red beds. Before that, however, Member 3 of the Lyaiol Formation (Figs 10,11) represents an infill interval (shelf margin fan, SMF) similar to but, in the area with which we are concerned, smaller than that of the Vetlasyan Formation. High sea level was interrupted by a short fall corresponding to the boundary between the Syrachoy and Ukhta Formations. In shallow-water shelf sections this event is marked by beds with red coloration; in the barrier reef massifs there was a short break and a replacement of stromatoporoid–algal bioherms by micrite mounds.

Evlanovo regional stage

The Evlanovo–Livny interval comprises three parts: early Ukhta (early Evlanovo ev_1); middle Ukhta (Sedju, Vanyu infill units, Keran and Vanyu reefs or bioherms, late Evlanovo, ev_2); late Ukhta (Tebuk–Viss or western Tebuk reef massif, latest Evlanovo–Livny, ev_3-liv).

The relative rise of sea level in Syrachoy to early Evlanovo time was less than in Domanik time and this resulted in a smaller area of starved basin and an eastward shift of the barrier reef (Figs 3,4). In early Evlanovo time the basin level was lower than in Syrachoy time and reefs continued along a similar line to those of the Syrachoy. Two transgressive pulses (TST) are recognizable in the early Evlanovo. There was then a major break and the development of the Sedyu Formation as a basin infill shelf margin wedge (SMW) (Figs 10,11). In places this is in excess of 300 m thick providing evidence of the contemporary depth of the early Evlanovo basins seaward from the reef front.

In the middle of Evlanovo time there was a cessation of reef formation and the formation in the fore-reef margin of the starved-depression argillites of the Sedyu Formation. This formed a large accumulation terrace which led to a shift to the east and southeast of the shallow-water shelf. The lower part of the Sedyu Formation may perhaps best be considered to represent a period of high sea level. The upper part of the Sedyu Formation, together with its age equivalents in the form of clay-rich beds and depression zones, can be related to shallow-water deposition.

The clay shelf margin wedge facies of the Sedyu Formation continues in the Vanyu infill sequence (Fig. 4); it is on the edge of the accumulation terrace of this unit that the youngest of the Frasnian reefs of the area developed. The Sedyu and Vanyu Formations are divided by a carbonate member which contains the small Keran and Vanyu biostromes. It is thought that the top of the Sedyu Formation corresponds to a shallowing period followed by the Keran limestones and Vanyu Formation developed as part of a subsequent transgressive phase.

Renewed reef development in the late Evlanovo was initiated on a flooding surface (MFS) formed by the base level of the Sedyu deposits; again the reef proper developed near the margin of the Sedyu wedge suggesting that seaward deposits were below base level and within an eastern Sedyu basin. The upper Evlanovo therefore represents a transgressive event (TST) with total increased depth given by maximum thicknesses of the reef which probably exceeded 200 m.

Livny regional stage

With latest Evlanovo and Livny time the reef front had migrated substantially eastward (Fig. 3) corresponding with a renewed period of transgression. The West Tebuk Barrier reef does not appear to reach the thicknesses of the Syrachoy reefs. Distinctive of Livny time is the development in the area of an extremely extensive lagoonal platform area in which, for the first time, evaporitic rocks are widespread, mostly in the form of gypsum and anhydrite and forming the Sulphate Member of the Ukhta Formation. It seems likely that climatic changes, introducing evaporitic conditions, have causal links with palaeoceanographic changes that led to the approximately synchronous and near-global Upper Kellwasser Event.

Post-Frasnian

The topography resulting from the latest Frasnian depositional relief is infilled by the basal beds of the Savinobor Formation which indicates a significant regression near the Frasnian/Famennian boundary. This terminal Frasnian regression is not particularly well dated, but is indicated in the sequence stratigraphic record by the nature of the infill surface of the succeeding, early Famennian, Savinobor Formation. The Livny contains the *linguiformis* Zone in the subsurface according to Ovnatanova *et al.* (1999*b*) so the Savinobor is probably in the Lower *triangularis* Zone.

International correlation

The standard for comparison of Euramerican Devonian sea-level changes has been the sequence in New York (House 1983; Johnson *et al.* 1985, 1986) and all interpretations have been heavily dependent upon the masterly summaries of the New York facies movements by Rickard (1975, 1981). Later attempts to review the New York sequence by House & Kirchgasser (1993) have provided correlations with international biostratigraphic scales which enable more precise correlation with distant areas. But the sea-level curves then given were local depth curves. Here Frasnian curves for New York are given in the more usual style of interpreted accommodation space (Fig. 12) and the scales enable comparison with the Timan–Pechora area in the light of the refined correlations given in the earlier sections. The European areas have been highly tectonized by Hercynian and other deformations and a similar synthesis has still to be achieved in that area for Frasnian sea-level changes. House (1983, 1985) suggested some eustatic parallels between New York and Europe; the more recent review by Johnson *et al.* (1986) is more precise but based wholly on the older conodont scale. The discussion of parallelisms and contrasts between the Timan-Pechora, Western Europe and eastern North American areas is treated here historically.

Late Givetian

The Taghanic Event represents the first period of palaeogeographic change in the late Givetian and a significant extinction event. The widespread effect of a major transgressive phase of the late Givetian was reviewed by House (1975; at that time defining the base of the Upper Devonian but since revised; see Klapper *et al.* 1987): this transgression had been named the Taghanic Onlap for North America by Johnson (1970). Details of the transgressive event are still largely unstudied but the Tully Limestone of New York (late Middle to Upper *varcus* Zone) is part of the eustatic complex and contains the last

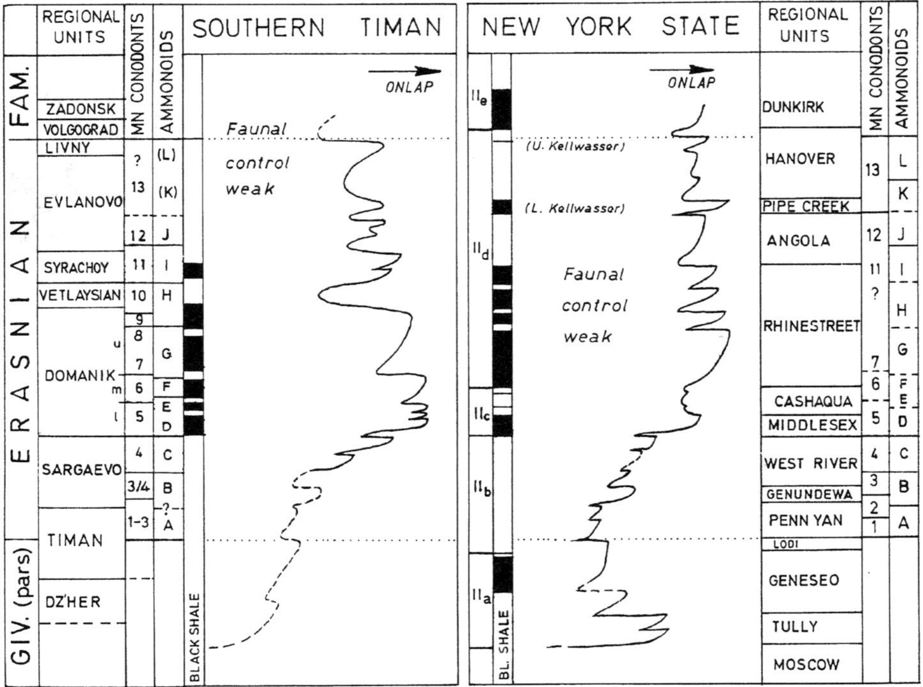

Fig. 12. Comparison of the sea-level curves deduced for the Frasnian of the Southern Timan based on the work reported here compared with that of New York (based largely on House & Kirchgasser 1993).

Devonian reef known in eastern North America. It has been suggested (House 1975) that the major transgressive pulse is recognizable in the Ardennes, with the Fromelennes Formation transgression, and as initiating the Frasnian sequence of the Timan–Pechora region. It is also thought to have initiated the carbonate ramp in the Canning Basin (Becker *et al.* 1993). Because of lack of a precise biostratigraphical tie, this pulse should be considered as tentative in the sea-level curve for the Timan–Pechora area (Fig. 12). Johnson *et al.* (1985) took the Tully Limestone as the start of their international Devonian Depophase IIa.

In New York the black Geneseo Shale, representing a significant anoxic/hypoxic event, follows the Tully Limestone. *Ponticeras* enters in the late Geneseo Shale (House 1961) at an earlier stage than in the Timan–Pechora region, before the thin Lodi Limestone, a level especially rich in the genus. The *norrisi* Zone is recognized at the top of the Lodi Limestone marking the top of the Givetian. The Lodi pulse has not been recognized in the Timan.

Early Frasnian

In New York the base of the Upper Devonian and Frasnian is taken at the regression level which initiates the Penn Yan Formation shales and silts (Kirchgasser 1996). From an evolutionary point of view, there is a break at this level characterized by the almost complete extinction of the Pharciceratidae and this was named by House (1985) the Frasnes Event, since it was thought to correspond with the base of the Assise de Frasnes and base of the Lower *asymmetrica* Zone. It has also been named as the *Manticoceras* Event (Walliser 1996) but that is quite incorrect since *Manticoceras* does not enter until very much later (in MN 3 or upper UD I-B). Since the record of pharciceratids is non-existent in the Timan and Belgium and the only representatives in New York are early simple-sutured forms within the Tully Formation (House 1961), this boundary event cannot be recognized with precision.

The New York Penn Yan Shale is a sequence of infilling shales and silts which introduces the genus *Koenenites* in the later part. The Penn Yan Shale is terminated by the Genundewa Limestone, a deeper-water facies interpreted as a transgressive pulse (House & Kirchgasser 1993); the Lower Genundewa Limestone is referred to Montagne Noire conodont zone MN 2 (Kirchgasser 1996) and UD I-B whilst the upper part belongs to MN 3. Below the Bluff Point Siltstone, material intermediate between *Koenenites* and *Hoeninghausia* occurs which serves to suggest that this interval must correlate with the entry of *Hoeninghausia* in the middle part of the Ust'-yarega in the Timan and is assigned to UD I-B. For these reasons the Genundewa/West River pulse is thought to be the equivalent of the early Middle Ust'yarega transgressive pulse which introduced goniatites into the Ukhta area (Fig. 12).

In the Timan the two pulses of first *Timanites*, and next *Komioceras*, do not find a faunal equivalent in New York in the West River Shale. But *Timanites* is known in the Maligne Formation of western Canada and in Australia in levels which seem to represent deepening pulses which are dated as UD I-C and MN Zone 4. This Timan Event (Becker & House 1997) is especially well developed as a widespread black styliolinite in southern Morocco (part of the 'Lower Kellwasser Member' of Wendt & Belka (1991) and Becker *et al.* (1997)).

The Middlesex Shale is a major transgressive tongue of black shale in the New York sequence; it contains *Sandbergeroceras* and so is referred to goniatite Zone UD I-D but is currently lacking critical conodonts (detailed work by J. Over, of Geneseo, New York, is in progress). MN 4 commences in the upper part of the West River and MN 5 can be recognized in the lower Cashaqua Shale. It seems reasonable to relate the transgressive pulse of the Middlesex which defines the start of the depophase IIc (Johnson *et al.* 1985) with that of the base of the Domanik Formation which (Figs 7–9) is dated as MN 5. On the Ardennes shelf of Belgium, Frasnian reefs seem to have grown during slowly rising relative sea level following regressive episodes which allowed the spread (progradation) of biostromal shallow-water carbonates. The eventual drowning of the earliest major (F2d) Frasnian reefs of the area (Carrière de l'Arche) falls in the local *Ancyrodella gigas* Zone (Vandelaer *et al.* 1989) which can be correlated with MN zones 5 and 6. Whilst the Domanik deepening appears to be recognizable in the southern Dinant Syncline reef complexes, it occurred on a different shelf topography and deeper-water setting. Therefore in that area only local mud mounds formed rather than a broad carbonate platform.

Mid-Frasnian

In New York the Cashaqua Shale succeeds the black Middlesex Shale and the shallowest facies appears to be developed in about the middle of the unit (House & Kirchgasser 1993). The onset of the black and very thick Rhinestreet Shale above is very sudden. For goniatites the entry of

Probeloceras, early, and *Prochorites*, later in the Cashaqua mark UD I-E and UD I-F respectively (Kirchgasser 1975; House & Kirchgasser 1993). UD I-G, characterized by *Naplesites*, is thought to lie around the Cashaqua/Rhinestreet boundary but the famous faunas of Clarke have not been located in place. The subsequent Rhinestreet Shale must represent a considerable period of time and it marks the thickest Devonian hypoxic shale of New York, and perhaps anywhere; it spans conodont zones from upper MN 6, with zonally diagnostic faunas of MN 6 and MN 7 in the lower Rhinestreet and an abundant fauna of MN Zone 11 at Relyea Creek near the top of the Rhinestreet. In general, this period encompasses the greatest extension of marine transgression in the Devonian (House 1983). As with the Middlesex black shale, however, the conodont record of the middle part of the Rhinestreet Shale is largely unsampled. Goniatite faunas within the Rhinestreet show no evidence of the beloceratids which characterize UD I-G/H and the later Frasnian but contain facies-restricted multilobed Triainoceratidae; these are mostly associated with several transgressive pulses (Sutton 1963) but precise dating is not available. Near the top of the Rhinestreet oxyconic *Carinoceras* enters in beds referred to UD I-I (House & Kirchgasser 1993).

The span of the Domanik Formation, from MN 5 to MN 10, covers a substantial part of this Rhinestreet anoxic interval of New York and is similarly a spectacular development of anoxic facies. The basal Rhinestreet transgression seems to correlate approximately with the Upper Domanik deepening (Figs 11,12). A major eustatic transgression (perhaps with a primary control in mid-ocean ridge activity) may have flooded critical epicontinental areas with nutrient-rich cold waters; this may have occurred in relation to the unusual ocean circulation of this time when the continental masses occupied one hemisphere and the other would have been mostly ocean. An additional influence from extremely high climatic and greenhouse temperatures seems likely.

For areas of Western Europe, biostratigraphic knowledge of this interval and documentation of facies movements is still rather vague. Although the Belgian F2h reefs date to this time (Sandberg *et al.* 1992), a transgression northward across the Old Red Sandstone is recognized here and in England. A similar transgression in the Canning Basin led not to anoxia, but to the introduction of oxic waters (Becker *et al.* 1993). In southern Morocco (Becker *et al.* 1997), the eustatic equivalent of the Rhinestreet deepening led to the deposition of nodular limestones, rich in gradually evolving beloceratids, above an unconformity or extreme condensation level with conglomerates.

At the end of the Domanik, the regression of the Vetlaysian, and the transgressive pulses giving the Syrachoy reefs (src_1 and src_2 of Fig. 4A) are approximately equivalent to the later Rhinestreet pulses recognized by Sutton (1963) but correlation is not sufficiently resolved to confirm this. As Lyaiol Member 2 and the Syrachoy Formation are referred to MN 11, this corresponds to the *Pa. semichatovae* transgression noted by Ziegler & Sandberg (1990) which caused the drowning of the Lion Mudmound in the Dinant Syncline. In NW Australia a significant transgression at the same time (MN 11) has been noted by Becker & House (1995, 1997). Depophase IId of Johnson *et al.* (1985) has been taken to be this eustatic pulse, but was first applied to the basal Rhinestreet pulse.

Most, if not all, of Lyaiol Member 4 is referred to MN Zone 12; the subsequent Sedyu infill, therefore, could correspond to a world-wide shallowing before the Lower Kellwasser Event but the possible record of a MN 13 index conodont in the upper Lyaiol 4 (Ovnatanova *et al.* 1999*a*,*b*) needs to be resolved before clear correlation can be made. Alternatively, the minor regression between Evlanovo ev_1 and ev_2 (Fig. 4A) may represent a regional expression of the pre-Lower Kellwasser eustatic sea-level fall. In New York the latter is shown also by the flood of clastics entering in the upper part of the Angola Shale. Piecha (1993) documented a corresponding important and widespread condensed marker unit from just below the Lower Kellwasser level of basinal sections in the Rhenish Slate Mountains. A shallowing episode and retreat of ammonoid biofacies is also indicated in the Canning Basin.

Late Frasnian

Taking the base of the Lower Kellwasser Limestone as a convenient marker, it has been suggested that this is represented in New York by the thin black shale unit of the Pipe Creek Shale (Johnson *et al.* 1986; House & Kirchgasser 1993). The Lower Kellwasser Limestone is referred to the base of goniatite zone UD I-K (Becker & House 1994*b*) and the whole of the subsequent late Frasnian belongs to MN 13, but an upper division with *Palmatolepis linguiformis* and *Ancyrognathus ubiquitus* is distinguishable. As with the other black shale tongues in the Upper Devonian of New York, the Pipe Creek Shale is interpreted as a transgressive unit (House & Kirchgasser 1993). Whilst both black limestone

Kellwasser horizons are transgressive and probably associated with climatic events that add to give their distinctive anoxic attributes (Becker & House 1994b), they are also limited by sharp regressive events. In the Timan the lower Kellwasser Event may correspond to the latest Evlanovo and the beginning of the deepening associated with the formation of the Tebuk–Viss reef massif.

The Belgian Frasnian 2j reefs of Belgium have been correlated with this transgression (Johnson *et al.* 1986) and include the spectacular mud mound at Beauchateau (Boulvain & Coen-Aubert 1992) which displays a shallowing episode within and subsequent drowning at about Lower Kellwasser time. According to Professor P. Bultynck (pers. comm.) the reddish mud mounds of the southern part of the Dinant Synclinorium are now referred more precisely to the Neuville Formation and to the Neuville and/or Valisettes Formations in the Philippeville Anticlinorium (Boulvain *et al.* 1999; Bultynck *et al.* 1998) which cover a broad time range. In this time span are the various terminal Frasnian reefs of the Timan–Pechora area (Vanyu and Keran reefs). Alternatively the equivalent would be the last Evlanovo reef (ev_2); the West Tebuk reefs seem to correspond to a later phase and are among the youngest Frasnian reefs of Euramerica. However, both conodont and goniatite resolution need to be improved to understand this more fully. Apart from latest Frasnian (with *P. linguiformis*) Domanik-type carbonates in the Tebuk–Viss region, there is no representation of an anoxic Upper Kellwasser level in the Timan–Pechora region: in this it corresponds with Western Australia (Becker *et al.* 1991).

Frasnian/Famennian boundary

The boundary between the end of MN 13 and the Lower *triangularis* Zone is well marked in all pelagic sequences. Boundary conodonts and goniatites are well constrained. In New York the base of the Famennian lies close to the top of the Hanover Shale (Over 1997*a*,*b*), at some distance above the last occurrence of *Crickites rickardi*; but the crickitids are not known at all in the Timan region and critical conodont faunas of the boundary interval have only been obtained from subsurface basinal facies (Klapper *et al.* 1996; Kuz'min *et al.* 1998; Ovnatanova *et al.* 1999*b*). As in other regions, the Frasnian/Famennian boundary seems to coincide in the Timan with a marked sequence boundary. There was a major and sharp regression which appears to have terminated the West Tebuk reefs and the Domanik-type sedimentation in the adjacent basin (Kuz'min *et al.* 1998).

Conclusions

This attempt to apply high resolution biostratigraphic and sequence stratigraphic methodology, with emphasis on conodonts and goniatites, to the famous Frasnian deposits in the Timan–Pechora region has resulted in a number of general conclusions.

1. It is clear that eustatic influences can be elucidated in the succession, and that some of the transgressive pulses identified in the sea-level curve produced here for the Timan–Pechora area show evidence of global effects on the broad scale. The interpretation favoured is illustrated in the compilation in Fig. 12.
2. There are distinctive regional attributes. Both the conodont and goniatite records have unusual characteristics with some endemism, and more work remains to document this. The major global transgression of the Taghanic Onlap, whilst it is thought to have initiated the Timan Frasnian sequence, is not precisely placed and the conodont/goniatite record, as a result, begins rather higher than in most other areas. The reef progradation (Fig. 4) through the Frasnian is somewhat distinctive, although it has an analogy in the westward progradation of clastics through the Frasnian in New York.
3. The many reef episodes during the Frasnian of the Timan–Pechora area were established during transgressive phases and are thought to have been terminated by shallowing events. This development is in contrast with many reefs of SW England, Belgium and Germany where growth terminated during tectonically or eustatically controlled drowning episodes in association with eutrophication. Timan reefs were not covered by basinal shales or cephalopod limestone caps with pelagic faunas.
4. The Timan–Pechora area is distinctive in the well developed reef complexes in the Late Frasnian where a complex history has to be further elucidated and dated. Other European regions are characterized by reef complexes which were mostly lost during earlier Frasnian transgressive peaks.
5. The formation of petroleum source rocks, typified by the Domanik facies, was related to the formation of rather deep (about 200 m) basins developed offshore and east of the reef fronts when these were mostly starved basins.

These have close similarities with the black shale developments of New York (especially the Middlesex, Rhinestreet and Pipe Creek Shales), but have their own distinctiveness in the abundant Radiolaria and the characteristic well bedded calcisilicate lithology. Environmental similarities are found with pelagic facies of the Lower and Upper Kellwasser Limestones where black limestones represent another anoxic facies.

6. The Domanik facies, as with the other anoxic facies, because of the general contemporaneity established, must reflect global causations. The possibility seems likely that the curious Devonian palaeogeography, with all continents virtually in one hemisphere, may have resulted periodically in cold, nutrient-rich waters flooding epicontinental areas in conjunction with upwelling systems. An association with high temperatures and a greenhouse-type regime leading to underflowing heavy saline bottom-waters may have led to particularly high plankton blooms resulting in eutrophication. In addition, the sea-floor anoxia may have been enhanced by general poor bottom circulation resulting from transgression over well planated areas.
7. The introduction of evaporitic facies in the terminal Frasnian seems likely to have been connected with the global climatic change that also introduced the Kellwasser Event which is the most spectacular mass extinction of the Devonian. Attention is drawn to how at least eight other extinction events in the Devonian (House 1985; Becker & House 1994*a*,*b*) are also associated with anoxic events and transgression–regression couplets. No discussion of any one of these events can ignore the evidence of the others and most share many common attributes.
8. The Timan–Pechora region, which has been a classic area for Frasnian faunas for over one and a half centuries, is shown to be an area of exemplary interest for reef carbonate and anoxic facies development. Also, the considerable evidence available from boreholes and outcrops should enable progress to be made in the quantification of the sea-level changes which have been documented here.

Acknowledgement is mainly due to support from INTAS Grant 93–750 which funded the work with M. R. H. as project coordinator: this covered the field and laboratory work of Russian collaborators and a visit to Moscow by M. R. H. and R. T. B.; these funds also supported a field excursion to Frasnian reef developments in Germany, France and Belgium which Professor P. Bultynck and Dr M. Coen-Aubert kindly organized. A visit to the Timan for field work by M. R. H. was funded by the Royal Society of London and a similar visit by R. T. B. was funded by the Deutsche Forschungsgemeinschaft. Particular thanks are due to the University of Southampton, especially J. Maha, for handling administrative and financial matters. Dr S. V. Yatskov gave considerable help with goniatite collections he had made in areas covered by this review and gave helpful comments in our discussions. We are especially indebted to Drs Yu. A. Yudina and M. N. Moskalenko who led an introductory field trip to the Ukhta region, and to Dr Bogatsky, Director of the Timan–Pechora Department of VNIGRI, for his help and that of his colleagues. We are all appreciative of the support from the directors and heads of our various institutes. Finally we wish to express gratitute to Professor M. A. Rzhonsnitskaya whose support for western and Russian scientists wishing to work in the Timan–Pechora area has been long continued and much appreciated.

References

AVCHIMOVITCH, V. I., TCHIBRIKOVA, E. V., OBUKHOVSKAYA, T. G., NAZARENKO A. M., UMNOVA, V. T., RASKATOVA, L. G., MANTSUROVA, V. N., LOBOZIAK, S. & STREEL, M. 1993. Middle and Upper Devonian miospore zonation of Eastern Europe. *Exploration et Production Elf-Aquitaine*, **17**(1), 79–147.

BECKER, R. T. 1992. Zur Kentniss von Hemberg-Stufe und *Annulata*-Schiefer im Nord-Sauerland (Oberdevon, Rheinisches Schiefergebirge, GK 4611 Hohenlimburg). *Berliner geowissenschaftliche Abhandlungen*, **E, 3**, 3–41.

—— & HOUSE, M. R. 1994*a*. International Devonian goniatite zonation, Emsian to Givetian, with new records from Morocco. *Courier Forschungsinstitut Senckenberg*, **169**, 79–135.

—— & —— 1994*b*. Kellwasser Events and goniatite successions in the Devonian of the Montagne Noire with comments on possible causations. *Courier Forschungsinstitut Senckenberg*, **169**, 45–77.

—— & —— 1995. High-resolution ammonoid biostratigraphy and the timing of facies shifts in the Upper Devonian of the Canning Basin. *First Australian Conodont Symposium (AUSCOS-1), Boucot Symposium, Abstracts and Programme.* MUCEP Special Publication, **1**, 12–13.

—— & —— 1997. Sea-level changes in the Upper Devonian of the Canning Basin, Western Australia. *Courier Forschungsinstitut Senckenberg*, **199**, 129–146.

——, —— & KIRCHGASSER, W. T. & PLAYFORD, P. E. 1991. Sedimentary and faunal changes across the Frasnian/Famennian boundary in the Canning Basin of Western Australia. *Historical Biology*, **5**, 183–196.

——, —— & —— 1993. Devonian goniatite biostratigraphy and timing of facies movements in the Frasnian of the Canning Basin, Western Australia. *In*: HAILWOOD, E. A. & KIDD, R. B. (eds) *High Resolution Stratigraphy*, Geological Society, London, Special Publication, **70**, 293–321.

——, —— & Klapper, G.– 1997. Event-sequence and sea level-changes in the Upper Givetian and Frasnian of the Tafilalt (Southern Morocco). *The Amadeus Grabau Symposium, International Meeting on Cyclicity and Bioevents in the Devonian System, Programme and Abstracts, University of Rochester, July 20–27, 1997*. 18.

——, ——, Menner, V. V., & Ovnatanova, N. S. 2000. Revision of ammonoid biostratigraphy in the Frasnian (Upper Devonian) of the Southern Timan (Northeastern Russian Platform). *Acta Geologica Polonica*, **50**, 67–97.

Bogoslovskiy, B. I. 1969. *Devonian ammonoids, I. Agoniatitida*. Trudy Paleontologicheskogo Instituta, AN SSSR, **124** (in Russian).

—— 1971. *Devonian ammonoids. II. Goniatitida*. Trudy Paleontologicheskogo Instituta, AN SSSR, **127** (in Russian).

Boulvain, F. & Coen-Aubert, M. 1992. La carrière de marbre rouge de Beauchateau: aperçu paléontologique, stratigraphique et sedimentologique. *Annales de la Societé Géologique de Belgique*, **115**, 19–22.

Boulvain, P., Bultynck, P., Coen, M., Coen-Aubert, M., Lacroix, D., Laloux, M., Casier, G.-J., Dejonghe, L., Dumoulin, V., Ghysel, P., Godefroid, J., Helsen, S., Mouravieff, M. A., Sartenaer, P., Tournier, F. & Vanguistaine, M. 1999. Les Formations du Frasnien de la Belgique. *Memoirs of the Geological Survey of Belgium*, **44–1999**, 1–125.

Bultynck, P., Helsen, S. & Hayduckiewich, J. 1998. Conodont succession and biofacies in upper Frasnian formations (Devonian) from the southern and central parts of the Dinant Synclinorium (Belgium) – (Timing of facies shifting and correlation with late Frasnian events). *Bulletin de l'Institut royal des Sciences naturelles de Belgique, Sciences de la Terre*, **68**, 25–75.

Fokin, N. A. 1975. Ostracods of the Lyayel suite in the Frasnian of the South Timan. *In*: *Geologia i poleznye iskopaemye Timano-Pechorskoy provinzii*, **3**, *Syktyvkar* (in Russian).

—— 1977. Ostracods and some problems of the stratigraphy of the Upper Frasnian substage Timan-Petchora province. *In*: *Geologia i neftegazonosnost severo-vostoka evropejskoj tchasti SSSR*, **7**, *Syktyvkar*, 107–115 (In Russian).

Holzapfel, E. 1899. *Die Cephalopodenkalke das Domanik im südlichen Timan*. Mémoires du Comité Géologique, St Petersburg, **10**.

House, M. R. 1961. Observations on the ammonoid succession of the North American Devonian. *Journal of Paleontology*, **36**, 247–284.

—— 1975. Facies and time in Devonian tropical areas. *Proceedings of the Yorkshire Geological Society*, **40**, 233–288.

—— 1983. Devonian eustatic events. *Proceedings of the Ussher Society*, **5**, 396–405.

—— 1985. Correlation of mid-Palaeozoic ammonoid evolutionary events with global sedimentary perturbations. *Nature*, **313**, 17–22.

—— 1996. The Middle Devonian Kačák Event. *Proceedings of the Ussher Society*, **9**, 79–84.

—— & Kirchgasser, W. T. 1993. Devonian goniatite biostratigraphy and timing of facies movements in the Frasnian of eastern North America. *In*: Hailwood, E. A. & Kidd, R. B. (eds) *High Resolution Stratigraphy*. Geological Society, London, Special Publications, **70**, 267–292.

Johnson, J. G. 1970. Taghanic Onlap and the end of North American Devonian provinciality. *Geological Society of America Bulletin*, **81**, 2077–2106.

——, Klapper, G. & Sandberg, C. A. 1985. Devonian eustatic fluctuations in Euramerica. *Geological Society of America Bulletin*, **96**, 567–587.

——, —— & —— 1986. Late Devonian eustatic cycles around margin of Old Red Continent. *Annales de la Société géologique de Belgique*, **103**, 141–147.

Keyserling, K. 1844. Beschreibung einiger Goniatiten aus dem Domanik-Schiefer. *Verhandlungen der Kaiserlichen-russischen mineralogischen Gesellschaft*, **1844**, 217–238.

—— 1846. *Wissenschaftliche Beobachtungen auf einer Reise in das Petschora-Land im Jahre 1843*. St Petersburg.

Khalymbadzha, V. G. 1981. *Upper Devonian conodonts of the East of Russian Platform, South Timan, Polar Ural and their stratigraphic significance*. Kazan, Izdatestvo Kazanskogo Universiteta (in Russian).

Kirchgasser, W. T. 1975. Revision of *Probeloceras* Clarke, 1898 and related ammonoids from the Upper Devonian of western New York. *Journal of Paleontology*, **49**, 58–90.

—— 1996. Evidence of distant transport (basinward) of reworked conodonts in a condensed sequence in the Upper Devonian (lower Frasnian) of western New York. *30th International Geological Congress, Beijing, China, 4–14 August 1996, Abstract Volume* **2**, 86.

——, Brett, C. E. & Baird, G. C. 1997. Sequences, cycles and events in the Devonian of New York State: an update and overview. *In*: Brett, C. E. & Ver Straeten, C. A. (eds) *Devonian cyclicity and sequence stratigraphy in New York State – Field Trip Guidebook*. The University of Rochester, New York, 5–21.

Klapper, G. 1989. The Montagne Noire Frasnian (Upper Devonian) conodont succession. *Canadian Society of Petroleum Geologists, Memoir*, **14**(3), 440–468.

—— 1997. Graphic correlation of Frasnian (Upper Devonian) sequences in Montagne Noire, France, and western Canada. *In*: Klapper, G., Murphy, M. A., & Talent, J. A. (eds) *Paleozoic Sequence Stratigraphy, Biostratigraphy, and Biogeography: Studies in Honor of J. Granville ('Jess') Johnson*. Geological Society of America, Special Paper, **321**, 113–129.

—— & Becker, R. T. 1999. Comparison of Frasnian (Upper Devonian) conodont zonations. *Bollettino della Società Paleontologica Italiana*, **37**(2–3), 339–348.

—— & Foster, C. T., Jr. 1993. Shape analysis of Frasnian species of the Late Devonian conodont genus *Palmatolepis*. *Paleontological Society Memoir*, **32** (*Journal of Palaeontology*, **67**(4), supplement).

——, Feist, R. & House, M. R. 1987. Decision on the

boundary stratotype for the Middle/Upper Devonian Series boundary. *Episodes*, **10**(2), 97–101.

——, KUZ'MIN, A. V. & OVNATANOVA, N. S. 1996. Upper Devonian conodonts from the Timan–Pechora region, Russia, and correlation with a Frasnian composite standard. *Journal of Paleontology*, **70**(1), 131–152.

KUSHNAREVA, T. I. 1959. The stratigraphy, lithology and oil of Devonian deposits. In: *The Geology and Oil of the Timan–Pechora region*. Trudy VNIGRI, Leningrad, **133**, 81–93 (in Russian).

—— & MATVIYEVSKAYA, N. D. 1966. Reef structure of the Pechora depression and their oil-gas prospects. *Geologiya Nefti i Gasa*, **8**, 30–33 (in Russian).

—— & —— 1973. Reef structure of the Pechora depression and their oil-gas prospects. *Petroleum Geology*, **10**, 349–352.

—— & RASKATOVA, L. G., 1980. Palynologic characteristics of the Lyaiol Formation of South Timan. *Doklady Akademii Nauk SSSR*, **253**(6), 1423–1428 (in Russian).

——, BUSIGINA, Y. N., FOKIN, N. A., CHALYMBADSHA, V. S. & YUDIN, V. S. 1979. New data on the stratotype section of the Lyaiol Suite, South Timan. *Izvestija Akademii Nauk SSSR, Seriya Geologitcheskaja*, **1974**(10), 133–139 (in Russian).

——, KHALAMBADZHA, V. G. & BUSIGNIA, Y. N. 1978. Biostratigraphic zonation of the Domanik Formation in a section of the stratotype. *Sovetskaya Geologiya*, **1**, 60–71 (in Russian).

KUZ'MIN, A. V. 1997. Aspects of the Frasnian conodont stratigraphy of the Timan–Pechora Province. *Ichthyolith Issues*, Special Publication, **21**, 22.

—— 1998. New species of Early Frasnian *Palmatolepis* (Conodonta) from southern Timan. *Paleontologischeskiy Zhurnal,* **1998** (2), 70–76 (in Russian).

—— & YATSKOV, S. V. 1997. Transgressive–regressive events and conodont and ammonoid assemblages in the Frasnian of the South Timan. *Courier Forschungsinstitut Senckenberg*, **199**, 25–36.

——, SHUVALOVA, G. A., OBUKHOVSKAIA, T. G., AVKHIMOVICH, V. I., YUDINA, Y. A. & MOSKALENKO, M. N. 1998. Frasnian/Famennian boundary in the Izhma-Pechora depression. *Bulleten Moskovskogo Obschestva Ispitateley Prirody, Odtel geologiya*, **73**(4), 27–38 (in Russian).

——, YATSKOV, S. V., ORLOV, A. N. & IVANOV, A. O. 1997. 'Domanik crisis' in the development of the fauna in the Frasnian marine basin of southern Timan. *Paleontologicheskiy Zhurnal*, **31**(3), 3–9 (in Russian).

LIKHAREV, B. K. 1931. Geological research in South Timan. Moscow – Leningrad. *Trudy Vsesoyusnogo geologorasvedotchnogo ob'edineniya, VSNCH SSSR*, **150**, 3–42 (in Russian).

LYASHENKO, A. I., 1956. Biostratigraphy of the Devonian deposits of the South Timan. *In*: *Questions of Stratigraphy, paleontology and lithology of the Paleozoic and Mesozoic regions of the European parts of the USSR. Moskow, Nauka, Trudy VNIGNI*, **7**, 4–31 (in Russian).

—— 1959. *Atlas of Brachiopods and Devonian Stratigraphy of the Russian Platform*. Moskva Gostoptehizdat (in Russian).

—— 1969. New species of the Lower Frasnian brachiopods of South Timan. *Trudy VNIGNI, Moskva Izdatelstvo Nedra*, **93**, 49–58 (in Russian).

—— 1973. Brachiopods and stratigraphy of the Lower Frasnian deposits of the South Timan and Volga-Ural oil-bearing Province. *Trudy VINGNI, Moskva Izdatelstvo Nedra*, **134** (in Russian).

—— 1985. New Upper Devonian brachiopods of South Timan. *In*: IL'IN, V. D. & LIPATOVA, V. V. (eds) *Stratigraficheskie Issledovaniya Prirodnyh Rezervyarov Nefti i Gaza. Moskva. VNIGRI*, 9–18.

LYASHENKO, G. P. 1957*a*. New species of Devonian goniatites. *Trudy VNIGNI,* **8**, 192–211 (in Russian).

—— 1957*b*. Novoye vidy Devonskikh goniatitov. *Trudy VNIGNI*, **6**, 192–211.

—— 1959. *Coniconchines from the Devonian of the central and northern districts of the Russian Platform*. VNIGNI (in Russian).

MAXIMOVA, S. V. 1970. *Ecological and facial peculiarities and conditions of formation of the Domanik Formation*. Nauka, Moscow (in Russian).

MEDYANIK, S. I. 1981. Palynological characteristics of the Frasnian sediments of the Timan–Pechora province. *Autoreferens Doctor work*. Moscow (in Russian).

—— & YATSKEVICH, B. A. 1981. Boundary of the Kynov and Sargaev horizons of the southern and central Timan using palynological characteristics. Moscow, *Izvestiya Akademii Nauk SSSR, Serija Geologicheskaja*, **8**, 132–136 (in Russian).

MENNER, V. V., ARKHANGELSKAYA, A. D., KUZ'MIN, A. V., MOSKALENKO, M. N., OBUCHOVSKAYA, T. G., OVNATANOVA, N. S., SHUVALOVA, G. A. & YATSKOV, S. V. 1992. Correlation of different facies sections of the Frasnian Stage in South Timan. *Bulletin MOIP, otdelenie geologii, Moskva*, **67**(6), 64–82 (in Russian).

MILLER, A. K. 1938. *Devonian Ammonoids of America*. Geological Society of America, Special Papers, **14**.

NALIVKIN, D. V. 1936. Srednepaleozoiskie fayn' bekhovyev Kolymy i Khandyti. *Materialy po Izycheniiu Okhotsko-Kolymskogo kraya, Seriya 1*, **4**, 1–28.

NEFIODOVA, M. I. 1955. The Devonian Brachiopods of Troizko-Pechora and Pechora areas. *Trudy VNIGRI, Novaya Seriya*, **88**, 419–455 (in Russian).

OVER, J. 1997*a*. Frasnian-Famennian boundary in Western New York: Dunkirk Beach and Buffalo Creek Valley. *In*: BRETT, C. E. & VER STRAETEN, A. (eds) *Devonian Cyclicity and Sequence Stratigraphy in New York State, Field Trip Guidebook*, 147–154.

—— 1997*b*. Conodont biostratigraphy of the Java Formation (Upper Devonian) and the Frasnian/Famennian boundary in western New York State. *In*: KLAPPER, G., MURPHY, M. A. & TALENT, J. A. (eds) *Paleozoic Sequence Stratigraphy, Biostratigraphy and Biogeography: Studies in Honor of J. Granville ('Jess') Johnson*. Geological Society of America, Special Paper, **321**, 161–177.

OVNATANOVA, N. S. 1976. New Late Devonian conodonts of the Russian Platform. *Paleontolocheskiy Zhurnal*, **10**(1), 210–219 (in Russian).

—— & KONONOVA, L. I. 1984. Korrelyatsiya verhknedevonskonizh neturneyskikh otlozheniy europeyskoy chasti SSSR po Konodontam. *Sovetskaya Geologiya*, **1984** (8), 32–42 (in Russian).

—— & KUZ'MIN, A. V. 1991. Conodonts of typical sections of the Domanik Formation in South Timan. *Izvestiya Akademii Nauk, SSSR, Seriya Geologicheskaya*, **3**, 37–50 (in Russian).

——, —— & MENNER, V. V. 1999*a*. The succession of conodont complexes in the type sections of the southern Timan–Pechora province. *Geologia i Mineral'nye Resursy Evropeiskogo Severo-Vostoka Rossiy*, **2**, 282–284 (in Russian).

——, —— & —— 1999*b*. The succession of Frasnian conodont assemblages in the type sections of the southern Timan–Pechora Province (Russia). *Bollettino della Società Paleontologica Italiana*, **37**(2–3), 349–360.

PARRISH, J. T. 1982. Upwelling and petroleum source beds, with reference to Paleozoic. *AAPG Bulletin*, **66**, 750–774.

PEDERSEN, T. F. & CALVERT, S. E. 1990. Anoxia vs. productivity: what controls the formation of organic-rich-sediments in sedimentary rocks. *AAPG Bulletin*, **74**, 454–466.

PIECHA, M. 1993. Stratigraphie, Fazies und Sedimentpetrographie der rhythmisch und zyklisch abgelagerten, tiefoberdevonisches Bechensedimente im Rechtrheinischen Schiefergebirge (Adorf-Bänderschiefer). *Courier Forschungsinstitut Senckenberg*, **163**, 1–151.

RASKATOVA, L. G. 1969. *Spores Complexes of Middle and Upper Devonian of South-east Part of the Central Devonian Field*. Voronezh (in Russian).

RICKARD, L. V. 1975. *Correlation of the Devonian Rocks in New York State*. New York State Museum Series, **4**.

—— 1981. The Devonian System in New York. *In*: OLIVER, W. A. & KLAPPER, G. (eds) *Devonian Biostratigraphy of New York, Part 1*. International Union of Geological Sciences, Subcommission on Devonian Stratigraphy, 5–22.

RUDWICK, M. J. S. 1985. *The Great Devonian Controversy: The shaping of scientific knowledge among Gentlemanly Specialists*. The University of Chicago Press, Chicago.

SANDBERG, C. A., ZIEGLER, W., DREESEN, R. & BUTLER, J. L. 1992. Conodont biochronology, biofacies and Event stratigraphy around Middle Frasnian Lion mudmound (F2h), Frasnes, Belgium. *Courier Forschungsinstitut Senckenberg*, **150**, 1–87.

SCHWARZKOPF, T. A. 1993. Model for prediction of organic carbon content in possible source rocks. *Marine and Petroleum Geology*, **1993**(10), 478–492.

Stratigraphic Dictionary of the USSR. Cambrian, Ordovician, Silurian, Devonian. 1975. Leningrad, Nedra, Leningradskoe otdelenie (in Russian).

SUTTON, J. I. 1963. Stratigraphy of the Naples Group (Late Devonian) in western New York. *New York State Science Service, Bulletin*, **380**.

TIKHONOVICH, N. N. 1930. Geology of Ukhta oil-bearing region. *Neftyanoe hozyaistvo*, **1930**, 8–9 (in Russian).

—— 1941. Geological structure of the Timan-Ural oil-bearing Province. *Sovietskaya Geologiya*, **1**, 43–60 (in Russian).

TSCHERNYSHEV, F. N. 1884. Materials on research into the Devonian deposits of Russia. *Trudy Geologicheskogo Komity*, **1**(3) (in Russian).

—— 1890. Timan works in 1889 (preliminary report). St Petersburg. *Izvestiya Geologitscheskogo Komity*, **9**, 41–84 (in Russian).

—— 1915. Orographic description of the Timan. St Petersburg. *Trudy Geologicheskogo Komity*, **12**(1) (in Russian).

TSZYU, Z. I. & KOSSOVOY, L. S. 1973. Timan–Pechora region. *In*: NALIVKIN, D. V., RZHONSNITSKAIA, M. A. & MARKOVSKII, B. P. (eds) *The stratigraphy of USSR. The Devonian System*, **1**, Nedra, Moscow, 145–165 (in Russian).

ULMISHEK, G. 1982. *Petroleum geology and resource assessment of the Timan–Pechora Basin, USSR, and the adjacent Barents-Northern Kara Shelf.* Argonne National Laboratory, Argonne, Illinois.

VANDELAER, E., VANDORMAEL, C. & BULTYNCK, P. 1989. Biofacies and refinement of conodont successions in the Lower Frasnian (Upper Devonian) of the type area (Frasnes-Nismes, Belgium). *Courier Forschungsinstitut Senckenberg*, **117**, 321–351.

WALLISER, O. H. 1996. Global events in the Devonian and Carboniferous *In*: WALLISER, O. H. (ed.) *Global Events and Event Stratigraphy*. Springer, Berlin, 225–250.

WENDT, J. & BELKA, Z. 1991. Age and depositional environment of Upper Devonian (Early Frasnian to Early Famennian) black shales and limestones (Kellwasser Facies) in the eastern Anti-Atlas, Morocco. *Facies*, **25**, 51–90.

YATSKOV, S. V. 1994. Devonian ammonoid zonation on Novaya Zemlya. *Newsletters on Stratigraphy*, **30**, 167–182.

—— & KUZ'MIN, A. V. 1992. Comparison of ammonoid and conodont assemblages in the Lower Frasnian of the South Timan. *Bulleten' Moskovskogo Obschestva Ispytatelei Prirody, Otdelenie geologitscheskoe*, **67**(1), 85–90 (in Russian).

YUDINA, YU. A. & MOSKALENKO, M. N. 1994. *Frasnian Key Sections of the Southern Timan*. Timan Department, All-Russian Petroleum Scientific Research Geological Exploration Institute (VNIGRI), Ukhta.

—— & —— 1998. *In*: M. RZHONSNITSKAYA (ed.) *Frasnian key sections of the Southern Timan (Field Guide book)*. Timan Department, All-Russian Petroleum Scientific Research Geologcal Exploration Institute, Ukhta.

ZAMYATIN, A. N. 1911. Lamellibranchiata from the Domanik Horizon of the South Timan. *St Petersburg Trudy Geologicheskogo Komiteta, Novaya Seriy*, **67** (in Russian).

ZIEGLER, W. 1971. Conodont stratigraphy of the European Devonian. In: SWEET, W. C. & BERGSTRÖM, M. (eds) *Symposium on Conodont Biostratigraphy*. Geological Society of America, Memoir, **127**, 227–284.

—— & SANDBERG, C. A. 1990. The Late Devonian standard conodont zonation. *Courier Forschungsinstitut Senckenberg*, **121**, 7–115.

Rudist lithosome development on the Maiella Carbonate Platform margin

IWAN STÖSSEL & DANIEL BERNOULLI
Geology Institute, ETH-Zürich, CH-8092 Zürich, Switzerland
(e-mail: iwan.s@pop.agri.ch)

Abstract: Cenomanian to mid-Campanian rudist lithosomes, exposed along a transect across the Maiella carbonate platform margin (central Apennines, Italy), are described in terms of faunal and matrix composition, geometry, and facies association. On the external platform, the lithosomes reveal a complex geometry and a comparatively high faunal diversity, whereas lithosomes of the inner platform are generally thinner and show a simple, sheet-like geometry. Based on the abundance of lithosomes and of the associated rudist-derived calcarenites, we propose that lithosome formation and, hence, sediment production preferentially occurred on the outermost platform, although the preservation potential of bioconstructions was low in these high-energy environments. Reworking of rudists led to sediment export both towards more internal areas as well as towards the adjacent basin. In contrast, on the inner platform, rudist lithosome formation was restricted by the lack of an adequate substrate and by higher sedimentation rates.

Benthic communities exert a major control on carbonate platform development. Their ecology determines to a significant extent the composition of the sediment, the sequential architecture and the lateral and vertical distribution of facies. Prediction of facies assemblages in the subsurface thus relies heavily on our knowledge of the palaeoecology of the carbonate-producing organisms, which for many benthic communities is poorly known. Therefore, models of fossil carbonate platforms are often based on the insights gained from the study of modern tropical shallow-water ecosystems. It is obvious that this actualistic approach may lead to an oversimplification in cases in which platforms were dominated by ecosystems not occurring today or in cases where environmental conditions during platform growth significantly differed from modern ones.

The Cretaceous was a time of extraordinary development of carbonate platforms and offshore banks (Simo *et al.* 1993). In contrast to modern tropical platforms, which are typically dominated by coral–algal or coral communities, many Cretaceous platforms were characterized by rudist and coral–rudist communities. The degree to which these communities differed from coral-dominated communities in terms of facies association and sequence architecture is a matter of controversy. As a consequence, the impact of rudists on carbonate platform development is poorly understood. In this paper, we document the abundance and the lateral and vertical distribution of various types of rudist accumulations, termed lithosomes, along a transect across the margin of an isolated, Bahamian-type carbonate platform exposed in Montagna della Maiella in central Italy. Following Gili (1993) we use the term 'rudist lithosome' instead of 'biostrome' for lithologies with significant autochthonous rudist accumulations, because it avoids any implications for shape, size or internal fabric of the sedimentary unit.

The Maiella carbonate platform margin

In the Montagna della Maiella, located in central Italy, the frontal anticline of the youngest and most external of the exposed thrust sheets of the southern Apennines fold-and-thrust belt offers kilometre-scale transects across a segment of the northern margin of the isolated Mesozoic to Tertiary Apulian carbonate platform (Fig. 1; Eberli *et al.* 1993; Vecsei *et al.* 1998). The axis of the anticline is perpendicular to the platform margin and, unlike the case of other Mesozoic carbonate platforms of the Apennines, no tectonic decoupling between the platform and the adjacent basin occurred. The transition from the internal platform to the proximal basin is thus preserved in physical continuity and exposed in spectacular seismic-scale outcrops parallel and perpendicular to the margin. The general stratigraphy, the overall depositional geometry and facies arrangement are known in detail (Bally 1954; Catenacci 1965; Crescenti *et al.* 1969) and the sequence-stratigraphic framework is well established (Accarie 1988; Vecsei 1991; Eberli *et al.* 1993; Sanders 1994, 1996; Mutti *et al.* 1996; Vecsei *et al.* 1998). Seven second-order

From: INSALACO, E., SKELTON, P. W. & PALMER, T. J. (eds) 2000. *Carbonate Platform Systems: components and interactions*. Geological Society, London, Special Publications, **178,** 177–190. 0305–8719/00/$15.00

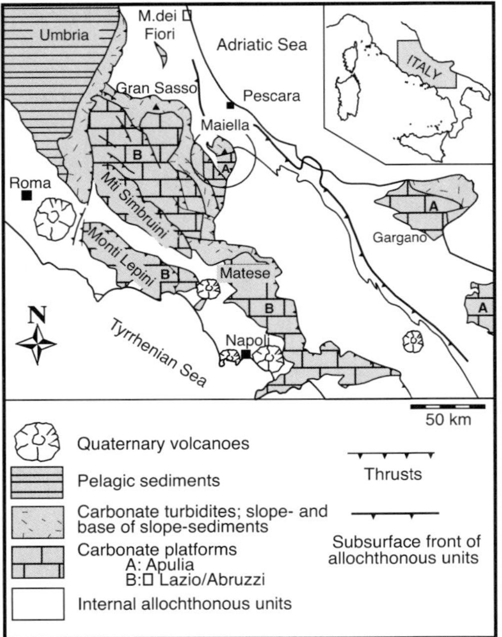

Fig. 1. Simplified geological map of central Italy and Upper Cretaceous facies distribution, modified from Eberli *et al.* (1993).

supersequences (*sensu* Van Wagoner *et al.* 1988) can be distinguished in the Upper Jurassic to Miocene succession (Fig. 2).

In this study, we concentrate on the entirely carbonate Cenomanian to mid-Campanian Cima delle Murelle Formation (Fig. 2). During the corresponding time interval, the platform margin of the Montagna della Maiella was, unlike many other Upper Cretaceous carbonate platforms and ramps, bounded by a steep, non-depositional escarpment, shaped by submarine erosion and gravitational collapse (Eberli *et al.* 1993).

Several sections along an outer to inner platfrom transect (Fig. 2) were studied in terms of cyclicity, lithosome abundance and composition (Stössel 1999). In order to visualize cycle stacking patterns, cycles were plotted as Fischer plot-type diagrams. However, as many of the cycles observed do not seem to have built up to sea level, no attempt was made to quantify the amplitude of the sea-level excursions, and the diagrams should not be mistaken as an approximation for sea-level curves.

On the outer platform, the Cima delle Murelle Formation is dominated by metre-scale cycles composed of coarse, cross-bedded bioclastic lime rudstones and grainstones with intercalated rudist lithosomes (Fig. 3). Peritidal lime mudstones to packstones are subordinate. The cycles show a shallowing-upward trend and are often capped by emersion surfaces, associated with meteoric and vadose diagenesis. Large-scale cross-bedding and coarse grain-size testify to deposition under high-energy conditions with high rates of sediment reworking and redistribution. Sedimentary structures indicate transport directions both towards the basin and towards more internal areas of the platform. Pervasive early marine cementation is ubiquitous. Especially on the outer platform, rudist lithosomes are preferentially preserved in intervals with thick cycles representing times of increased creation of accommodation space.

The equivalent succession of the inner platform is characterized by a gradual change from lime mudstone-dominated peritidal cycles in the lower part of the succession to subtidal lagoonal cycles which are composed of fine-grained bioclastic lime grainstone to packstone in the upper part (Fig. 3). Rudist lithosomes are restricted to the lagoonal cycles which are only rarely capped by emersion surfaces. Early marine cements were hardly observed. The facies transition from the outer to the inner platform appears to be gradual, but outcrops are not physically continuous.

No generally applicable model of rudist cycle formation can be developed. Nevertheless, there seem to be some common features which might relate to rudist palaeoecology. The succession from poorly sorted, mud-rich sediment through rudist lithosomes to well sorted and cross-bedded grainstones was reported from many Late Cretaceous open-shelf settings (e.g. Breyer 1991; Skelton *et al.* 1995; Sanders & Baron-Szabò 1997). This facies succession might reflect the inability of rudist lithosomes to withstand high levels of hydrodynamic energy.

Cycle stacking patterns reveal at least three orders of cyclicity, which are most clearly developed in the internal platform (Fig. 2; Stössel 1999). From the inner to the outer platform, cycles increase in thickness and significantly decrease in number. This relationship suggests that the preservational potential of individual cycles was lower and/or that the environment was less sensitive to variations in relative sea level on the outer platform margin. Fischer plots allow recognition of two third-order sequence boundaries in the Cima delle Murelle Formation (Fig. 2).

Facies associations

Figure 4 documents the contribution of various lithofacies to total stratigraphic thickness. On

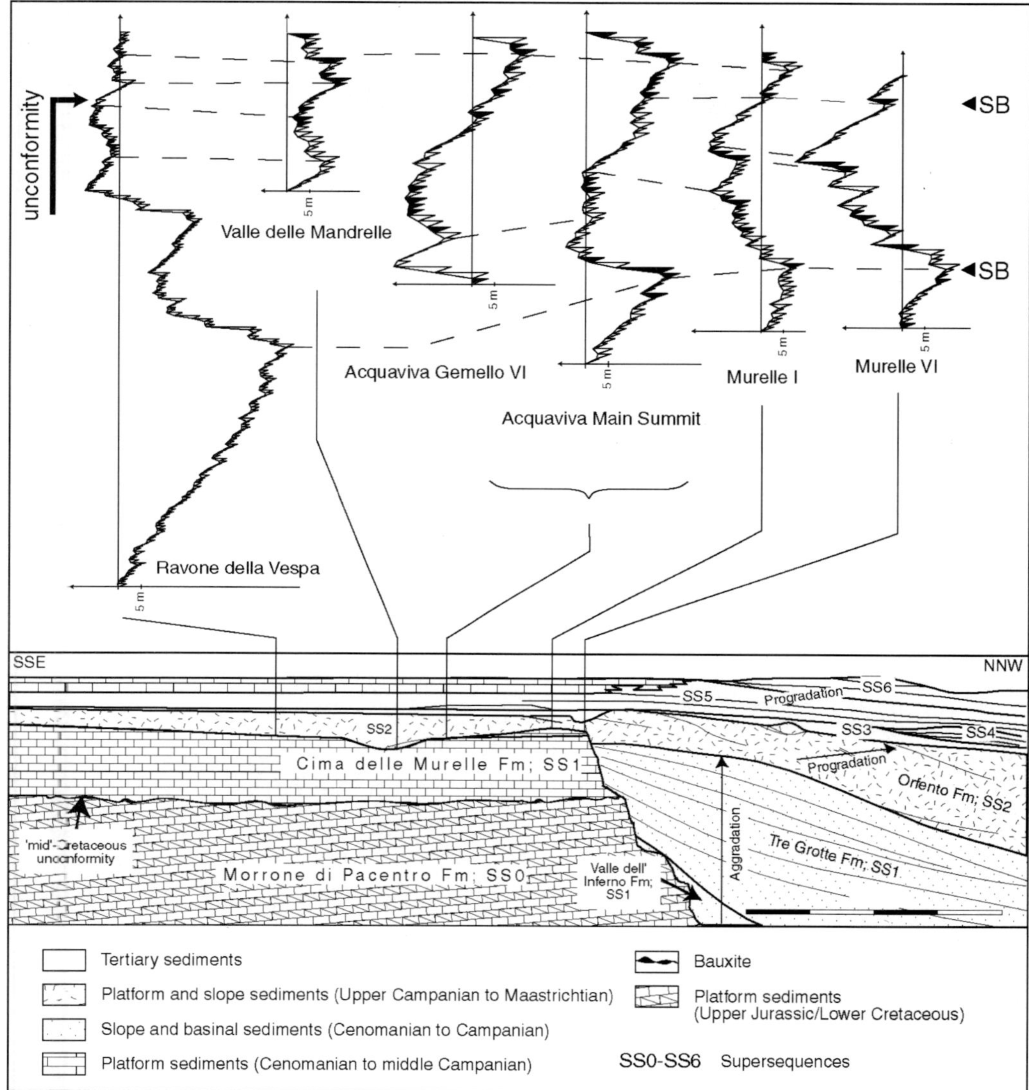

Fig. 2. Schematic cross-section of the Maiella platform margin; modified from Eberli *et al.* (1993) and Mutti *et al.* (1996). Fischer plots of several sections within the Cima delle Murelle Formation in the upper part of the figure show the cyclic stacking pattern, but should not be used as approximation for sea-level curves. Cycles containing rudist lithosomes are marked in black. Arrows labelled 'SB' mark the position of two sequence boundaries identified in the field. Correlations shown by dashed lines are based on cyclic stacking patterns, on lithology, and on the interpretation of photo mosaics.

the most external platform margin, the facies assemblage of the Cima delle Murelle Formation is dominated by coarse, cross-bedded lime grainstones and rudstones which are interbedded with rudist lithosomes. Despite the strong predominance of rudist fragments in all the coarse-grained sediments, rudist lithosomes with rudists in growth position contribute only up to 20% of total stratigraphic thickness. Sedimentary structures indicate sediment export both towards the basin as well as towards more internal settings.

Towards the platform interior, the contribution of grainstone-dominated lithologies to total stratigraphic thickness decreases rapidly but gradually. Also the rudist lithosomes show a

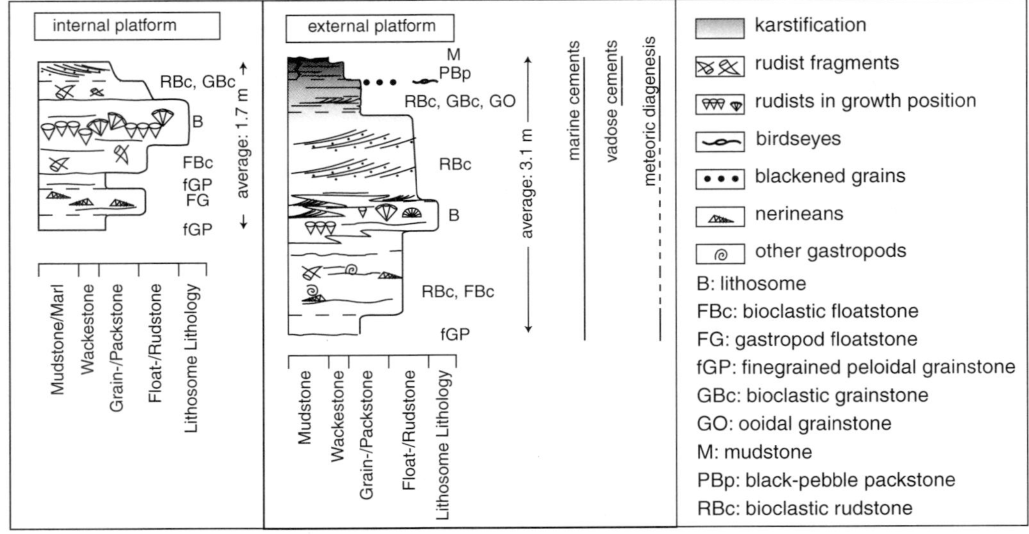

Fig. 3. Cycle architecture and composition along the inner to the outer platform transect. The drawings represent an idealized synthesis of the numerous cycles observed. Especially on the outer platform margin, the variability in cycle composition is considerable, which makes cycle recognition often difficult.

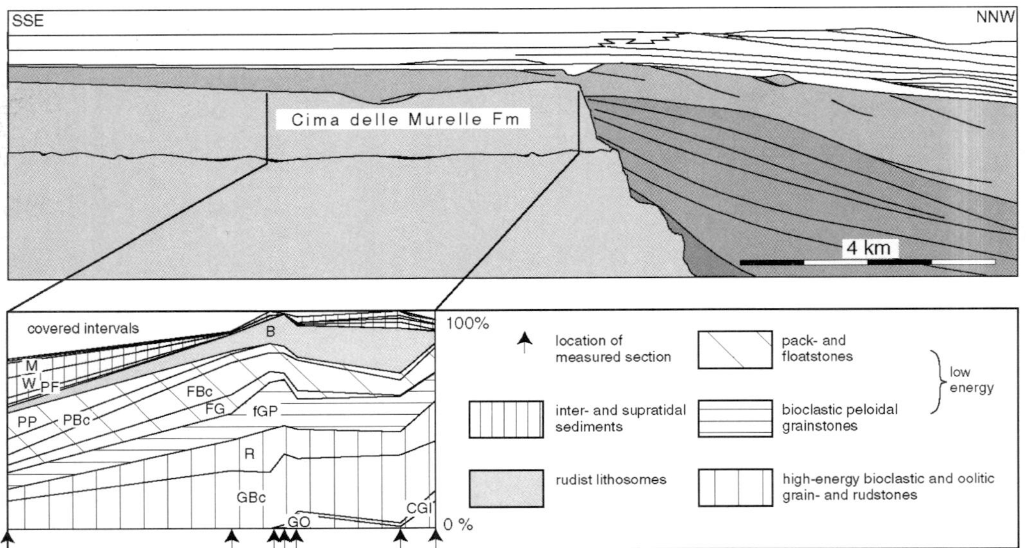

Fig. 4. Percentage distribution of lithologies in relation to total stratigraphic thickness of the measured sections. This plot averages facies abundance over the whole stratigraphic interval from Cenomanian to Lower Campanian. Note the absence of a sharp facies transition from the external to the internal part of the platform. Abbreviations: B, rudist lithosome lithology; CGl, conglomeratic rudstone; FBc, bioclastic floatstone; FG, gastropod floatstone; fGP, fine-grained peloidal to bioclastic grainstone; GBc, bioclastic grainstone; GO, oolitic to bioclastic grainstone; M, lime mudstone; PBc, bioclastic packstone; PF, fenestral packstone; PP, peloidal packstone; R, bioclastic rudstone; W, wackestone.

decrease in total thickness, while lagoonal sediments including fine-grained peloidal and bioclastic lime grainstones, packstones and floatstones become more important. The change in facies association is gradual, and there is no evidence for a narrow hydrodynamic barrier as found in modern walled reef complexes (Wilson 1975).

Lithosome types

Rudist lithosomes occurring along the outer to inner platform transect show marked variations in faunal composition, geometry and internal organization. In order to document these variations, Sanders (1996) subdivided the rudist lithosomes of the Maiella Platform margin into four types (radiolitid biostrome, hippuritid biostrome, caprinid biostrome and oyster biostrome). This classification scheme proved often difficult to apply, as transitional forms are extremely common. In order to better account for these transitional forms, we suggest a refinement of this scheme. However, our scheme, as all classifications, also represents an artificial subdivision. The main characteristics of the various lithosome types are listed in Fig. 5.

Type A: caprinid lithosome

The lithosomes of type A are dominated by recumbent rudists (*sensu* Skelton & Gili scheme, described in Skelton, 1991), i.e. caprinids, although radiolitids, a few corals, hydrozoans and encrusting algae were also observed and indicate a comparatively high diversity. They are up to 6 m thick (Fig. 6); their geometry is poorly defined, but where recognizable it is sheet-like. The matrix consists of poorly sorted bioclastic grainstone to packstone or floatstone with an arenitic matrix. The lithosomes gradually develop from the underlying bioclastic grainstones, and are overlain in turn by cross-bedded bioclastic grainstone to floatstone. Apart from well sorted and cross-bedded bioclastic lenses, no internal organization can be recognized, and neither biomorphs nor bioclasts show a preferred orientation or concentration. Associated sedimentary facies as well as matrix and texture of the lithosomes indicate high rates of reworking and redispersal of sediment.

Similar lithosomes have been described from numerous localities in Italy, including various outcrops in Friuli (e.g. Sartorio 1987; Swinburne & Noacco 1993), Rocca di Cave (Carbone *et al.* 1971; Carbone & Sirna 1981) and Matese (Accordi *et al.* 1990), and elsewhere (Gili *et al.* 1995*a* and references therein). Recumbent morphotypes are not known during the time interval from Turonian to late Campanian (Ross &

Lithosome Type	Dominant Fauna	Diversity	Matrix	Base	Packing	Geometry	Schematic Cross-Section
Type A Caprinid Lithosome	caprinids, radiolitids, corals, hydrozoans	high	grain-packstone to floatstone	gradual	open	sheet-like (poorly defined)	
Type B Grainstone Lithosome	radiolitids, hippuritids	high	grain- to packstone	gradual	very open	complex inter-fingering pattern with grainstones	
Type C High Diversity Lithosome	radiolitids, hippuritids, caprinids, corals	very high	pack- to floatstone	gradual	moderate to dense	lateral extensive, but variations in thickness	
Type D Hippuritid-Coral Lithosome	hippuritids, corals, radiolitids	moderate	grainstone	gradual	moderate	sheet-like	
Type E Dense Hippuritid Lithosome	hippuritids, corals	low	floatstone (only little matrix)	gradual or sharp	very dense	probably sheet-like poorly defined	
Type F *Distefanella* Lithosome	*Distefanella*	low	mud-rich floatstone	sharp	moderate to dense	thin and sheet-like	
Type G Thin Radiolitid-Hippuritid Lithosome	hippuritids, radiolitids	high	grain- to packstone	rapid transition or sharp	moderate to dense	thin and sheet-like	
Type H Thick Radiolitid Lithosome	radiolitids	moderate	mud-rich floatstone	gradual or sharp	moderate	undefined	
Type I Thin Radiolitid Lithosome	radiolitids (mainly *Milovanovicia*)	low	mud-rich floatstone	sharp	moderate to dense	thin and sheet-like	
Type K Osteid Lithosome	this lithosome type was not observed in this study, but see Sanders (1996)						

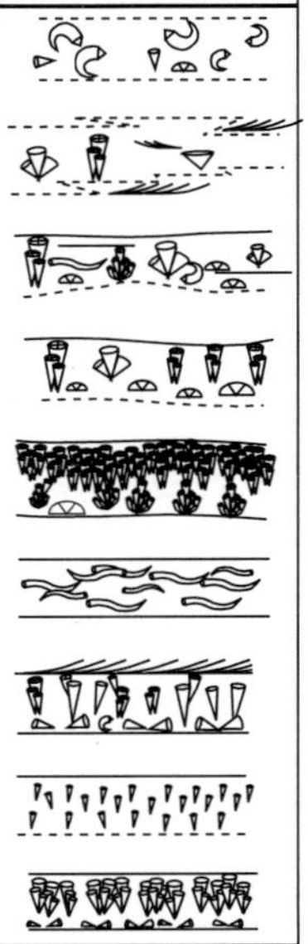

Key to symbols: Hippuritid; Radiolitid; Caprinid *s.l.*; *Distefanella*; Coral; gradual base; sharp base; cross bedding

Fig. 5. Characterization of the lithosome types defined in the Maiella platform margin. Diversity and packing density are based on qualitative visual estimates.

Fig. 6. Caprinid lithosome (lithosome type A). Large and complete specimens of a variety of rudists (*Caprina schiosensis, Caprina carinata, Caprina* sp., *Schiosia* sp., *Sphaerucaprina* sp., *Sauvagesia* sp.) as well as gastropods float in a poorly sorted packstone to floatstone matrix. The aragonitic layers of the shells were dissolved and are preserved as moulds. Neither preferred orientation of fossils nor bedding can be recognized (bedding is approximately horizontal; length of walking stick: approximately 1.2 m), Cima delle Murelle, Maiella. (from Stössel 1999, fig. 75).

Skelton 1993). Correspondingly, within the Cima delle Murelle Formation, type A lithosomes are restricted to the Cenomanian.

Type B: grainstone lithosome

Grainstone lithosomes are characterized by small isolated elevator rudists (mainly radiolitids) or small clusters of radiolitids (*Radiolites* sp., *Sauvagesia* sp., *Gorjanovicia* sp. and many others) and hippuritids which may be concentrated along lenses and layers, forming an open fabric. A high proportion of the observed rudist shells is not in growth position, but unbroken. This type of lithosome is up to 14 m thick. The matrix normally is a bioclastic lime grainstone to packstone (which also forms the surrounding facies). Lime mud content increases with increasing packing. Lenses of well sorted and rounded, cross-bedded grainstone to rudstone are intercalated. The base usually is irregular and gradual. Mapping of the boundaries of these lithosomes, which is often arbitrary in detail, reveals a complex pattern of interfingering with the cross-bedded grainstone facies, resulting in a complex and irregular geometry (Fig. 7).

The close relationship of this lithosome type with cross-bedded grainstones and rudstones and its close affinity to the platform margin indicates a depositional environment with intense reworking and redistribution of sediment. Layers and lenses of rudist congregations probably testify to intervals or areas of relative quiescence, whereas abundant high-energy events prevented the formation of denser or thicker congregations and account for the abundance of toppled and reworked rudist shells.

Type C: high-diversity rudist lithosome with a complex internal organization

Lithosome type C is characterized by an exceptionally high diversity (*Radiolites, Praeradiolites, Distefanella, Plagioptychus, Hippurites, Vaccinites, Hippuritella,* other rudist genera, and corals) and a complex internal organization (Fig. 8). This type of lithosome is up to 5 m thick, and laterally very persistent (possibly for several kilometres). Internally, these lithosomes often consist of distinct beds with a different faunal composition and varying amounts of matrix. The matrix is heterogeneous and consists of bioclastic packstone to floatstone, locally rich in lime mud; however, lenses of coarse-grained bioclastic grainstone are abundant. Typically, these internal fabrics cannot be traced laterally over a distance of more than a couple of metres.

The lithosomes commonly overlie cross-bedded grainstones to rudstones. The basal contact is transitional or sharp, whereas the top is usually sharp, and sometimes erosive. Their commonly greater thickness in comparison with most other lithosome types, the high faunal diversity and the generally large size of the rudists indicate

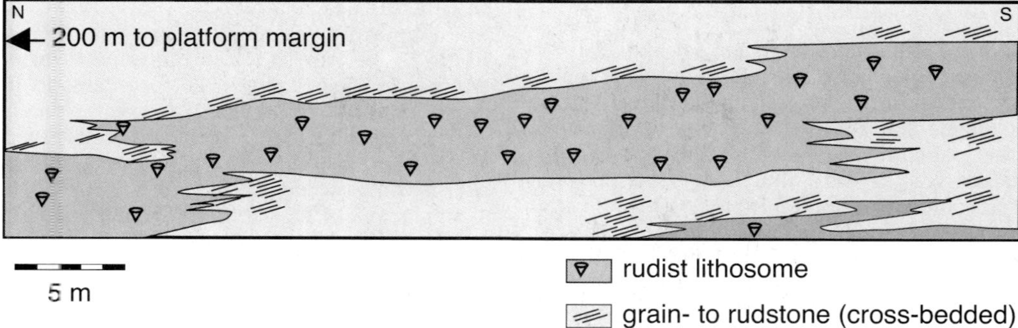

Fig. 7. Schematic drawing of a type B lithosome. The lithosome is dominated by radiolitids, most of them not in growth position. Isolated clusters of hippuritids occur as well. Because of the interfingering with the adjoining cross-bedded grainstone to rudstone units, the lithosome reveals a complex geometry, yet no indications for a positive topography can be recognized. The interfingering of the grainstone units and the lithosome documents the dynamic interplay between rudist growth and migration of carbonate sand bodies. Lithosome geometry seems to be controlled by the hydrodynamic environment rather than by rudist palaeoecology. (from Stössel 1999, fig. 77).

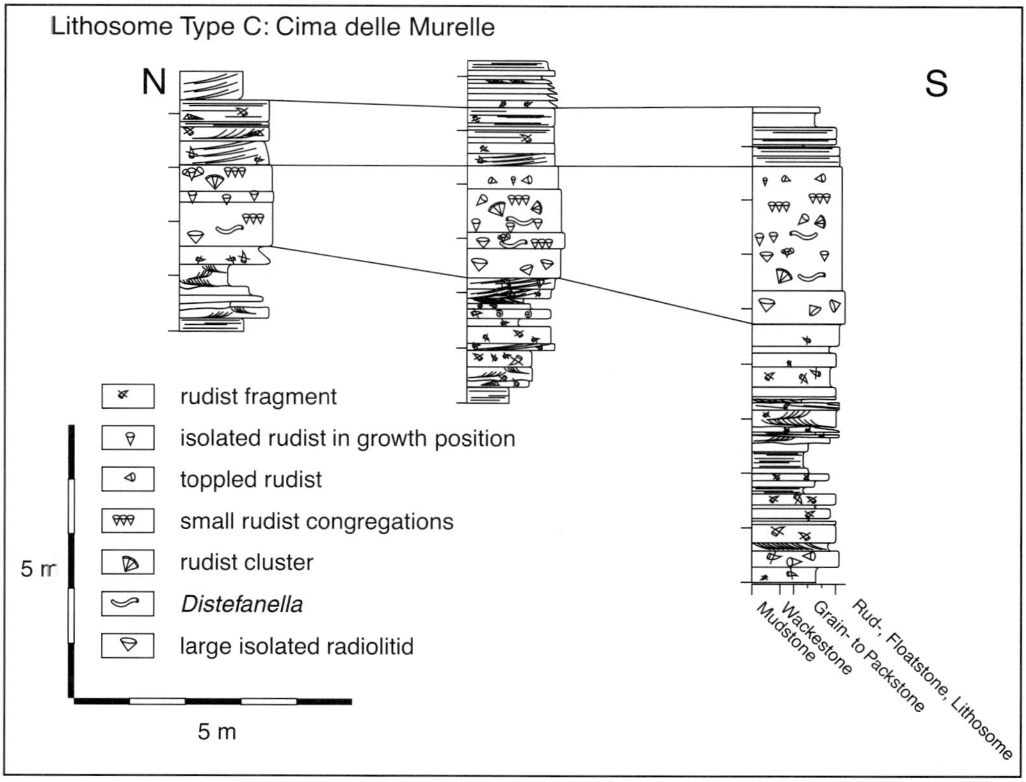

Fig. 8. Detailed logs of type C lithosomes. In each section, there is a distinctive subhorizontal zonation and type of bedding. However, it is impossible to follow the internal structure over a distance of even a few metres. (from Stössel 1999, fig. 79).

that during formation of these lithosomes, conditions were ideal for rudist growth. Matrix and intercalated lenses of bioclastic grainstone suggest elevated hydrodynamic energy, which was, however, not sufficient to destroy the lithosomes.

Type D: hippuritid-coral lithosome

Type D lithosomes have a sheet-like geometry and are up to 2.5 m thick. Both base and top are typically transitional. The matrix is very homogeneous and consists of bioclastic grainstone; the population density is moderate. Corals, which may be very abundant, occur preferentially, but not exclusively at the base of the lithosome. The rudist fauna is dominated by large clusters of gregarious hippuritids (with abundant occurrences of the genus *Hippuritella*). Isolated but abundant radiolitids are interspersed. Corals are very frequent along certain intervals. Hydrozoans occur as a subordinate faunal element.

Type D lithosomes combine the faunal diversity of type E lithosomes with the open fabric and the grainstone-dominated facies of type B lithosomes. Like type B lithosomes, they show a strong affinity to the outer platform margin, and their rudists apparently were able to thrive in a comparatively high-energy environment.

Type E: dense hippuritid lithosome

Lithosomes of this type are among the most impressive bioconstructions of Montagna della Maiella. They are characterized by very dense, paucispecific assemblages of gregarious hippuritids (*Hippurites colliciatus* and related species, interspersed specimens of *Vaccinites sulcatus*, *Vaccinites gosaviensis* and others; corals occur at the base; Fig. 9). These assemblages can often be subdivided into clusters. The lithosomes are up to some 6 m thick. The matrix, where present, is a mud-rich packstone with micritized and bioeroded rudist fragments. The hippuritids are often broken due to compaction. The base of these lithosomes can be either gradual or sharp. The top seems to be erosive and flat, but is rarely exposed. No indications of a positive topography were observed. Within the dense clusters, still denser clusters of rudists can be recognized.

Fig. 9. Close-up of a hippuritid cluster. There is very little space left between individual shells, which are in close contact with each other. These lithosomes are made of almost monospecific rudist assemblages (diameter of coin at bottom, centre: 2 cm). Valle delle Mandrelle (northern slope of Cima dell'Altare), Maiella. (from Stössel 1999, fig. 81).

Similar dense hippuritid congregations have been described from many Upper Cretaceous platforms (e.g. Höfling 1985; Grosheny & Philip 1989; Skelton *et al.* 1995; Schumann 1995; Sanders & Baron-Szabò 1997). However, in contrast to these occurrences, a preferential orientation of the hippuritids was not observed. The environmental interpretation of this lithosome type, however, remains ambiguous. While Sanders & Baron-Szabò (1997) interpreted them as having formed under relatively high-energy conditions (compared to coral–rudist limestones and open radiolitid biostromes), Skelton *et al.* (1995) interpreted similarly dense hippuritid congregations from the southern Pyrenees as having formed in 'usually quiet, restricted water of the platform top, where they were sporadically disturbed, however, by storm surges'. On the Maiella platform margin, these dense hippuritid congregations are rare, but show a wide geographic distribution. They were, however, not recorded on the outermost platform seemingly suggesting an affinity with more protected environments. On the other hand, they are usually associated with coarse bioclastic material, indicating slightly elevated energetic conditions.

Type F: Distefanella lithosome

Type F lithosomes are sheet-like and usually less than 1 m thick. They reveal monospecific assemblages of the radiolitid genus *Distefanella*. These accumulations form dense and laterally very persistent layers (at least for more than 300 m) of complete *Distefanella* specimens. Within individual layers, all specimens of *Distefanella* have a similar size, sometimes with both valves articulated. Nevertheless, in most cases, they seem to represent para-autochthonous accumulations, as all the specimens are toppled (Fig. 10). The matrix consists of medium- to coarse-grained bioclastic grainstone to packstone. Although the shells do not seem to be in growth position and are associated with rather coarse-grained sediments, it seems unlikely that they were assembled by hydrodynamic activity. Instead, it is suggested that these *Distefanella* individuals grew in an environment with rapid sedimentation. Their thin shell and cylindrical shape

Fig. 10. *Distefanella* layers in a succession of bioclastic grainstones to rudstones. These layers are interpreted as para-autochthonous equivalents of type F lithosomes (length of hammer: 50 cm). Monte Acquaviva, main summit, Maiella.

probably enabled them to compete with elevated sediment accumulation rates. The tabulae which are thought to reflect seasonal growth increments (Cestari & Sartorio 1995) are widely spaced, suggesting rapid growth. Subsequent and repeated removal of the sediment by current activity (which, however, was not strong enough to remove the heavy *Distefanella* shells) destabilized the subvertical shells and caused their burial in a subhorizontal position. The density of the assemblage might therefore be enhanced by condensation. This interpretation is in agreement with the coarse-grained bioclastic matrix.

Type G: thin radiolitid–hippuritid lithosome

Type G lithosomes are characterized by a sheet-like geometry (less than 1 m thick) and by a high faunal diversity. They comprise very few generations of rudists in growth position (large radiolitics and, subordinately, slender, gregarious hippuritids). Rudists are typically arranged in small clusters, but isolated specimens occur as well. The matrix consists of poorly sorted, fine-grained grainstone or packstone. These lithosomes can be traced over more than 300 m without evident changes in thickness or fauna. The base is often a rapid transition from rudist floatstone to lithosome lithology. They either grade into a rudist floatstone, or are sharply overlain by cross-bedded bioclastic grainstones to rudstones (Fig. 11).

As it combines relatively high diversity with a thin, sheet-like geometry, this lithosome type is considered to represent a transitional form between lithosome type I of the inner platform and the lithosomes showing a higher diversity of the outer platform (types B and C). The long growth form of the rudists, and the large vertical distance between successive rudist congregations suggest comparatively high rates of sediment accumulation. Like lithosomes of type I, lithosomes of type G probably reflect rapid, but short-lived colonizations of large areas.

Type H: thick radiolitid lithosome

Type H lithosomes are up to several metres thick and, in that case, include many generations of rudists. Their overall geometry is poorly defined. The fauna is dominated by radiolitids (mainly *Milovanovicia*, but also *Durania* and other radiolitids; Fig. 12), and rare hippuritids in growth position. The rudists are isolated from each other or form small clusters, attached to radiolitids in growth position or large rudist fragments. The spacing of the radiolitids in growth position is often surprisingly regular. The matrix is mud-rich with large fragments of radiolitids. The base may be both gradual (evolving from floatstone lithologies) or sharp. Most of the type H lithosomes grade into radiolitid floatstones, which are capped by well sorted cross-bedded grainstones and rudstones.

Type H lithosomes are very similar to type I lithosomes in terms of radiolitid dominance and matrix composition. However, net population density, diversity and thickness are similar to type B lithosomes. The high content of lime mud in the matrix suggests rather low-energy conditions. The sometimes significant thickness and the homogeneity of these lithosomes suggest ecological equilibrium during prolonged intervals of time: the rudists must have been ideally adapted to the environment. The regular spacing which is often observed might be explained by some sort of biotic interaction of adjacent individuals.

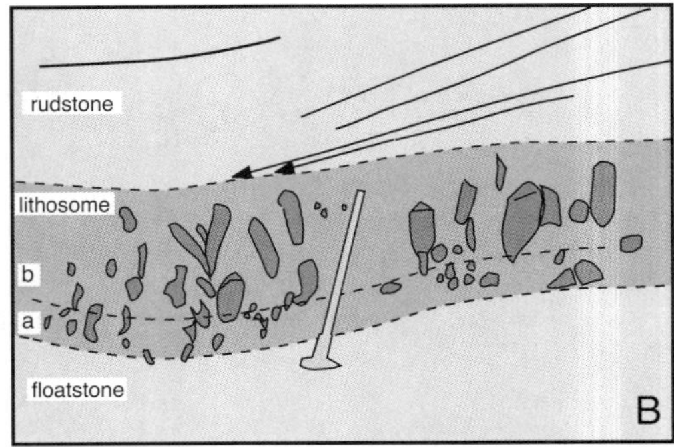

Fig. 11. Photograph of a type G lithosome at M. Acquaviva Gemello (**A**) and corresponding line drawing (**B**). The base of the lithosome consists of a layer of toppled, but almost intact rudists (a). This layer probably represents an initial, though 'unsuccessful' phase of colonization. However, the toppled rudists served as a substrate for the subsequent colonization by radiolitids. The second phase of colonization resulted in the growth of large individuals, which are either isolated or part of small clusters, preserved in the second layer of the lithosome (b). Population density seems to decrease slightly up-section. Finally, the lithosome was covered by bioclastic rudstone with inclined bedding (length of hammer: 50 cm). Monte Acquaviva Gemello, Maiella. (from Stössel 1999, fig. 85).

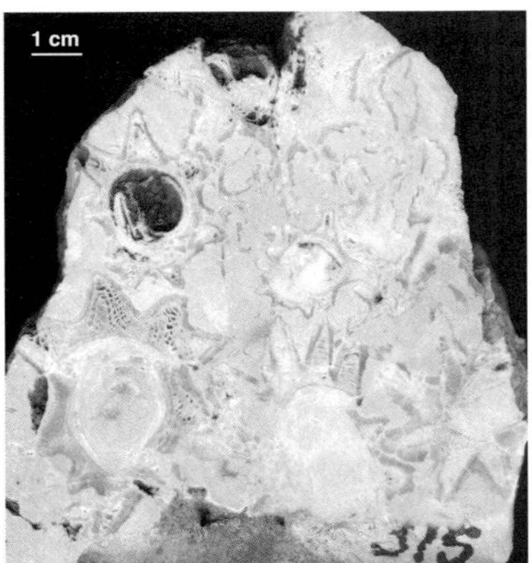

Fig. 12. Cluster of *Milovanovicia*, which are broken due to compaction. The section is parallel to bedding. The individuals are in contact with each other, but because of their 'spiny outline' (i.e. because of their folded outer shell layer) there is a lot of space between the shells. This space is filled with fine-grained wackestone to packstone. Rudists are more closely packed than average in type H lithosomes (Sample 315). Base of Monte Acquaviva Gemello, Maiella. (from Stössel 1999, fig. 87).

Type I: thin radiolitid lithosome

Type I lithosome is characterized by a sheet-like geometry (often less than 50 cm thick) and a moderate to high population density (Fig. 13). These lithosomes are dominated by one or a few generations of *Milovanovicia* and related radiolitids, usually in growth position. As an exception, requieniid fragments were observed. Sanders (1996) also reported the occurrence of ostreids. In general, however, species diversity seems to be very low. Type I lithosomes are associated with fine-grained lithologies: mainly fine-grained peloidal and bioclastic grainstones to packstones. Their matrix consists of bioclastic packstone and wackestone with abundant benthic foraminifera as well as large, bioeroded and angular radiolitid and gastropod fragments. Both base and top are usually sharp and well defined.

Fig. 13. Detail of a type I lithosome. Note the grainstone to floatstone lithology at the base of the lithosome; in this lithology, large fragments of rudists are abundant and it is interpreted as the relic of an earlier phase of colonization, possibly destroyed by a high-energy event. This layer made colonization by the subsequent generation possible. Numbers refer to sample numbers (length of hammer: 50 cm), Ravone della Vespa, Maiella. (from Stössel 1999, fig. 88).

The sedimentary facies suggests low-energy conditions. The floatstone to packstone which is often observed at the base is interpreted as a deposit of a high-energy event, which provided the coarse substrate required for rudist colonization in an environment of otherwise fine-grained sedimentation. The flat base and the laterally very extensive, sheet-like geometry suggest very rapid, but short-lived colonization, which indicates that conditions were favourable for rudist growth only during short periods of time.

Type K: ostreid lithosome

In the area investigated, no ostreid lithosome occurs. Sanders (1996), however, described lithosomes dominated by ostreids, requieniids and monopleurids from other parts of Montagna della Maiella (representing very internal facies).

Discussion

Figure 14 plots the percentage distribution of the described lithosome types versus distance to platform margin. From this diagram and from the descriptions above, the following trends can be read:

- average lithosome thickness decreases away from the platform margin (from 2.2 m on the outermost platform to 45 cm on the inner platform);
- whereas lithosomes of the outer platform reveal a complex geometry, the inner platform is characterized by lithosomes with simple, sheet-like geometries;
- diversity within lithosome units also decreases away from the platform margin: on the outer platform, hippuritids, various radiolitids, caprinids and a variety of corals are abundant; on the inner platform, there exists a strong

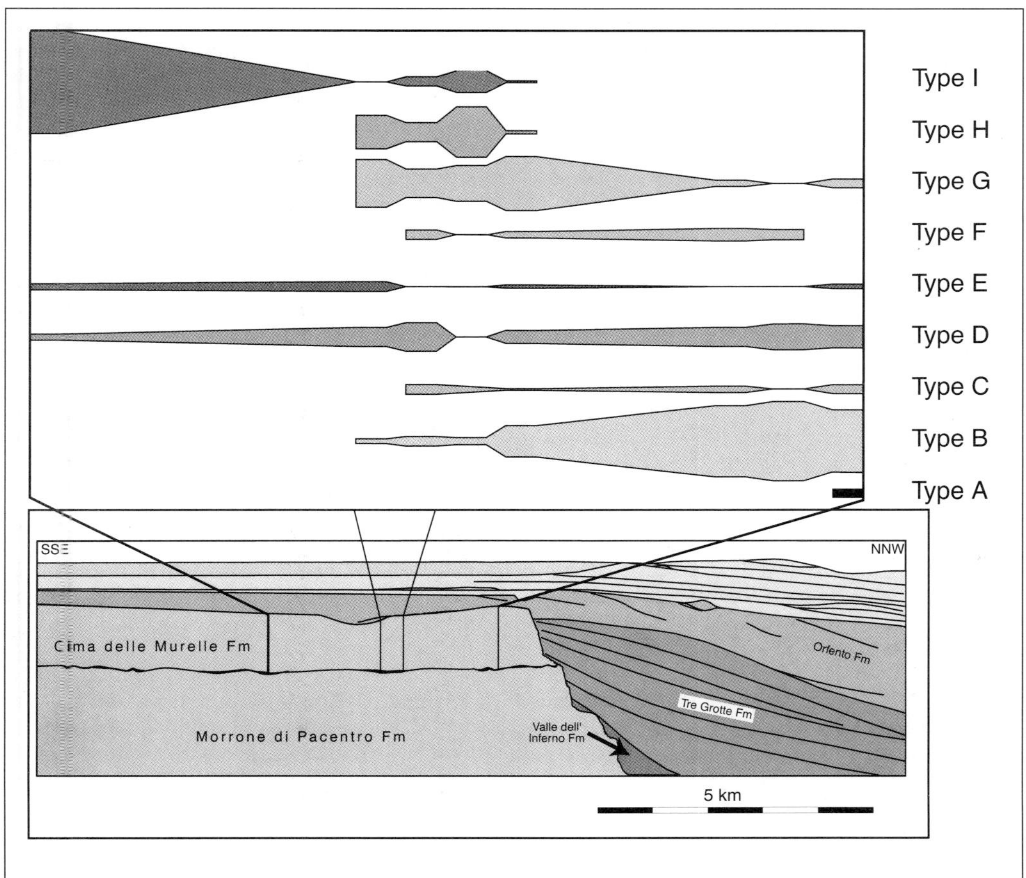

Fig. 14. Abundance of the various lithosome types along the Maiella platform margin.

predominance of Biradiolitinae (especially *Milovanovicia* and related taxa), and hippuritids are much less common.

In none of the lithosomes studied was any indication of a significant syndepositional topographic relief observed. Moreover, there is no positive evidence for a wave-resistant biogenic framework (neither primary nor secondary, with the possible exception of type H lithosomes) which could have formed a hydrodynamic barrier. The absence of a barrier is also suggested by the gradual change in terms of both facies associations and lithosome characteristics along the outer to inner platform transect.

Textural or faunal successions within individual lithosomes were rarely observed. In many cases, the faunal composition is either uniform from base to top, or vertical variations in faunal composition cannot be traced laterally. Only in a few cases may a succession either from coral-dominated to rudist (mainly hippuritid)-dominated assemblages (type D and E lithosomes) or from open-frame radiolitid assemblages to denser hippuritid assemblages be observed (type B lithosome; see Gili *et al.* 1995*b* and references therein for similar successions). Whether this shift represents an ecological succession (i.e. driven by autogenic processes; Etter 1994) or a sedimentological succession (driven by allogenic processes, including taphonomic feedback; cf. Kidwell & Jablonski 1983; sensu Gili *et al.* 1995*b*) is difficult to ascertain. Yet, the high variability of lithosomes suggests that lithosome formation did not follow an idealized 'Bauplan', but resulted from a complex interplay of depositional and biological processes, and autogenic processes played a subordinate role only.

Bounding surfaces of lithosomes may help to constrain boundary conditions of lithosome development. Both upper and lower surfaces are generally flat and document very little or no syndepositional topography on outcrop scale. No colonization of hardgrounds, erosional or emersion surfaces was observed, and rudist colonization on the Maiella platform margin appears to have occurred exclusively on unconsolidated sediment. Bases of lithosomes occurring on the outermost platform are preferentially gradual transitions from bioclastic floatstone and rudstone to lithosome lithologies. In contrast, in the inner platform area, lithosome bases are usually sharp and well defined, and overlie fine-grained grainstone to packstone. However, also here, the lowermost part of the lithosomes is usually made of a thin layer of floatstone, representing the record of high-energy events. This pattern reflects the need for adequate (i.e. coarse-grained) substrate for rudist colonization.

Lithosome tops are usually rapid transitions from lithosome lithologies either to floatstones without evident changes in matrix composition, or to cross-bedded grainstones. In these cases, increasing sedimentation rates may have caused the demise of the rudist lithosomes, although other factors (such as variations in nutrient concentrations or salinity) cannot be excluded. Lithosomes may also be overlain with a sharp contact by well sorted, cross-bedded grainstone which indicates choking by migrating sandwaves or dunes. In some cases, lithosomes are cut by erosive or even emersion surfaces. As these surfaces represent an unknown amount of time, it is impossible in this case to establish a genetic link between lithosome and overlying lithology.

Conclusions

The dominance of rudist fragments in the bioclastic portion of the Cima delle Murelle Formation, the generally high ratio of rudist fragments to rudists in growth position, and the minor contribution of cumulative lithosome thickness to total stratigraphic thickness (rarely more than 20%) indicate a low preservational potential for rudist lithosomes in general. Yet, the highest percentage of lithosomes occurs on the outermost platform margin, where grain-size distribution and bioclastic components indicate the lowest preservational potential. This suggests that lithosome formation and, hence, sediment production preferentially occurred in this area, although frequent phases of destruction and recolonization obviously controlled lithosome formation. The dynamic nature of this environment resulted in lithosomes with complex geometries. According to the non-equilibrium hypothesis (Krebs 1985), a constant level of disturbance might account for a higher diversity. The production of bioclastic sediment obviously outpaced the creation of accommodation space on the outer platform margin, which resulted in sediment export both to the proximal basin as well as to more internal areas of the platform.

In the more internal settings of the platform, however, the lack of adequate substrates must have limited the establishment of rudist lithosomes. This may (partly) explain the decrease of lithosome abundance from the outer to the inner platform. Furthermore, high sedimentation rates which are documented by the long and slender growth forms of rudists, or the large vertical spacing between individual colonization episodes (type G lithosomes, for example) probably prevented or restricted the growth of rudist lithosomes.

We thank M. Mutti and J.-H. van Konijnenburg for many inspiring discussions and comments, as well as for cooperation in the field. E. Gili, F. van Buchem and P. Skelton carefully reviewed an earlier version of the paper. P. Skelton is also thanked for his constant encouragement, and the big editorial effort, which he put into the paper. Financial support was provided by the Swiss National Science Foundation (grants 20-35907.92 and 20-45131.95 to D. Bernoulli and M. Mutti).

References

ACCARIE, H. 1988. *Dynamique Sédimentaire et Structurale au Passage Plate-forme/Bassin. Les Faciès Carbonatés Crétacés et Tertiaires: Massif de la Maiella (Abruzzes, Italie).* Ecole des Mines de Paris, Mémoires des Sciences de la Terre, **5**.

ACCORDI, G., CARBONE, F., CESTARI, R., REALI, S. & SIRNA, G. 1990. Cretaceous rudist colonization in north-eastern Matese. *In:* ACCORDI, G., CARBONE, F. & SIRNA, G. (eds) *Rudist Communities and Substratum in Matese Mounts, Molise, Italy.* Field trip guide of the 2nd International Conference on Rudists, Rome, 1990, Centro di Studio per la Geologia dell'Italia Centrale, Dipartimento di Scienze della Terra, Univeristà 'La Sapienza', Roma, 19–43.

BALLY, A. 1954. *Geologische Untersuchungen in den SE-Abruzzen.* PhD Thesis, University of Zurich.

BREYER, R. 1991. Das Coniac der nördlichen Provence ('Provence rhodanienne')- Stratigraphie, Rudistenfazies und geodynamische Entwicklung. *Frankfurter Geowissenschaftliche Arbeiten*, Frankfurt, **9**.

CARBONE, F. & SIRNA, G. 1981. Upper Cretaceous reef models from Rocca di Cave and adjacent areas in Latium, Central Italy. *In:* TOOMEY, D. F. (ed.) *European Fossil Reef Models.* Society of Economic Paleontologists and Mineralogists, Special Publication, **30**, 427–445.

——, PRATURLON, A. & SIRNA, G. 1971. The Cenomanian shelf-edge facies of Rocca di Cave (Prenestini Mts., Latium). *Geologica Romana*, **10**, 131–198.

CATENACCI, E. 1965. Resoconto sommario delle osservazioni stratigrafiche compiute sulla Maiella (Appennino Abruzzese). *Bolletino del Servizio Geologico d'Italia*, **86**, 17–25.

CESTARI, R. & SARTORIO, D. 1995. *Rudists and Facies of the Periadriatic Domain.* Agip, Milano.

CRESCENTI, U., CROSTELLA, A., DONZELLI, G. & RAFFI, G. 1969. Stratigrafia della serie calcarea dal Lias al Miocene nella regione Marchigiano-Abruzzese (Parte II-Litostratigrafia, Biostratigrafia, Paleogeografia). *Memorie della Società Geologica Italiana*, **8**, 343–420.

EBERLI, G. P., BERNOULLI, D., SANDERS, D. & VECSEI, A. 1993. From aggradation to progradation: The Maiella Platform, Abruzzi, Italy. *In:* SIMO, J. A., SCOTT, R. W. & MASSE, J.-P. (eds) *Cretaceous Carbonate Platforms.* AAPG, Memoir, **56**, 213–232.

ETTER, W. 1994. *Palökologie, Eine methodische Einführung.* Birkhäuser, Basel.

GILI, E. 1993. Facies and geometry of les Collades de Basturs Carbonate Platform, Upper Cretaceous, South-Central Pyrenees. *In:* SIMO, J. A., SCOTT, R. W. & MASSE, J.-P. (eds) *Cretaceous Carbonate Platforms.* AAPG, Memoir, **56**, 343–352.

——, MASSE, J.-P. & SKELTON, P. W. 1995*a*. Rudists as gregarious sediment-dwellers, not reef-builders, on Cretaceous carbonate platforms. *Palaeogeography, Palaeoclimatology, Palaeoecology*, **118**, 245–267.

——, SKELTON, P. W., VICENS, E. & OBRADOR, A. 1995*b*. Corals to rudists – an environmentally induced assemblage succession. *Palaeogeography, Palaeoclimatology, Palaeoecology*, **119**, 127–136.

GROSHENY, D. & PHILIP, J. 1989. Dynamique biosédimentaire de bancs à rudistes dans un environnement péridéltaïque: la formation de la Cadière d'Azur (Santonien, SE France). *Bulletin de la Société Géologique de France*, **6**, 1253–1269.

HÖFLING, R. 1985. Faziesverteilung und Fossilvergesellschaftungen im karbonatischen Flachwasser-Milieu der alpinen Oberkreide (Gosau Formation). *Münchner Geowissenschaftliche Abhandlungen, Reihe A, Geologie und Paläontologie*, **3**, 1–241.

KIDWELL, S. M. & JABLONSKI, D. 1983. Taphonomic feedback: ecological consequences of shell accumulation. *In:* TEVESZ, M. J. S. & MCCALL, P. L. (eds) *Biotic Interactions in Recent and Fossil Benthic Communities.* Plenum, New York, 195–248.

KREBS, C. J. 1985. *Ecology: The Experimental Analysis of Distribution and Abundance.* Harper & Row, New York.

MUTTI, M., BERNOULLI, D., EBERLI, G. P. & VECSEI, A. 1996. Depositional geometries and facies associations in an Upper Cretaceous prograding carbonate platform margin (Orfento Supersequence, Maiella, Italy). *Journal of Sedimentary Research,* **66(B)**, 749–765.

ROSS, D. J. & SKELTON, P. W. 1993. Rudist formations of the Cretaceous: a palaeoecological, sedimentological and stratigraphical review. *In:* WRIGHT, V. P. (ed.) *Sedimentological Review*, **1**, Blackwell, Oxford, 73–91.

SANDERS, D. 1994. *Carbonate platform growth and erosion: the Cretaceous to Tertiary of Montagna della Maiella, Italy.* PhD Thesis, Swiss Federal Insitute of Technology, Zürich.

—— 1996. Rudist biostromes on the margin of an isolated carbonate platform: The Upper Cretaceous of Montagna della Maiella, Italy. *Eclogae Geologicae Helvetiae*, **89**, 845–871.

—— & BARON-SZABÒ, R. C. 1997. Coral-rudist bioconstructions in the Upper Cretaceous Haidach Section (Gosau Group; Northern Calcareous Alps, Austria). *Facies,* **36**, 69–90.

SARTORIO, D. 1987. Reef and open episodes on a carbonate platform margin from Malm to Cenomanian: the Cansiglio example (Southern Alps). *Memorie della Società Geologica Italiana,* **40**, 91–97.

SCHUMANN, D. 1995. Upper Cretaceous rudist and stromatoporoid associations of Central Oman (Arabian Peninsula). *Facies,* **32**, 189–202.

SIMO, J. A., SCOTT, R. W. & MASSE, J.-P. 1993. Cretaceous carbonate platforms: an overview. *In:* SIMO, J. A., SCOTT, R. W. & MASSE, J.-P. (eds) *Cretaceous Carbonate Platforms.* AAPG, Memoir, **56**, 1–14.

SKELTON, P. W. 1991. Morphogenetic versus environmental clues for adaptive radiation. *In:* SCHMIDT-KITTLER, N. & VOGEL, K. (eds) *Constructional Morphology and Evolution.* Springer, Berlin, 375–388.

——, GILI, E., VICENS, E. & OBRADOR, A. 1995. The growth fabric of gregarious rudist elevators (hippuritids) in a Santonian carbonate platform in the southern Central Pyrenee. *Palaeogeography, Palaeoclimatology, Palaeoecology*, **119**, 107–126.

STÖSSEL, I. 1999. Rudists and carbonate platform evolution: The Late Cretaceous Maiella Carbonate Platform margin, Abruzzi, Italy. *Memorie di Scienze Geologiche (Padova)*, **51**, 333–413.

SWINBURNE, N. H. M. & NOACCO, A. 1993. The platform carbonates of Monte Jouf, Maniago, and the Cretaceous stratigraphy of the Italian Carnian Prealps. *Geologica Croatica*, **46**, 25–40.

VAN WAGONER, J. C., POSAMENTIER, H. W., MITCHUM, R. M., VAIL, P. R., SARG, J. F., LOUTIT, T. S. & HARDENBOL, J. 1988. An overview of the fundamentals of sequence stratigraphy and key definitions. *In:* WILGUS, C. K., HASTINGS, S., POSAMENTIER, H., VAN WAGONER, J., ROSS, C. A., KENDALL, G. ST. C. (eds) *Sea Level Changes: An Integrated Approach.* Society of Economic Paleontologists and Mineralogists, Special Publication, **42**, 39–45.

VECSEI, A. 1991. Aggradation und Progradation eines Karbonatplattform-Randes: Kreide bis Mittleres Tertiär der Montagna della Maiella, Abruzzen: *Eidgenössische Technische Hochschule und Universiät Zürich, Geologisches Institut, Mitteilungen, Neue Folge*, **294**.

——, SANDERS, D. G. K., BERNOULLI, D., EBERLI, G. P. & PIGNATTI, S. 1998. Cretaceous to Miocene sequence stratigraphy and evolution of the Maiella carbonate platform margin. *In:* DE GRACIANSKY, P.-C., HARDENBOL, J., JACQUIN, T., VAIL, P. R. & FARLEY, M. B. (eds) *Mesozoic and Cenozoic Sequence Stratigraphy of European Basins.* Society of Economic Paleontologists and Mineralogists, Special Publications, **60**, 53–74.

WILSON, J. L. 1975. *Carbonate Facies in Geologic History.* Springer, Berlin.

Fluctuations in the carbonate production of Phanerozoic reefs

WOLFGANG KIESSLING[1], ERIK FLÜGEL[2] & JAN GOLONKA[3]

[1]*Museum für Naturkunde, Humboldt-Universität, Invalidenstrasse 43, D-10115 Berlin, Germany. Present address: Department of Geophysical Sciences, University of Chicago, 5734 South Ellis Avenue, Chicago, IL 60637, USA*

[2]*Institut für Paläontologie, Universität Erlangen, Loewenichstrasse 28, D-91054 Erlangen, Germany*

[3] *Institute of Geological Sciences, Jagiellonian University, U. Oleandry 2a, 30–093 Krakow, Poland*

Abstract: A comprehensive database on Phanerozoic reefs is used to evaluate the carbonate production of reefs through time. Net, gross and export carbonate productions of 2760 Phanerozoic reefs are calculated and the cumulative production for 32 time slices is evaluated. The total amount of carbonate produced in the reef ecosystem in a given time slice is a function of global reef abundance, average reef size and the relative amount of carbonate exported from the reefs. Carbonate production of reefs is usually low, but characterized by prominent peaks in the mid-Silurian Givetian–Frasnian, the Late Triassic, the Late Jurassic, the mid-Cretaceous and the Neogene.

The determinants of reefal carbonate production are correlated with a variety of intrinsic and extrinsic parameters such as palaeogeographic setting, dominant biota, reef type, bioerosion, petrographic composition, eustatic sea level, oceanic crust production rates, atmospheric CO_2 concentrations, and global nutrient level. The calculated carbonate production, however, is rarely correlated with particular Earth system parameters. This implies that either the controls on reefal carbonate production are too complex to allow reliable predictions, or biotic factors represent more important controls than physico-chemical parameters. The constructed curve of Phanerozoic reefal carbonate export production is also poorly correlated with proposed curves of global shallow-water carbonate production suggesting that reefs rarely contributed in a quantitatively significant way to the global carbonate budget.

Reefs interact with their environment in diverse and complex ways. The quantitatively most important feedback is given by the reefal carbonate factory. The calcium carbonate production of modern coral reefs is known to play a substantial role in the global sedimentary carbonate budget, although the relative magnitude of reefal carbonate production is still poorly known owing to large uncertainties in the quantification of carbonate production in non-reefal environments (Milliman & Droxler 1996).

The relative role of reefal carbonate production in ancient reefs is even less well understood. Many authors commonly equate global carbonate production with reef prosperity (Bosscher & Schlager 1993; Weissert and Mohr 1996; Morrow *et al.* 1996; Copper 1997). There is no doubt that reefs may have a strong influence on the geometry and sedimentary characteristics of carbonate depositional systems, especially when they occur at the margins of carbonate platforms (e.g. Scaturo *et al.* 1989). However, we demonstrate in this contribution that reef prosperity is not always accompanied by high carbonate sedimentation rates, and that reef decline only rarely coincides with the demise of shallow-water carbonate production.

Methods

Database

A comprehensive database on ancient reefs was evaluated for this contribution. The database structure, its advantages and shortcomings have been discussed in previous papers (Flügel *et al.* 1996; Kiessling *et al.* 1999) and is only summarized herein. Our reef database is a locality/palaeolocality-based collection of mostly Phanerozoic (a few Vendian reefs are also included) and exclusively pre-Quaternary reefs containing information on their geometrical, stratigraphical, palaeontological and petrographical features. The database currently contains 2760 entries. Each dataset summarizes data of an area of around 315 km^2 owing to the 20 km distance requirement for reef 'sites' to be included separately (Kiessling *et al.* 1999). A rather wide definition of reefs was applied for the database. Four requirements are needed for

From: INSALACO, E., SKELTON, P. W. & PALMER, T. J. (eds) 2000. *Carbonate Platform Systems: components and interactions*. Geological Society, London, Special Publications, **178,** 191–215. 0305–8719/00/$15.00

a bioconstruction to be included in the database: (a) control on the formation by sessile benthic organisms (biological control); (b) lateral restriction of the structure; (c) (inferred) rigidity of the structure; (d) minimum diameter of structure larger than 0.5 m.

The age of a reef was determined as precisely as possible, but a stratigraphic resolution better than stage level is rarely achieved for most of the Phanerozoic (Kiessling *et al.* 1999). As our maps represent time slices that correspond to supersequences, we emphasize reef attributes within supersequences (Table 1). The supersequences are constrained by second-order unconformities. The Phanerozoic is divided into 32 supersequences for our evaluation. The names of supersequences are based on the megasequence names of Sloss (1963), but stage and series names are applied in the text for an easier reading. Chronostratigraphic ages refer to the time scale of Gradstein and Ogg (1996) with modifications in the Palaeozoic (especially in the Cambro-Ordovician and Permian). Although the supersequence stratigraphic resolution is rather coarse, it does not bias our results, since reefal carbonate production volumes rather than rates are analysed and we want to evaluate long-term rather than short-term fluctuations.

The database fields relevant for this paper are reef dimensions (thickness, length), age and debris potential. Reef dimensions were only stored in two dimensions, that is thickness and length or width. Owing to often imprecise measures or poor exposure, a four-interval classification for reef dimensions was chosen: 1 = less than 10 m thick, less than 20 m wide or long; 2 = 10–100 m thick, 20–100 m wide or long; 3 = 100–500 m thick or wide or long; 4 = more than 500 m thick or long.

The debris potential of reefs is defined as the relative amount of allochthonous reef sediment produced by the reef organisms. Debris potential determines the physical impact of reefs on their environment and is in turn controlled by a variety of environmental variables. We use the term 'debris potential', because the actual debris production is determined by reef size. In modern reefs, bioerosion is commonly thought to be the most important control on debris potential (Hutchings 1986; Hubbard *et al.* 1990) and significant downslope transport of reef debris was observed without any storm activity (Hughes 1999). However, water energy, especially when considerably enhanced during storms, can strongly increase the debris production of reefs (Macintyre *et al.* 1987; Scoffin 1993; Dollar & Tribble 1993). In Palaeozoic reefs, bioerosion is rarely evident and significantly less intense than in Mesozoic and Cenozoic reefs (Bertling 1997; Kiessling *et al.* 1999). Therefore, the action of storms and tectonics (e.g. earthquakes; Stoddart 1972) may have been relatively more important in the past. Intense storms act directly through erosion but can also affect the diagenetic regime of reefal carbonates, thereby enhancing carbonate dissolution and karstification (Jan 1998). It is well known that the fossil record contains many records with abundant reef-building organisms but little evidence of in-situ framework. This observation often caused problems in the classification or recognition of ancient reefs. However, even Holocene reefs are often characterized by poor framework preservation in the subsurface (Hubbard *et al.* 1998) and thus the preservation of in-situ framework is not part of our reef definition. On the other hand, there are common mounds which appear completely separated from surrounding sediments and obviously did not shed any rubble in their environment. This applies to many mud mounds, some of them even growing on a shaly substratum (e.g. Ruppel & Kerans 1987), but there are also examples for reef mounds (e.g. Davies 1989) and biostromes (examples in Höfling 1997).

To document these strong differences in debris production of different reefs and reef types, we introduced a field that classifies the debris potential in three intervals: 1 = low, 2 = moderate, 3 = high debris potential. Low debris potential is supposed for reefs with a dominance of in-situ framework and/or lacking debris aprons (Fig. 1a). Reefs that contain a high amount of bioclastic material, internally or at their flanks, were also considered as having a low debris potential, if the material is devoid of reef builders and therefore not produced by the reef. A moderate debris potential was assigned to reefs with a moderately developed debris apron and modest percentage of toppled reef builders in the reef body (Fig. 1b). High debris potential was assumed for reefs with a distinct prevalence of toppled reef builders, plenty of reworked reef carbonate in the reef edifice and/or pronounced debris aprons (Fig. 1c). The absolute amount of debris production is not relevant for this field. Thick reefs can have a low debris potential (e.g. most Waulsortian mounds), whereas small patch reefs may exhibit a high debris potential (e.g. the small Upper Jurassic Laisacker coral reef in Germany; Flügel *et al.* 1993).

The philosophy behind all database entries was to get a reliable quantification of data that are often described only qualitatively in published literature. This is the reason for choosing interval rather than metric or percentage values. Still, some interpretations of figured examples

Table 1. *Definition of supersequences as used in this paper*

Name of supersequence	Number	Trivial name	Included 'stages'	Age of maximum trangression (Ma)	Duration of time slice (Ma)
Sauk I	1	Early Cambrian	Nemakit/Daldynian–Toyonian	520	33
Sauk II	2	Middle Cambrian	Middle Cambrian–Dresbachian	502	14
Sauk III	3	Tremadocian	Franconian–Tremadocian	488	15
Sauk IV	4	Middle Ordovician	late Early–early Middle Ordovician	472	21
Tippecanoe I	5	Late Ordovician	Darriwilian–Ashgillian	452	21
Tippecanoe II	6	Llandoverian	Llandoverian	435	15
Tippecanoe III	7	Wenlockian–Ludlovian	Wenlockian–early Pridolian	425	10
Tippecanoe IV	8	Lochkovian	middle Pridolian–middle Pragian	412	16
Kaskaskia I	9	Emsian–Eifelian	upper Pragian–Eifelian	396	22
Kaskaskia II	10	Givetian–Fransian	Givetian–lower Famennian	368	20
Kaskaskia III	11	Tournaisian	upper Famennian–lower Visean	348	22
Kaskaskia IV	12	Visean/Serpukhovian	middle Visean–Serpukhovian	328	15
Lower Absaroka I	13	Moscovian	Bashkirian–Kasimovian	302	27
Lower Absaroka II	14	Asselian	Gzhelian–Asselian	287	11
Lower Absaroka III	15	Artinskian	Sakmarian–Kungurian	277	17
Lower Absaroka IV	16	Guadalupian	Roadian–Changhsingian	255	20
Upper Absaroka I	17	Middle Triassic	Induan–lower Carnian	232	24
Upper Absaroka II	18	Late Triassic	upper Carnian–middle Hettangian	218	21
Upper Absaroka III	19	Early Jurassic	upper Hettangian–lower Aalenian	195	24
Lower Zuni I	20	Middle Jurassic	middle Aalenian–middle Bathonian	169	12
Lower Zuni II	21	Late Jurassic	upper Bathonian–middle Tithonian	152	20
Lower Zuni III	22	Berriasian	upper Tithonian–lower Valanginian	140	12
Upper Zuni I	23	Barremian	upper Valanginian–lower Aptian	126	18
Upper Zuni II	24	Albian	upper Aptian–middle Cenomanian	105	23
Upper Zuni III	25	Turonian	upper Cenomanian–lower Campanian	90	13
Upper Zuni IV	26	Campanian	middle Campanian–Danian/Selandian	76	23
Lower Tejas I	27	Ypresian	Thanetian–Ypresian	53	9
Lower Tejas II	28	Lutetian	Lutetian–Bartonian	45	12
Lower Tejas III	29	Rupelian	Priabonian–Rupelian	33	8.5
Upper Tejas I	30	Aquitanian	Chattian–Aquitanian	22	8
Upper Tejas II	31	Serravallian	Burdigalian–Serravallian	14	9.5
Upper Tejas III	32	Messinian	Tortonian–Gelasian	6	9

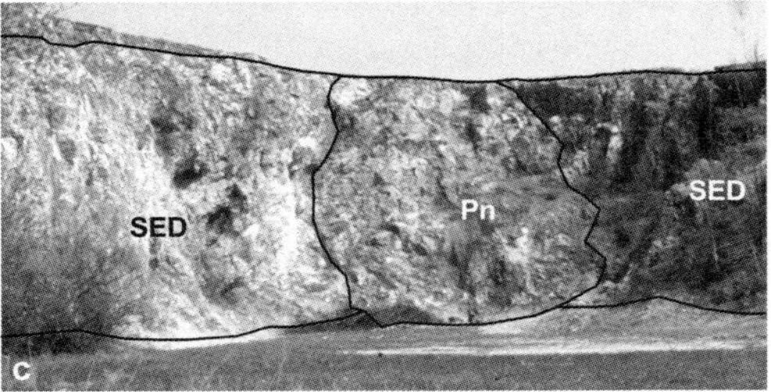

Fig. 1. Field examples of reef carbonate production and debris potential. **(a)** Microbial mound in Upper Cambrian Wilberns Formation, Llano River, Texas. The reef stands isolated in shaly substratum. Nearly all net production (P_n) is trapped in the reef body (SED = 0; $P_g = P_n$). Debris potential in database = 1. **(b)** Coral reef in Upper Albian Mural Limestone, Paul Spur, Arizona. The autochthonous reef body (P_n) and the facies with reworked reefal debris (SED) are of about equal volume. Debris potential in database = 2. **(c)** Coral reef in the lower Tithonian of Laisacker, Bavaria, Germany. Reefal debris (SED) is very pronounced and even within the autochthonous reef edifice (P_n) reef builders are rarely in life position. Debris potential in database = 3.

and interpolations were occasionally necessary, especially for the debris potential. However, if a value could not be assigned with some confidence from reading between the lines or outcrop photographs, the database field remained blank.

Calculation of carbonate production

The calculation of carbonate production applied the basic principles elaborated by Chave *et al.* (1972), Land (1979) and Hubbard *et al.* (1990) for modern reefs. However, owing to the limited stratigraphic resolution for the majority of ancient reefs, we cannot directly provide carbonate production rates, but are limited to the evaluation of carbonate volumes produced in a given time slice. Carbonate net production is the amount of $CaCO_3$ retained by the reef system and is thus equivalent to the volume of the preserved reef edifice. Net production consists of both in-situ reef carbonates and reworked reefal sediment that did not leave the reef body. Carbonate export production is the volume of $CaCO_3$ that is produced by reef organisms but shed from the reef body to its surroundings. Export production is equivalent to sediment export of Hubbard *et al.* (1990). Carbonate gross production, finally, is the total amount of carbonate produced by the reef organisms.

For simplicity we assumed a tabular form for all reefs in the database. Thus the volume (V) of a preserved reef body can be calculated by:

$$V = Th \times L \times Wd \qquad (1)$$

where Th is the thickness, L is the length and Wd represents the width of the reef in metres. The volume of a given reef represents the sum of autochthonous carbonate productions and the retained allochthonous carbonate production and is thus equal to the net carbonate production (P_n) sensu Hubbard *et al.* (1990).

As the database contains interval rather than metric scales, the interval units had to be translated into metric values. Additionally, the width of a reef had to be estimated in most cases owing to poor original data. The transformations for the calculation of reef volumes are indicated in Table 2. Missing values were replaced by moderate values for the calculation (equivalent to interval 2 in the database). A larger-than-observed value for reef length was taken if the reef occurs in a continuous reef tract where not all reefs are separately recorded in the database.

The debris potential field is crucial for the definition of partitioning between autochthonous and allochthonous carbonate production. As mentioned before, the volume of debris produced by a reef (debris production) depends on its size and debris potential. Debris production is partly retained within the reef body and partly exported. Hubbard *et al.* (1990) concluded that about equal proportions of the detrital reef sediment are retained and exported, whereas Land (1979) found that retained carbonate production may be much less than exported production. However, in all cases reported thus far, there is a linear relationship between retained and exported carbonate production. We therefore assume that the debris production retained in the reef is generally proportional to the production exported from the reef. Thus the debris potential field of the database provides a

Table 2. *Transformation of interval classes in the database to metric values*

Interval	Thickness (m)	Length (m)	Width (m)
1	5	10	10
2	50	60	60
3	250	250	200
4	650	5000	2000
4 (biostromes)		1500	1500
3 (in reef tract)		15 000	2000
4 (in reef tract)		20 000	2000

Note different transformations for different reef types

measure of carbonate export production, even if it was derived solely from the proportion of reworked reef builders within the reef. This is especially important because the total volume of exported carbonate can rarely be evaluated when exposure is not perfect. Based on observations of Silurian, Triassic and Holocene reefs (Ingels 1963; Hubbard *et al.* 1990, 1998; Zankl 1977; Harris 1994) we assumed that high debris potential is equivalent to 80% of the gross carbonate production (P_g) being exported from the autochthonous reef body. For moderate debris potential a value of 50% was taken (based on modern examples; Scoffin *et al.* 1980; Hubbard *et al.* 1990), whereas 10% was postulated for low debris potential. Hence the carbonate export production, equivalent to the sediment production (SED) of Land (1979) and Hubbard *et al.* (1990), can be directly determined from the reef's volume and its debris potential. This can be calculated by:

$$\mathrm{SED} = P_n \times D_p \qquad (2)$$

where D_p is the debris potential transformed into the relative amount of carbonate net production (P_n) being exported (8, 1 or 0.1, respectively). Again, missing values for debris potential (27% of all reef sites) were replaced by moderate values for the calculation of carbonate production. With the determination of P_n and SED, the gross production (P_g) can be calculated in a straightforward way:

$$P_g = P_n + \mathrm{SED} \qquad (3)$$

The definitions of variables used for the calculation of carbonate export are illustrated in Figs 1 and 2.

Two sample calculations may exemplify the huge possible range of export production by reefs. A small mound of 5 m thickness, 10 m length, 10 m width and a low debris potential is calculated to have a net production (P_n) of 500 m^3, an export production (SED) of 50 m^3 and a gross production (P_g) of 550 m^3. In contrast, a 650 m thick, 5000 m long and 2000 m wide reef complex with high debris potential has the parameters: $P_n = 6.5 \times 10^9$ m^3, SED $= 5.2 \times 10^{10}$ m^3, $P_g = 5.85 \times 10^{10}$ m^3.

Although some mounds evidently received substantial amounts of carbonate from external sources (Bosence *et al.* 1985), imported carbonate volume is considered of minor quantitative importance and is thus not discussed here. The role of carbonate dissolution is also not considered quantitatively, although carbonate dissolution may be important in reef (Tudhope & Risk 1985) as well as in the forereef slope and basin environment (Droxler *et al.* 1988). Our net production values represent volumes that are actually preserved, whereas export and gross production are calculated maximum values. A substantial volume of exported reef carbonate may never have entered the sedimentary record.

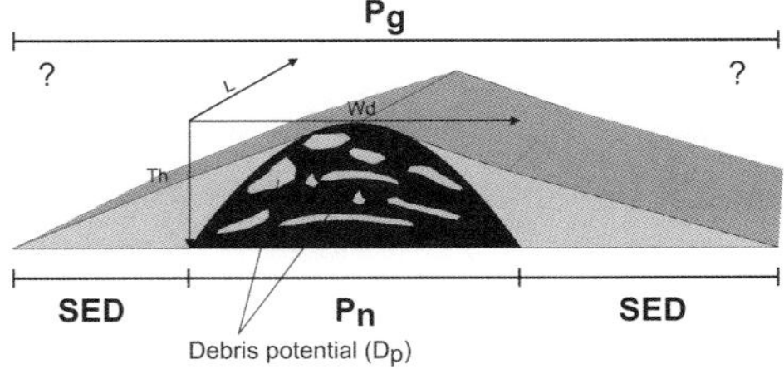

Fig. 2. Definition of variables used to calculate the carbonate production of the database reefs. The terminology of P_g (gross production), P_n (net production = autochthonous carbonate production + retained allochthonous carbonate production) and SED (sediment production = carbonate export production) was taken from Land (1979) and Hubbard *et al.* (1990). Additionally, a variable termed 'debris potential' (D_p) was introduced to allow an estimation of SED which is often not directly observed owing to limited exposure. It is assumed that the debris potential can be evaluated from the amount of reworked material in the reef body and that debris potential is proportional to SED. Reef dimensions: Th = thickness; Wd = width; L = length.

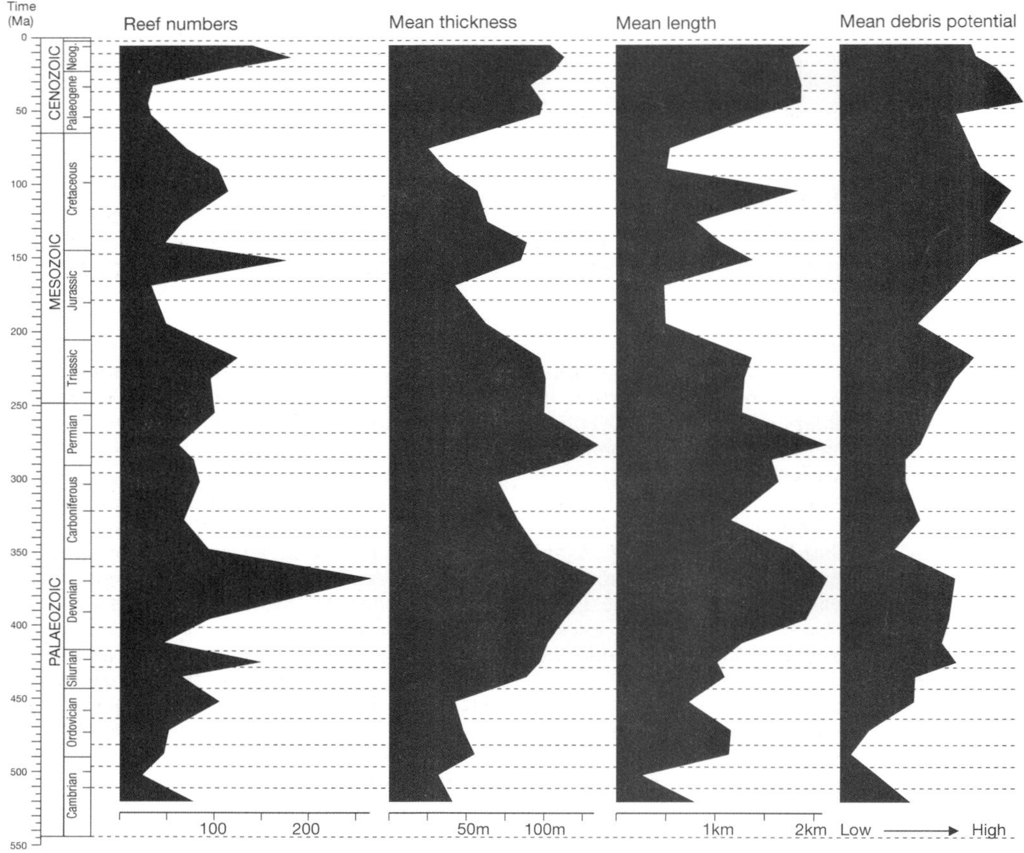

Fig. 3. Secular variations in the main determinants of Phanerozoic reefal carbonate export production, calculated for supersequences (indicated by stippled horizontal lines). The mean extension and especially the mean debris potential tend to increase through time, whereas there is no significant trend for the number of reef sites and mean thickness.

Reef carbonate production through time

Fluctuations of determinants

The major determinants of reef carbonate production vary considerably throughout the Phanerozoic (Fig. 3). The number of reef sites fluctuates most strongly, varying between 23 in the Middle Cambrian supersequence and 257 in the Givetian–Frasnian supersequence. Distinct peaks are evident in the Late Ordovician, Wenlockian–Ludlovian, Givetian–Frasnian, Late Triassic, Late Jurassic, middle Cretaceous and the Neogene. The Givetian–Frasnian supersequence is clearly the Phanerozoic acme of reef expansion. This is especially noteworthy, since reef numbers in Fig. 3 were not corrected for carbonate cycling (cf. Wilkinson & Walker 1989) and therefore Palaeozoic numbers should be considerably greater than indicated. Other time intervals such as the Middle Cambrian to Tremadocian, the Lochkovian, the Early and Middle Jurassic and most of the Palaeogene exhibit few reef sites.

Reef dimensions show less pronounced but still significant variations through time. Reef thickness peaks in the Devonian, the Early Permian, the Late Jurassic and the Cenozoic. Reef lengths parallel the fluctuations in thickness except for an additional peak in the Albian time slice. Averaged for the whole Phanerozoic, reef dimensions are significantly correlated with reef numbers ($r = 0.39$ for thickness, $r = 0.38$ for length, $p = 0.03$ for both), but thickness peaks may occur during times of only moderate reef abundance (e.g. Early Permian).

The mean debris potential of reefs exhibits a significant increase through time (Fig. 3).

Palaeozoic reefs (especially Cambrian to Middle Ordovician and Carboniferous to Permian) contain significantly less reworked reefal debris than Mesozoic and Cenozoic reefs. Debris potential is not correlated with reef numbers, mean thickness or mean length.

Fluctuation of carbonate production

The reefal carbonate production in a particular time slice is determined by the cumulative gross carbonate production of all reef sites (Fig. 4). Owing to the admittedly incomplete dataset and to the fact that reef sites rather than individual reefs were considered, the resulting values are much lower than actual values. For instance, the global gross carbonate production rate of Recent reefs is estimated as 9×10^8 tons per year (Milliman 1993; Milliman & Droxler 1996), while the calculated gross production for the whole Serravallian supersequence is 1945 km^3, equivalent to 5.25×10^{12} tons of $CaCO_3$ (density of 2.7 g cm^{-3} assumed). Considering that the Serravallian supersequence sums up reef data of 9.5 Ma (Table 1), the calculated annual carbonate production is roughly 5.52×10^5 tons, that is 1650 times less than the production rate of modern reefs. Since Miocene reefs exhibit a wider geographical range than Recent reefs it is likely that our approach produces values that are at least 2000 times below true mass values, if the estimations for modern reefs are correct. Estimations of modern reef production vary strongly and may often be exaggerated as recently shown by a new evaluation of global reef area (Spalding & Grenfell 1997). Additionally, reef carbonate production may vary significantly even on short time scales such as the late Quaternary (Kleypas 1997).

However, even the lower estimate of global reef area (255×10^3 km^2; Spalding & Grenfell 1997) translates into a global mass production rate of 3.8×10^8 tons per year assuming an average production of 1.5 kg m^{-2} a^{-1} (Ware *et al.* 1992). This value is still almost 700 times above our calculation for the Serravallian time slice. Thus our calculated values cannot be taken as actual values. A compensation factor covering almost three orders of magnitude could be introduced but would not enhance the reliability of our values. The discrepancy is most likely due to the incomplete database which in turn is due to (1) unexplored areas/unknown reef sites, (2) incomplete coverage of published data (e.g. some Russian and Chinese references), (3) problems with reef definitions and (4) complete destruction of reefs by carbonate cycling with increasing age (loss of carbonate due to erosion or subduction). For the first two factors an almost equal bias for all time slices can be assumed (Kiessling *et al.* 1999). Problems with reef definition were overcome by the application of a very broad definition for the database. The complete destruction of older reefs by carbonate cycling can be balanced by numerical reconstruction. Following the approach of many

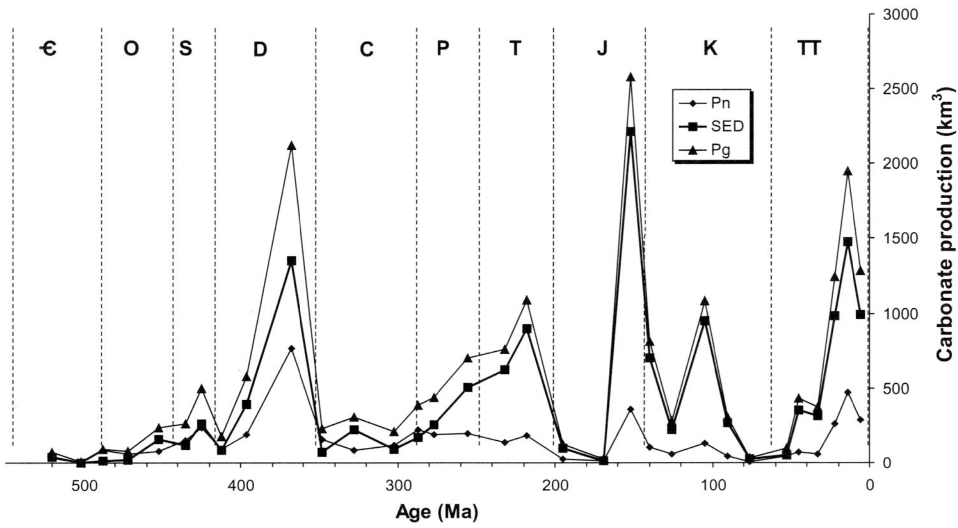

Fig. 4. Cumulative reefal carbonate production in supersequences as calculated from the database. P_n = net production determined by the sum of preserved reef dimensions; SED = sum of sediment produced by the reefs and exported from the reef body; P_g = gross production, sum of P_n and SED.

previous authors (e.g. Gregor 1985; Wilkinson & Walker 1989; Wold & Hay 1990) an exponential decay for reef carbonates was assumed. The exponential model fit to Phanerozoic carbonate flux data is quite poor (Wilkinson & Walker 1989; Mackenzie & Morse 1992), owing to the strong actual fluctuations in neritic carbonate production and the linear component due to changing first-order sea level (Mackenzie & Morse 1992). The strongly fluctuating primary signal in the dataset is also the reason for the different decay curves calculated for gross, net and export production, respectively. Nevertheless, an exponential model is still the most reasonable approach to carbonate cycling. We applied the decay constant of 0.0025 Ma^{-1} (Wilkinson & Walker 1989), although this may be a minimum value not completely balancing the almost complete destruction of oceanic atoll reefs in the Palaeozoic and early Mesozoic. Our approach to balance carbonate cycling is given by:

$$\Sigma[P_g, P_n, \text{SED (reconstructed volume)}] = \Sigma[P_g, P_n, \text{SED (database volume)}] \times e^{t \times 0.0025} \quad (4)$$

where t is the mean age (in Ma) of the time slice considered (time of maximum transgression in the case of supersequences). Although our reconstructed values cannot be directly compared to the Recent, we conclude that actual fluctuations in reef carbonate production volumes are reflected by the database calculations.

In order to achieve carbonate production rates rather than total masses for time slices, the duration of the supersequence has to be accounted for. The total carbonate production is simply divided by the duration of the supersequence in millions of years. We are aware of the limitations of this approach (Schlager 1999), but currently see no other way to roughly estimate carbonate production rates from the database. The production rates can be reconstructed in the same way as volumes (Equation 4). The

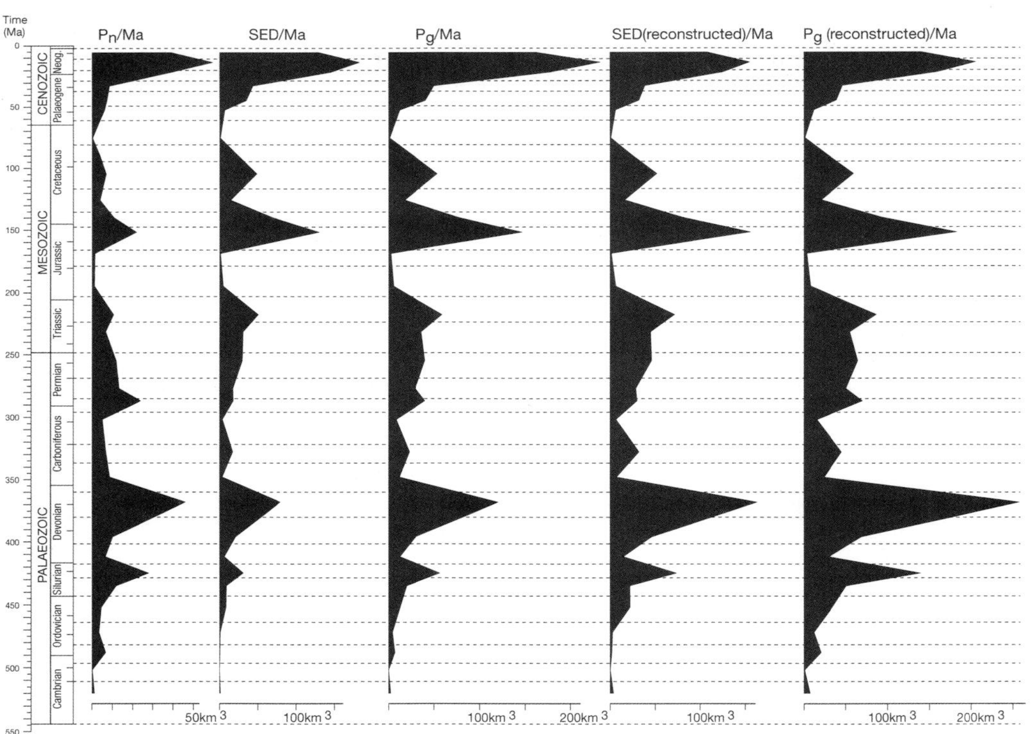

Fig. 5. Phanerozoic reefal carbonate production rates. P_n/Ma = total net carbonate production rate in a supersequence; SED/Ma = total export sediment production rate in a supersequence; P_g/Ma = total gross production rate in a supersequence; SED (reconstructed)/Ma = export sediment production rate balanced for carbonate cycling processes; P_g (reconstructed)/Ma = gross production rate balanced for carbonate cycling processes. Reconstructed values are based on the assumption of an exponential decay constant of 0.0025 following Wilkinson & Walker (1989).

calculated original and reconstructed carbonate production rates are indicated in Fig. 5. All curves exhibit very strong variations and show a similar pattern. No significant trend with time is evident. Only six distinct peaks are observed for reconstructed Phanerozoic gross carbonate production rates: (1) Wenlockian–Ludlovian, (2) Givetian–Frasnian, (3) Late Triassic, (4) Late Jurassic, (5) mid-Cretaceous, (6) Neogene.

Cambrian to Middle Ordovician reefal carbonate production is negligible on the original as well as on the reconstructed curves. Carbonate production rates rise slightly in the Late Ordovician and the Llandovery but a pronounced peak is not evident before the Wenlock. The Wenlockian–Ludlovian represents only a minor peak, although reefs were abundant (Copper & Brunton 1991) and debris potential was moderate. This is due to the small size of most Silurian reefs. Carbonate production rates declined in the latest Silurian and most of the Early Devonian. After a slight rise in the Emsian–Eifelian supersequence, the Phanerozoic maximum of total reconstructed production rate was reached in the Givetian–Frasnian supersequence. Although the debris potential was moderate as compared to most Mesozoic and Cenozoic supersequences, the sheer number of reefs and their great dimensions (e.g. Franke 1973; Ulmishek 1988) were sufficient to produce probably the largest amount of reefal carbonate in the Phanerozoic (Fig. 5). The mass extinction events at the Frasnian–Famennian and Devonian–Carboniferous boundaries reduced the reef carbonate production profoundly. A slight rise is evident already in the Visean/Serpukhovian supersequence but another significant decline occurred in the Moscovian. Reconstructed gross carbonate production rate increased again in the Asselian supersequence and a plateau of moderate carbonate production up until the Late Triassic is evident. Reconstructed sediment production, however, increased more or less constantly from the Moscovian to the Late Triassic. The Late Triassic is the first time slice for which export production is significantly higher than net production. Throughout the Early and Middle Jurassic both export and gross production were very low, but expanded profoundly in the Late Jurassic. Although Middle Jurassic reefs have been shown to have fairly high carbonate production rates (Geister 1989), the limited number of reefs in the pre-Oxfordian Jurassic did not produce a significant cumulative carbonate volume. The major peak in the Late Jurassic supersequence (Figs 4 and 5) is only partly due to increased net production; the majority of reef limestone production was exported. Although many European reefs are deeper-water mud mounds with low debris potential (e.g. Hammes 1995), large reef complexes with high debris potential are known from Europe and central Asia (Michailova 1968; Fortunatova *et al.* 1986). After a significant decline in the earliest Cretaceous, carbonate export and gross production rose significantly in the Albian supersequence. The Albian peak is caused by moderately thick but laterally extensive reefs with high debris potential in the Gulf of Mexico region, Israel and Pacific guyots (Adams 1985; Bein 1976; Rougerie & Fagerstrom 1994). The decline of carbonate production in the Turonian and Campanian supersequences is due to the near-absence of true reefs (Gili *et al.* 1995; Skelton *et al.* 1997) and the moderate debris potential of the prevailing biostromes. Although reef production increased in the Palaeogene, a significant rise is not evident before the Aquitanian supersequence. The Neogene supersequences are characterized by an acme in reef construction in a short time interval. Although the debris potential of Miocene reefs is moderate or only poorly known, the wealth of thick reef complexes in the Mediterranean and especially in southeast Asia (Vincelette & Soeparjadi 1976; Stewart & Durkee 1985; Sun & Esteban 1994) makes the Miocene reefs a very important carbonate factory.

Controls on carbonate production

Intrinsic controls

All curves of reefal carbonate production, whether original or reconstructed, indicate that their Phanerozoic fluctuation is larger than for any other reef attribute or physico-chemical Earth system parameter. A positive feedback mechanism of multiple factors is likely during peaks of reef carbonate production, whereas a negative feedback is responsible for the strong depressions. Since carbonate export production is defined by (1) debris potential, (2) reef dimensions and (3) reef abundance, it is worthwhile analysing the possible controls on these variables separately. A comprehensive correlation matrix on all fields in the database shows that significant ($p < 0.01$) correlations are common (Table 3).

(1) The mean debris potential in a supersequence exhibits most correlations suggesting that this measure is controlled by many factors. Supersequences with an enhanced concentration of mud mounds, microbial

Table 3. *Simplified correlation matrix of the determinants of export production with other fields in the database*

	Reef abundance	Mean thickness	Mean length	Mean debris potential	Export production	Gross production
Reef abundance					+++	+++
Mean thickness	+				+	++
Mean length	+	+++			+	++
Mean debris potential			++		++	+
Mean palaeolatitude				++		
Percent inferior reefs		−−−	−−−		−−	−−
Percent margin reefs	+	+++	+++	+++	+++	+++
Percent framework reefs		+		+++	+++	++
Percent mud mounds				−−−	−	
Percent biostromes		−	−−			
Percent microbial reefs			−−	−−−		
Percent coralline sponge reefs	+					
Percent scleractinian reefs				+++	++	+
Percent pelecypod reefs		−−	−			
Percent constructor guild				+++	+	+
Percent binder guild				−−−	−	
Percent reefs with bioerosion				+++	+	+
Mean diversity	+	+		+++	++	++
Tropical reef diversity*	++	++	+	+++	+++	+++
Mean succession	++	++			+	+
Mean lateral zonation	+++	++	+	++	+++	+++
Mean micrite content				−−−	−−	−−
Mean sparite content				−−−		

Plus sign indicates a positive correlation; minus sign indicates a negative correlation. Three signs: correlation coefficient greater than 0.6; two signs: correlation coefficient of less than 0.6 and strongly significant correlations ($p < 0.01$); one sign: significance of $p < 0.05$

* Less than 30° palaeolatitude

reefs, the binder guild, and high content of micrite and/or sparite tend to have low overall debris potential. Since microbial reefs are mostly mud-dominated mounds, it is possibly their abundance that ultimately triggers the low debris potential of many supersequences. On the other hand, the mean debris potential is significantly enhanced in supersequences with a high concentration of shelf/platform margin reefs, laterally zoned reefs, framework-dominated reefs, scleractinian coral reefs, the constructor guild, reefs with evidence of bioerosion and/or a high mean diversity. Some of these correlations were expected, whereas others are surprising and difficult to explain. The most straightforward relations with debris potential are the percentage of margin reefs and the percentage of reefs affected by bioerosion. Reefs on the edge of carbonate platforms are prone to permanent wave action and tend to be strongly affected by severe storms (Scoffin 1993). Bioerosion is seen as the major agent in modern reefs to produce reefal debris of all sizes (Hutchings 1986; Hubbard *et al.* 1990; Glynn 1997; Perry 1998). The significant increase of bioerosion since the Mesozoic (Kiessling *et al.* 1999) may thus be largely responsible for the coeval increase in debris potential. The fact that scleractinian reefs have a significantly enhanced debris potential in comparison to, for example, coralline sponge reefs may be explained by their comparatively low mechanical resistance (Schumacher & Plewka 1981) and by the usually intense bioerosion in scleractinian coral reefs. It is less easy to explain the strong positive correlation of debris potential and the abundance of framework reefs. The observation that framework-dominated reefs have a higher debris potential than binder-dominated reefs is logical, but what about baffler-dominated mounds? Because of their often more delicate skeletons, we would expect bafflers to be more easily broken and reworked than massive constructors. The solution probably lies in the preferred environmental setting of baffler-dominated reefs. They occur most commonly in intrashelf/intraplatform or deeper-water settings and are thus more protected from hydrodynamic influence. The strong positive correlation of debris potential with mean diversity is a strong argument in favour of the intermediate disturbance hypothesis formulated by Connell (1978). Reefs that are frequently disturbed by factors promoting debris production tend to be more diverse than reefs in protected settings (see also Bertling & Insalaco 1998).

(2) Reef dimensions exhibit less common correlations with other reef attributes. The most significant correlation is the larger average size in margin settings as compared with other environments. Although this relation has often been observed (e.g. Antoshkina 1998), the statistical significance for the whole Phanerozoic is noteworthy. On the contrary, supersequences with a high percentage of intrashelf/intraplatform reefs tend to have smaller reefs on average. The percentage of microbial reefs in a supersequence shifts down the mean length, whereas a high percentage of bivalve reefs tends to reduce average thickness (e.g. many thin rudist biostromes in Cretaceous supersequences). Diversity, succession and lateral zonation of reefs are significantly correlated with mean thickness.

(3) The number of reef sites (reef abundance) in a supersequence is related to only a few other attributes in the database. The most significant correlation is evident between reef abundance and mean lateral zonation in supersequences; that is, reefs tend to be most abundant when they exhibit the highest geometrical complexity. The mean ecological succession and the diversity of low-latitude reefs also correlate positively with reef abundance, suggesting a strong biological control on reef abundance. Weak correlations also exist between reef abundance and reef dimensions, the percentage of margin reefs and the percentage of coralline sponge reefs. The fact that no other correlations are evident is disappointing and complicates the interpretation, because reef abundance is the major control of global reefal carbonate production (Table 3).

Nevertheless, the correlation of reefal carbonate production (original data) with other database fields is not so bad. Apart from their determinants, export production and gross production in supersequences are positively correlated with the percentage of shelf/platform margin reefs, the percentage of true reefs, the mean diversity and the degree of lateral zonation. Thus the factors favouring reef growth at the shelf/platform margin, reef diversity and lateral zonation also favour the carbonate production of reefs. Weak ($0.01 > p < 0.05$) positive correlations also exist between gross production and the percentage of scleractinian reefs, mean succession and the percentage of reefs with evidence of bioerosion. Although bioerosion is strongly correlated with debris potential, the debris production (export production) thus appears only weakly controlled by bioerosion intensity.

Extrinsic controls

Numerous extrinsic controls on global reef growth were suggested. Sea level (Neumann & MacIntyre 1985; Webster *et al.* 1998; Bernecker *et al.* 1999), plate tectonics (Leinfelder 1994; Stanley & Hardie 1998), palaeoclimate (Webby 1984; Flügel 1994), oceanography (Johnson *et al.* 1996; Grigg 1997), nutrients (Hallock & Schlager 1986; Hallock 1988; Hallock *et al.* 1988) and tectono-sedimentary setting (Ahr 1989; Weidlich & Fagerstrom 1999) were proposed as major controls on the growth and demise of reefs. All these Earth system parameters are strongly related and it is thus difficult to isolate any particular factor as a dominant control. Additionally, not all possible controls can be evaluated quantitatively from published literature. The analysis of the influence of Earth system parameters on all reef attributes will be discussed elsewhere (Kiessling, submitted). Here we limit ourselves to the evaluation of the relation between the major determinants of carbonate production and extrinsic controls. Only long-term and global changes in Earth system parameters were analysed. The quantitative evaluation may thus differ from short-term or regional controls. The secular variations of major Earth system parameters were collected from the literature, averaged to agree with our time slice definitions and tested for correlations with the indicated reef attributes (Table 4).

Three sea-level curves (Vail *et al.* 1977; Hallam 1984; and a sea level curve derived from our palaeogeographic maps) were tested for correlations. Only the Hallam curve shows a significant correlation with debris potential and export production. The correlation is negative, that is, high eustatic sea-level seems to diminish debris potential and export production of reefs. These correlations obviously oppose the view of highstand shedding as a dominant process in carbonate sequence stratigraphy (Schlager *et al.* 1994). However, highstand shedding appears to be an important process in relation to low-order (third or fourth order) cycles (Droxler & Schlager 1985; Goldhammer & Harris 1989; Glaser & Droxler 1991), while high-order sea-level highs tend to reduce the debris production of reefs. It should be noted, though, that lowstand shedding can also be an important process at higher stratigraphic resolutions (Becker & House 1997; Vecsei & Sanders 1997; Hansen 1999).

Oceanic crust production rates were derived by Gaffin (1987) from the Vail sea-level curve. Owing to the phase lag between sea-floor generation and eustatic sea-level change, the Gaffin curve differs considerably from the Vail curve (Gaffin 1987). The calculated sea-floor generation rates are negatively correlated with mean reef thickness, reef length and export production rates, whereas no such relation is evident with any sea-level curve. This could imply that factors more intimately related to oceanic crust production (CO_2 degassing, change of water chemistry) influence reef dimensions. The same relationship may be responsible for the positive correlation between the reconstructed Mg/Ca mole ratio in sea water (Hardie 1996; Stanley & Hardie 1998) and reef dimensions.

An elevated partial pressure of CO_2 (pCO_2, equivalent to RCO_2 of Berner 1994) coincides with a significantly lower debris potential. The correlation may be arbitrary since pCO_2 and debris potential exhibit a significant but opposing trend through time. However, the link between reef growth and pCO_2 is commonly accepted (Berger 1982; Buddemeier & Fautin 1996; Kleypass *et al.* 1999). Recent evidence proved a non-linear increase of coral growth with increasing aragonite saturation (Gattuso *et al.* 1998). Since carbonate saturation is basically inversely proportional to atmospheric pCO_2 (Najjar 1992; Kleypass *et al.* 1999), we would expect less robust reef builders during times of elevated pCO_2. However, the obvious calcification in many Silurian–Devonian reef builders (during times of strongly elevated pCO_2) opposes this view. Furthermore, the correlation of pCO_2 and debris potential is negative. Thus the effect of pCO_2 on debris potential is probably more indirect. The strong negative correlation of reconstructed pCO_2 and the percentage of margin reefs in a supersequence ($r = -0.65$, $p < 0.01$) suggests that pCO_2 may control the preferred environmental setting of reefs and thereby also the global debris potential. A possible mechanism could be the influence of pCO_2 on atmospheric circulation and storm frequency.

The direct control of palaeoclimate on reefal carbonate production appears insignificant, though. We have tested various Phanerozoic temperature curves (Berner 1994; Frakes *et al.* 1992; Worsely *et al.* 1994; Veevers 1990) for their correlation with reefal carbonate production determinants, but only the idealized curve of Veevers (1990) exhibits significant correlations. The correlation suggests that average reef dimensions are greater during icehouse intervals. The mean global precipitation as taken from Frakes (1979) is negatively correlated with the debris potential. Continental runoff as modelled

Table 4. *Simplified correlation matrix of the determinants of export production with Earth system parameters*

	Reefs/Ma	Mean thickness	Mean length	Mean debris potential	Export production	Export production rate[10]	Gross production	Gross production rate[10]
Eustatic sea level[1]				– –	–	– –/O		–/O
Oceanic crust production rate[2]		– –	– –			–/–		–/O
Mg/Ca in seawater[3]		+						
pCO_2[4]				– – –		–/O		–/O
Greenhouse intensity[5]		– – –	– –			–/O		–/O
Precipitation[6]				– –				
Continental runoff[7]		– –	–	–	–	–/–	–	–/O
$^{87}Sr/^{86}Sr$[8]				–				
Nutrient level[9]	+		++	+++	+	++/+	+	+/O

Legend as for Table 3.
[1] Hallam (1984); [2] Gaffin (1987); [3] Stanley & Hardie (1998); [4] Berner (1994); [5] Veevers (1990); [6] Frakes (1979); [7] Tardy *et al.* (1989); [8] Veizer *et al.* (1997); [9] Martin (1996); [10] Original and reconstructed values, respectively

by Tardy *et al.* (1989) is inversely correlated with mean reef dimensions, export production and gross production in a supersequence. This implies that supersequences with an elevated continental erosion rate tend to contain smaller reefs and a lower global carbonate production on average. However, an indirect measure of continental erosion, the $^{87}Sr/^{86}Sr$ ratio in marine carbonates (Veizer *et al.* 1997), exhibits only a weak negative correlation with debris potential and none with reef carbonate production.

The sensitivity of the modern reef ecosystem to nutrient concentrations is well known (Hallock & Schlager 1986) and nutrient availability may have partially controlled the history of reef building (Wood 1993). The secular variations of nutrient concentrations are difficult to quantify owing to the impossibility of direct measurement and limitations in indirect deduction. A rather speculative and rough estimation of global nutrient levels by Bambach (1993) and Martin (1995, 1996) is all that can be used for statistical testing. Martin (1995) suggested a principal increase of Phanerozoic nutrient levels from superoligotrophic to mesotrophic/eutrophic conditions, with temporarily elevated levels during icehouse periods. Based on an interval quantification of the suggested nutrient levels, a significant positive correlation with mean length, mean debris potential and carbonate production in a supersequence is evident. Although, again, this may be an arbitrary correlation due to the parallel trend of debris potential and nutrient levels through time, there is a straightforward explanation for this correlation: modern reefs are well known to be more strongly affected by bioerosion with increased nutrient loads (Hallock 1988) and bioerosion is among the major triggers of debris potential in modern reefs. Thus the parallel trend of bioerosion, debris potential and nutrient levels through time may actually indicate a control of global nutrient availability.

Other geological and geochemical parameters such as the $\delta^{13}C$ record in sedimentary carbonates, the area covered by volcanics, the area covered by evaporites, granite abundance, calcite/aragonite ratio in ooids, oolite abundance, $\delta^{34}S$ in apatite and carbonate and continental dispersal, have been tested for correlations with reefal carbonate production determinants. None produced highly significant correlations with the determinants of reef carbonate production.

The common correlations of Earth system parameters with reef attributes are adverted by generally weak correlations with the calculated carbonate production of reefs (Table 4). The correlations are weak, because none of the geological factors is significantly correlated with reef abundance, the major determinant of global carbonate production by reefs. More significant correlations exist between Earth system parameters and carbonate production rates. Significant ($p < 0.05$) correlations are evident between export/gross production rate and the Hallam sea-level curve, oceanic crust production rate, pCO_2, palaeoclimate, continental runoff and nutrient level. We suggest that eustatic sea level and nutrient availability are the major ultimate controls of reefal carbonate production. High first-order sea-level is correlated with low debris potential and export production rates, whereas nutrient availability tends to increase with reef lengths, debris potential and export production rates. In contrast, the influence of global palaeoclimate (mean temperature and precipitation) on export production is considered to be minor.

The quantity and strength of storms, however, are likely to be major controls on the carbonate export production, although recent experimental work has clearly shown that substantial export production can occur without major storms (Hughes 1999). Periodic storms not only enhance the debris potential of reefs but can also favour reef calcification and growth (Highsmith *et al.* 1980). Thus the frequency of storms should be positively correlated with both carbonate export and gross production. Unfortunately, no detailed study on the fluctuation of storm intensity through time is currently available and statistical tests are not possible. Overall wind energy should rise during cool periods owing to enhanced temperature gradients, but the generation of tropical storms (hurricanes) requires heat and should be enhanced during global warmth. However, the intensity of storms cannot simply be derived from global palaeoclimate (Barron 1989). A qualitative comparison of reefal debris potential with the proposed Palaeozoic and Mesozoic hurricane pathways (Marsaglia & Klein 1983) suggests a relation, but no such dependence is evident for export production. Interestingly, a recent compilation of post-Carboniferous storm deposits and their palaeolatitudinal distribution (PSUCLIM 1999) suggests that those deposits are more frequent in the northern hemisphere, especially in time slices when prolific reef growth is also concentrated in the northern hemisphere (e.g. Late Jurassic and middle Cretaceous; Kiessling *et al.* 1999). This underlines the importance of storms for both reef growth and reef destruction and thus sediment production.

Reefs and carbonate sedimentation patterns

Size of Phanerozoic carbonate platforms

A total of 32 detailed palaeogeographic lithofacies maps were constructed by one of the authors (Golonka). The maps depict the global palaeogeographic distribution of Phanerozoic sediments and sedimentary environments. The time slices used for the map constructions agree with the definition of supersequences used in this paper, except for an additional map for the Precambrian–Cambrian boundary (545 Ma) and a missing map for the Artinskian (277 Ma). The maps were presented at international conferences (e.g. Golonka & Ford 1997; Golonka *et al.* 1997) and a sample map for the Norian western Tethys is shown in Kiessling *et al.* (1999, fig. 9), but the complete set of maps is still unpublished.

Based on the lithofacies maps, the area covered by carbonate platforms and shallow-water limestones was measured with GIS software. No distinction could be made between platforms and ramps in the analysis. Mixed carbonate–siliciclastic deposits, chalks and deep-water carbonates were not measured. The measurements thus basically reflect the expanse of warm-water carbonate platforms and shallow-water limestones (collectively termed 'carbonate platforms' hereafter) through time. The measured area of carbonate platforms for each supersequence is depicted in Fig. 6. No correction for carbonate cycling was done in the construction of the graph. The lowest size of carbonate platforms is evident for the Precambrian–Cambrian boundary with an area of roughly 3.5×10^6 km^2. The 520 Ma reconstruction agreeing with our Early Cambrian supersequence exhibits a significant rise with nearly 10×10^6 km^2 occupied by carbonate platforms. A stepwise increase of platform size until the Late Ordovician leads to the Palaeozoic maximum area of 14.4×10^6 km^2 covered by carbonate platforms. After a strong drop in the Late Silurian and Early Devonian (6.9×10^6 km^2), a continuous rise until the Late Devonian and Tournaisian is evident. The decline in platform carbonates from the Visean–Serpukhovian to the Middle Triassic is only interrupted by a minor peak in the Moscovian supersequence. The Late Triassic sees a significant rise in carbonate platform area. A minor drop in the Early Jurassic is followed by a slight increase until the Late Jurassic. The Cretaceous starts with a reduction of carbonate platform area until the Barremian, followed by a very pronounced rise in the Late Cretaceous. The Campanian supersequence appears to represent the Phanerozoic maximum of carbonate platform expansion covering an area of 16.5×10^6 km^2. The carbonate platform area shrinks considerably in the first Cenozoic supersequence. The

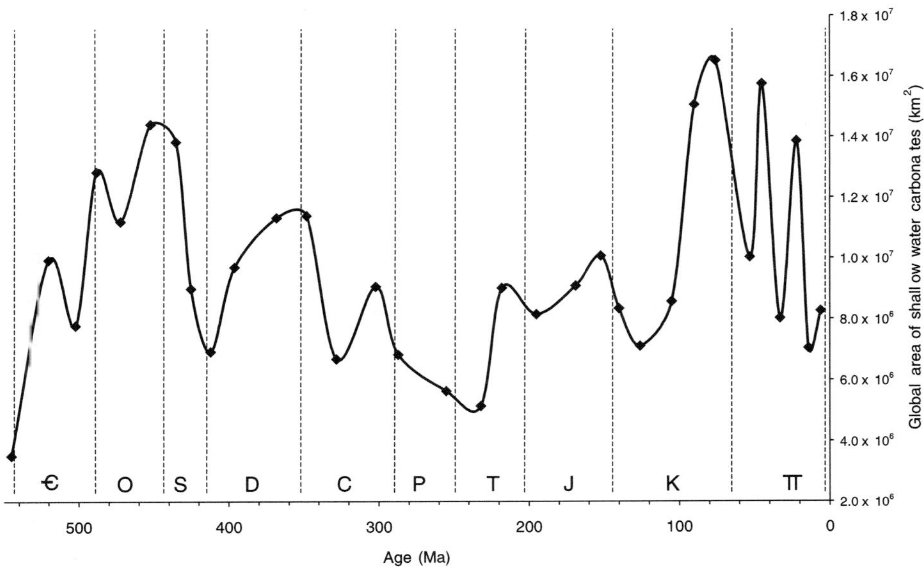

Fig. 6. Measured area of shallow-water carbonate areas (excluding chalk) through time based on 32 detailed lithofacies maps.

Cenozoic is generally characterized by strongly fluctuating platform areas between time slices. Prominent peaks are evident in the Lutetian and Aquitanian supersequences, whereas the Ypresian, Rupelian and Serravallian supersequences represent minima of platform expansion.

The observed pattern of global carbonate platform areas through time differs markedly from the record of Tethyan carbonate platforms as analysed by Philip *et al.* (1995) based on the palaeoenvironmental maps of Dercourt *et al.* (1993). The differences are most pronounced in the Late Triassic and Late Cretaceous, when peaks in our curve are opposed by depressions in the curve of Philip *et al.* (1995). The shapes of the curves agree well in the Jurassic and Cenozoic. Although the differences between curves are partially cleared up by carbonate platforms outside the western Tethys (e.g. Burma-Malaya in the Late Triassic), some differences are fundamental and cannot be explained in a straightforward way. The very large size of late Murghabian to late Anisian platforms given by Philip *et al.* (1995) for the Tethys alone exceeds by far the contemporaneous extent of global carbonate platforms in our survey ($>10 \times 10^6$ km^2 opposed to $< 6 \times 10^6$ km^2). A more detailed comparison and a detailed evaluation of platform development through time are beyond the scope of this paper but will be a worthwhile project in the near future.

Global carbonate sedimentation

The area of carbonate platforms through time may serve as a measure of productivity in the shallow-water carbonate factory. A variety of additional curves was proposed for global carbonate production through time. Schematic, intuitively produced curves (Kazmierczak *et al.* 1985; Morrow *et al.* 1996) as well as quantitative data (Wilkinson & Walker 1989; Bluth & Kump 1991) were evaluated for this paper. The compilation of total carbonate productions (Fig. 7)

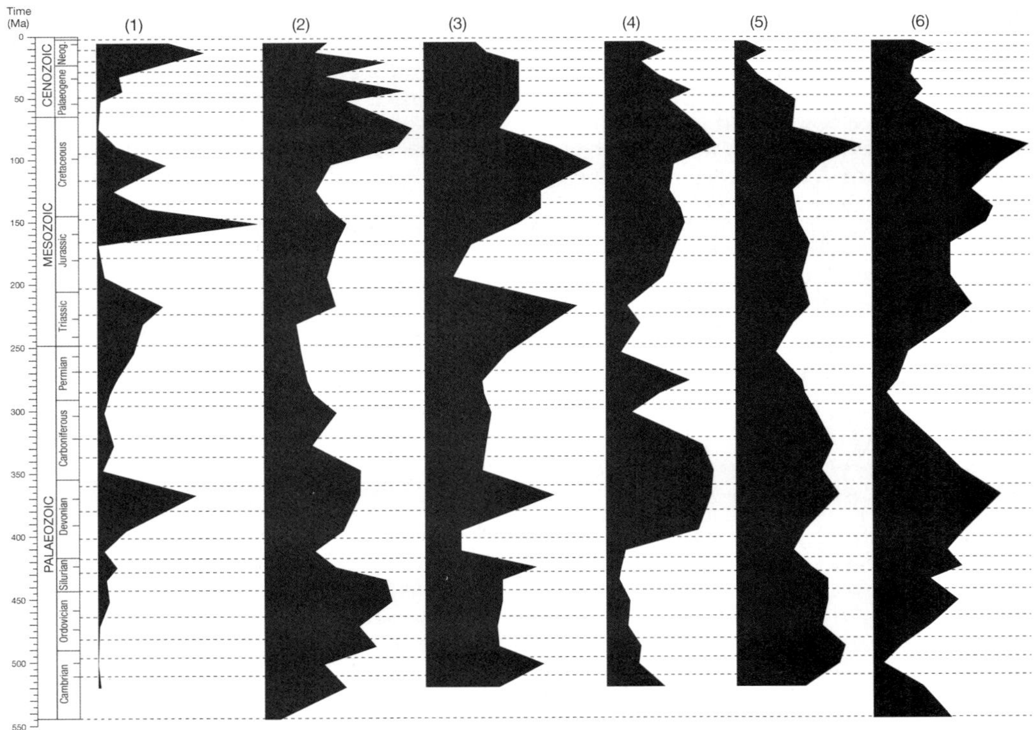

Fig. 7. Comparison of reefal carbonate export with the size of carbonate platforms, maximum carbonate platform accumulation rates and curves for global carbonate production. (1) Reefal carbonate export volume as calculated form the database (raw data); (2) area occupied by carbonate platforms as measured from the lithofacies maps (raw data); (3) maximum carbonate platform accumulation rate (Bosscher & Schlager 1993); (4) remaining mass of continental carbonates (Wilkinson & Walker 1989); (5) total carbonate sedimentation area (Bluth & Kump 1991); (6) total carbonates (Kazmierczak *et al.* 1985).

shows that there is little agreement. This is explained by the variety of applied methods and differences in analysed sedimentary environments. Kazmierczak *et al.* (1985), Morrow *et al.* (1996) and Bluth & Kump (1991) refer to total carbonates, whereas Wilkinson & Walker (1989) separated carbonate mass of oceanic and continental carbonates. Since carbonates are deposited in a variety of environments it is no surprise that the fit of carbonate platform area and the other curves is poor. For the post-Late Jurassic this may be largely explained by the rising importance of the pelagic carbonate factory. Even when excluding oceanic carbonates, however, there is no correlation between our measured carbonate platform size and normalized carbonate mass. Since all quantitative datasets on Phanerozoic carbonate sedimentation are based on various papers of Ronov and colleagues (Ronov *et al.* 1980 and references cited in Wilkinson & Walker 1989; Bluth & Kump 1991), we conclude that there are major inconsistencies between the Ronov data and our maps. The discrepancies are probably due to the fact that Ronov *et al.* (e.g. 1984, 1989) depicted lithologies on present-day base maps, while our maps are based on used palinspastic reconstructions.

An independent measure of shallow-water carbonate productivity is provided by the Bosscher & Schlager (1993) compilation of platform carbonate accumulation rates. Although their dataset is quite heterogeneous, a curve of maximum accumulation rates for given time slices may hint at actual fluctuations in the shallow-water carbonate factory (Fig. 7, curve 3). Nevertheless, the curve based on the Bosscher & Schlager (1993) data does not fit better to our carbonate platform curve than the previous datasets.

Links between reef growth and global carbonate sedimentation patterns?

If Phanerozoic reef growth had a major impact on global carbonate sedimentation, we would expect that the calculated carbonate export volume correlates significantly with carbonate sediment area, volume or mass. This would be expected even though much of the reefal export production is probably transported off the shallow-water environment (e.g. Burne 1974; Land 1979) where it is more likely to be dissolved (Droxler *et al.* 1988). The statistical tests, however, failed to produce significant correlations between reefal export production and measure of shallow-water carbonate sedimentation. The carbonate depositional area (Bluth & Kump 1991) even correlates negatively with the carbonate export production rates of reefs. The absence of a correlation between reefal carbonate export and shallow-water carbonate sedimentation (especially platform carbonates) suggests that averaged for the whole Phanerozoic, reefs had a minor impact on the global carbonate budget.

There are many examples for the decoupling of reef and carbonate platform growth. Very extensive early Palaeozoic carbonate platforms existed in North America and parts of Siberia. Although reefs were locally abundant on these platforms, they were small and had a low debris potential. Thus they contributed little to carbonate platform growth. In the Late Ordovician, for example, the majority of large debris-shedding reef complexes is located outside carbonate platform areas (Kazachstan, Urals) and reefs on carbonate platforms were mostly small (Fig. 8). The Middle to Late Devonian rise in carbonate platform area after the Late Silurian–Early Devonian collapse is paralleled by an increase in reefal carbonate export. For the Late Devonian we can assume that reefs actually played a major role in carbonate platform development (cf. Copper 1997).

However, the Frasnian–Famennian reef collapse is not reflected by decreased carbonate platform size. Extensive late Famennian–early Visean platform and ramp carbonates are known from the SW United States, and many parts of Russia (Ronov *et al.* 1984; Vinogradov 1968; Pol'ster *et al.* 1985). Reefs in the Tournaisian supersequence are mostly deeper-water Waulsortian mounds with very little debris potential. The Bosscher & Schlager (1993) data in this case agree well with the reefal carbonate production data. They indicate a major decline in the accumulation rate of carbonate platforms in the Famennian–Tournaisian. The remainder of the Carboniferous and the Early Permian see little sediment from reefs. Carbonate platforms are not very widespread but significant shallow-water limestones occur in the Arctic region (e.g. Barents sea) and South China (Enos 1995). The accumulation rate of carbonate platforms is continuously low.

A rise in reefal carbonate export in the Middle Permian to Middle Triassic is opposed by continuously decreasing carbonate platform areas. However, the accumulation rate of carbonate platforms increases significantly in this interval. Note that the Early Triassic collapse of both reefal carbonate production and carbonate platform growth is not mirrored in the curves owing to the lumping of Early and Middle Triassic in our time slice definitions. The prominent peak in

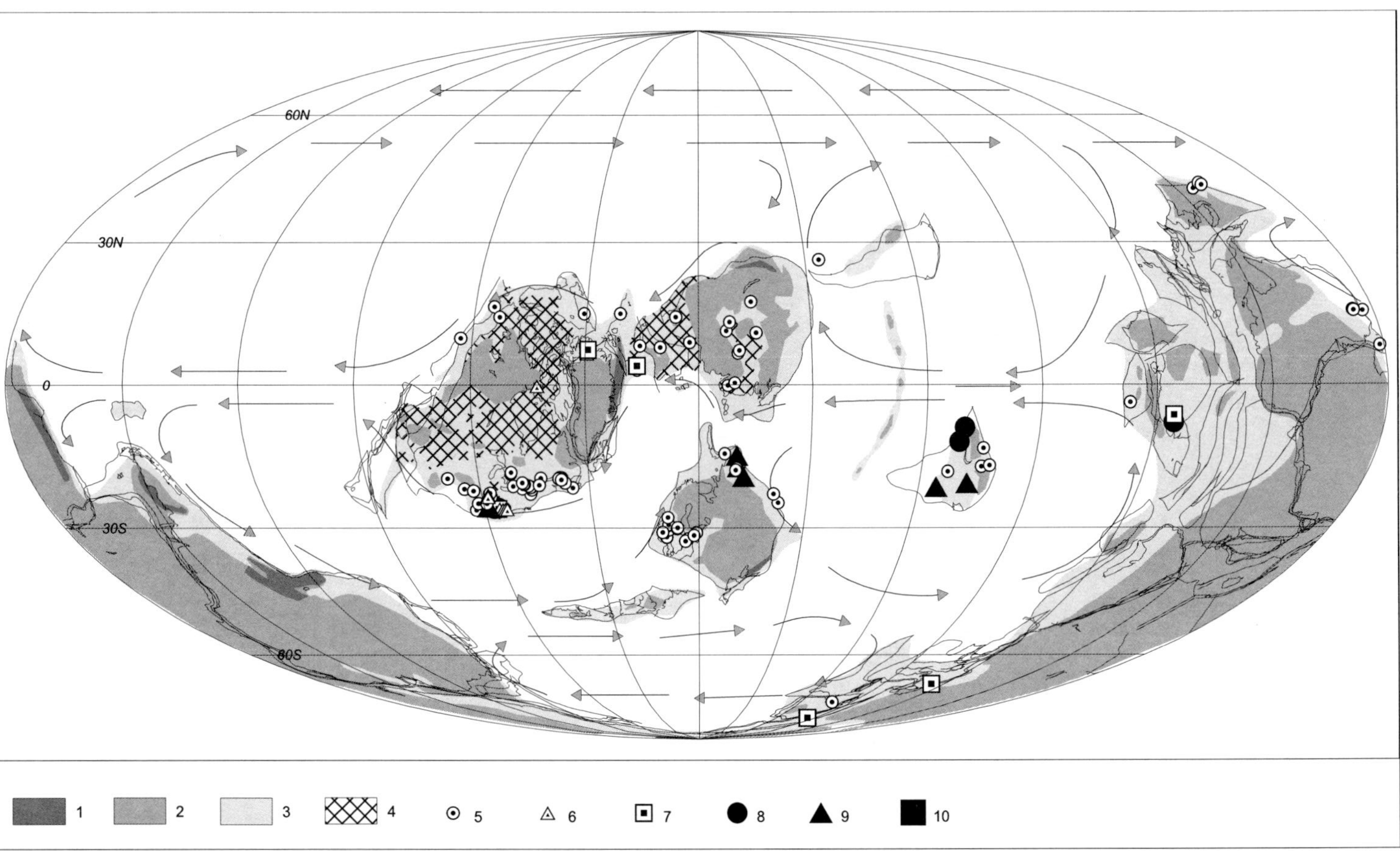

Fig. 8. Palaeogeographic map of the Late Ordovician supersequence as an example of poor fit between reefal carbonate export production and carbonate platform development. 1, mountains; 2, land; 3, shelf; 4, carbonate platforms; 5–10 carbonate export production: 5 = ($25 - 5 \times 10^7$ m^3); 6 = ($5 \times 10^7 - 2 \times 10^8$ m^3); 7 = (2×10^8 m^3 – 10^9 m^3); 8 = ($10^9 - 6.5 \times 10^9$ m^3); 9 = ($6.5 \times 10^9 - 2.25 \times 10^{10}$ m^3); 10 = ($> 2.25 \times 10^{10}$ m^3).

the Late Triassic reefal carbonate export is well reflected in most carbonate production curves, in the size of carbonate platforms and in the tremendous increase in carbonate platform accumulation rates (Fig. 7). Thus the Late Triassic may be the second supersequence when reefs contributed significantly to the growth of carbonate platforms. This is also indicated by the close spatial association of debris-shedding reefs and carbonate platforms (Fig. 9). However, even for moderately sized Late Triassic reefs (moderate debris potential) no reefal components are to be found in a distance of some 10–20 m off the reef (Schäfer 1979).

The Early Jurassic decline of the reef ecosystem corresponds to an extremely reduced carbonate export production and a breakdown of carbonate platform accumulation rates. However, the size of carbonate platforms and the total carbonate sedimentation areas are only slightly diminished. While the export production is even further reduced in the Middle Jurassic, all other carbonate production measures increase during this epoch. The pronounced Late Jurassic peak is mirrored by a minor increase in carbonate platform area and slightly elevated accumulation rates of platform carbonates. The subsequent Early Cretaceous reduction of reefal carbonate production agrees with a reduction of carbonate platform area but global carbonate production rates are almost unchanged.

The Albian supersequence sees a prominent rise in reefal carbonate export and a profound increase of platform carbonate accumulation rates. The size of carbonate platforms and global carbonate masses and sedimentation areas is only slightly elevated. The Late Cretaceous breakdown of the reefal carbonate factory corresponds to a decline in platform carbonate accumulation rates but is opposed by strongly expanded shallow carbonate depositional areas, and an increased global carbonate sedimentation. For the Ypresian supersequence the results are contradictory as well. The reefal carbonate export rises insignificantly, the accumulation rate of platform carbonates increases considerably and the carbonate platform area shrinks dramatically. The strong Lutetian expansion and Rupelian retreat of carbonate platform size is only reflected by minor changes in reefal carbonate export and virtually constant platform accumulation rates. Although the Aquitanian expansion of reefal carbonate export is mirrored by an increasing size of carbonate platform area, other carbonate production curves indicate a decrease, and accumulation rates remained unchanged. The Cenozoic peak in reefal carbonate production (net, export and gross production) in the Middle Miocene supersequence is opposed by a pronounced depression in carbonate platform size and platform carbonate accumulation rates. The subsequent decline in carbonate export is traced by most other curves except for carbonate platform area which exhibits a slight rise.

In summary, reefal carbonate production and carbonate platform evolution appear to be largely decoupled throughout the Phanerozoic. A parallel rise and decline of reefal carbonate export, carbonate platform area and platform carbonate accumulation rates is only evident in the Middle and Late Devonian, in the Triassic to Early Jurassic and in the middle Cretaceous. A significant contribution of reefal carbonates to platform growth is only likely in the Givetian–Frasnian, the Late Triassic, the Late Jurassic and the Neogene. A bias in these results is given by the lumping of carbonate platforms and ramps. Burchette & Wright (1992) indicated that the evolution of reefs and carbonate ramp complexes is contrary. It is reasonable that the development of flat platforms depends strongly on marginal barriers. Thus the growth of shelf/platform margin reefs must have an influence on the development of true carbonate platforms. Probably the correlation between platform and reef evolution would be better if ramps and true platforms could be separated quantitatively in the analyses. An additional bias may be present in the lumping of onshore and offshore carbonate transport from reefs (cf. Fagerstrom 1987, pp. 130–133). A high percentage of carbonate produced by shelf/platform margin reefs is shed in an offshore direction. This carbonate will preferentially dissolve or accumulate in oceanic environments that are prone to subduction and is therefore not represented in most global carbonate production curves. Nevertheless, the role of reefs in the global carbonate budget appears to be largely limited to an indirect constructional control and direct quantitative contributions are of minor importance.

Conclusions

Reefal carbonate production fluctuates strongly throughout the Phanerozoic and exhibits significant peaks in the mid-Siluvian, the Givetian–Frasnian, the Late Triassic, the Late Jurassic, the middle Cretaceous and the Neogene. The determinants of the total amount of carbonate produced by reefs in a given time slice (gross production) are reef abundance, preserved reef dimensions and the debris potential of reefs. All determinants vary through time but not as

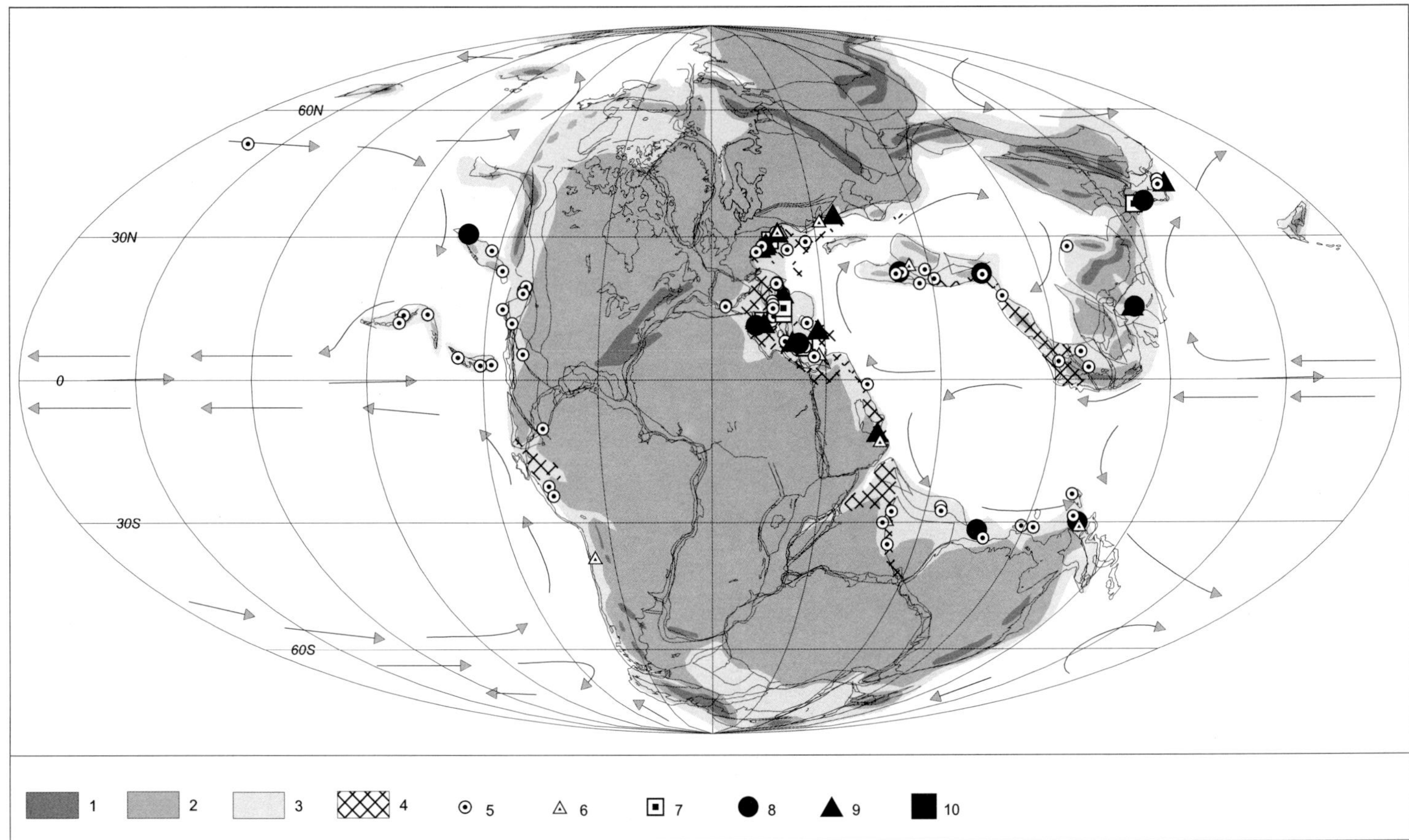

Fig. 9. Palaeogeographic map of the Late Triassic supersequence as an example of good fit between reefal carbonate export production and carbonate platform development. See Fig. 8 for legend.

significantly as carbonate production, suggesting that peaks and lows of the determinants overlap to produce either prominent peaks or profound lows in the reef carbonate curve.

The determinants of reefal carbonate production are in turn controlled by a variety of extrinsic and intrinsic factors. The most important intrinsic factors are the palaeogeographic position, the reef type, the dominant biota, the intensity of bioerosion and the petrographic composition. Gross carbonate production is most significantly enhanced in shelf/platform margin reefs and in reefs with a high biodiversity. Extrinsic controls on the determinants of reefal carbonate production are eustatic sea level, oceanic crust production rate, atmospheric CO_2 concentrations, general climatic state (greenhouse versus icehouse), mean global precipitation, continental runoff and global nutrient levels. Of these controls only continental runoff and nutrient level appear to have a direct impact on the total gross carbonate production in a given time slice, although gross production rates are also correlated with eustatic sea level, oceanic crust production rates, pCO_2 and palaeoclimate.

Reefal carbonate export production is not significantly correlated with the measured area of carbonate platforms, the mass of cratonic carbonates, the total carbonate sedimentation area and carbonate volume, or with the maximum accumulation rates of carbonate platforms. Sediment produced by reef complexes is only likely to contribute significantly to the formation of carbonate platforms in the Givetian–Frasnian, the Triassic, the Late Jurassic and the Neogene. Although reefs are important carbonate factories throughout the Phanerozoic, they obviously had little direct influence on the size and growth potential of carbonate platforms. The Phanerozoic history of carbonate platforms appears to be largely decoupled from the history of reef building.

We gratefully acknowledge the critical remarks of P. Skelton (Milton Keynes), W. Schlager (Amsterdam), O. Weidlich (Berlin) and an anonymous reviewer. Important contributions to the reef database were provided by P. Copper (Sudbury), V. Kuznetsov (Moscow), G. Webb (Brisbane) and B. Webby (Sydney).

References

ADAMS, S. 1985. Lithofacies of the middle Glen Rose reef buildup, Lower Cretaceous shelf margin, east Texas and Louisiana. *In:* BEBOUT, D. & RATCLIFF, D (eds) *Lower Cretaceous Depositional Environments from Shoreline to Slope.* Austin, Texas, 13–22.

AHR, W. M. 1989. Sedimentary and tectonic controls on the development of an Early Mississippian carbonate ramp, Sacramento Mountains area, New Mexico. *In:* CREVELLO, P. D., WILSON, J. L., SARG, F. & READ, J. F. (eds) *Controls on Carbonate Platform and Basin Development.* SEPM, Tulsa, Special Publication, **44**, 203–212.

ANTOSHKINA, A. I. 1998. Organic buildups and reefs on the Paleozoic carbonate platform margin, Pechora Urals, Russia. *Sedimentary Geology*, **118**, 187–211.

BAMBACH, R. K. 1993. Seafood through time: changes in biomass, energetics, and productivity in the marine ecosystem. *Paleobiology*, **19**, 372–397.

BARRON, E. J. 1989. Severe storms during Earth history. *Geological Society of America Bulletin*, **101**, 601–612.

BECKER, R. T. & HOUSE, M. R. 1997. Sea-level changes in the Upper Devonian of the Canning Basin, Western Australia. *Courier Forschungs-Institut Senckenberg*, **199**, 129–146.

BERNECKER, M., WEIDLICH, O. & FLÜGEL, E. 1999. Response of Triassic reef coral communities to sea-level fluctuations, storms and sedimentation: Evidence from a spectacular outcrop (Adnet, Austria). *Facies*, **40**, 229–280.

BEIN, A. 1976. Rudist fringing reefs of Cretaceous shallow carbonate platform of Israel. *AAPG Bulletin*, **60**, 258–272.

BERGER, W. H. 1982. Increase of carbon dioxide in the atmosphere during deglaciation: the coral reef hypothesis. *Naturwissenschaften*, **69**, 87–88.

BERNER, R. A. 1994. Geocarb II: A revised model of atmospheric CO_2 over Phanerozoic time. *American Journal of Science*, **294**, 56–91.

BERTLING, M. 1997. Bioerosion of Late Jurassic reef corals – implications for reef evolution. *In*: *Proceedings of the 8th International Coral Reef Symposium*, **2**, 1663–1668.

—— & INSALACO, E. 1998. Late Jurassic coral/microbial reefs from the northern Paris Basin – facies, paleoecology and paleobiogeography. *Palaeogeography, Palaeoclimatology, Palaeoecology*, **139**, 139–175.

BLUTH, G. J. S. & KUMP, L. R. 1991. Phanerozoic paleogeology. *American Journal of Science*, **291**, 284–308.

BOSENCE, D. W. J., ROWLANDS, R. L. & QUINE, M. L. 1985. Sedimentology and budget of a Recent carbonate mound, Florida Keys. *Sedimentology*, **32**, 317–343.

BOSSCHER, H. & SCHLAGER, W. 1993. Accumulation rates of carbonate platforms. *Journal of Geology*, **101**, 345–355.

BUDDEMEIER, R. W. & FAUTIN, D. G. 1996. Global CO_2 and evolution among the Scleractinia. *Bulletin de l'Institut Oceanographique (Monaco), Special Issue*, **14**, 33–38.

BURCHETTE, T. P. & WRIGHT, V. P. 1992. Carbonate ramp depositional systems. *Sedimentary Geology*, **79**, 3–57.

BURNE, R. V. 1974. The deposition of reef-derived sediment upon a bathyal slope: the deep off-reef envrionment, north of Discovery Bay, Jamaica. *Marine Geology*, **16**, 1–19.

CHAVE, K. E., SMITH, S. V. & ROY, K. J. 1972. Carbonate production of coral reefs. *Marine Geology*, **12**, 123–140.

CONNELL, J. H. 1978. Diversity in tropical rain forests and coral reefs. *Science*, **199**, 1302–1310.

COPPER, P. 1997. Reefs and carbonate productivity: Cambrian through Devonian. *In*: *Proceedings of the 8th International Coral Reef Symposium*, **2**, 1623–1630.

—— & BRUNTON, F. 1991. A global review of Silurian reefs. *Special Papers in Palaeontology*, **44**, 225–259.

DAVIES, G. R. 1989. Lower Permian palaeoaplysinid mound, northern Yukon, Canada. *In:* GELDSETZER, H. H. J., JAMES, N. P. & TEBBUTT, G. E. (eds) *Reefs – Canada and Adjacent Areas.* Canadian Society of Petroleum Geologists, Calgary, Memoir, **13**, 638–642.

DERCOUT, J., RICOU, L. E. & VRIELYNCK, B. (eds) 1993. *Atlas of Tethys Palaeoenvironmental Maps. 1: 20 000 000.* Gauthier-Villars, Paris.

DOLLAR, S. J. & TRIBBLE, G. W. 1993. Recurrent storm disturbance and recovery: A long-term study of coral communities in Hawaii. *Coral Reefs*, **12**, 223–233.

DROXLER, A. W. & SCHLAGER, W. 1985. Glacial versus interglacial sedimentation rates and turbidite frequency in the Bahamas. *Geology*, **13**, 799–802.

——, MORSE, J. W. & KORNICKER, W. A. 1988. Controls on carbonate mineral accumulation in Bahamian basins and adjacent Atlantic Ocean sediments. *Journal of Sedimentary Petrology*, **58**, 120–130.

ENOS, P. 1995. The Permian of China. *In:* SCHOLLE, P. T. M. AND ULMER-SCHOLLE, D. S. (eds) *The Permian of Northern Pangea, Vol. 2: Sedimentary Basins and Economic Resources.* Springer, Berlin, 225–256..

FAGERSTROM, J. A. 1987. *The Evolution of Reef Communities.* Wiley, New York.

FLÜGEL, E. 1994. Pangean shelf carbonates: Controls and paleoclimatic significance of Permian and Triassic reefs. *In:* KLEIN, G. D. (ed.) *Pangea: Paleoclimate Tectonics and Sedimentation During Accretion Zenith and Breakup of a Supercontinent.* Geological Society of America, Boulder, Special Paper, **288**, 247–266.

——, ALT, T., JOACHIMSKI, M. M., RIEMANN, V. & SCHELLER, J. 1993. Korallenriff-Kalke im oberen Malm (Unter-Tithon) der Südlichen Frankenalb (Laisacker, Marching): Mikrofazies-Merkmale und Fazies-Interpretation. *Geologische Blätter von Nordost-Bayern*, **43**, 33–56.

——, KIESSLING, W. & GOLONKA, J. 1996. Phanerozoic reef patterns: data survey, distribution maps and interpretation. *Göttinger Arbeiten zur Geologie und Paläontologie*, **SB2**, 391–396.

FORTUNATOVA, N. K., MICHEEV, I. G., IBRAGIMOV, A. G., FARBIROVICH, V. P. & SHVEZ-TENETA-GURII, A. G. 1986. Metodi vydeleniya rifovych fazii v verchnejurskich karbonatych otlozheniyach Yzhnogo Uzbekistana. *In:* SOKOLOV, B. S. (ed.) *Fanerosoiskie rifi i koralli SSSR.*. Nauka, Moskow, 149–161.

FRAKES, L. A. 1979. *Climates Through Geologic Time.* Elsevier, Amsterdam.

——, FRANCIS, J. E. & SYKTUS, J. I. 1992. *Climate Modes of the Phanerozoic: the history of the Earth's climate over the past 600 million years.* Cambridge University, Cambridge.

FRANKE, W. 1973. Fazies, Bau und Entwicklungsgeschichte des Iberger Riffes (Mitteldevon und Unterkarbon III, NW-Harz, W-Deutschland). *Geologisches Jahrbuch*, **A11**, 3–127.

GAFFIN, S. 1987. Ridge volume dependence on seafloor generation rate and inversion using long term sealevel change. *American Journal of Science*, **287**, 596–611.

GATTUSO, J.-P., FRANKIGNOULLE, M., BOURGE, I., ROMAINE, S. & BUDDEMEIER, R. W. 1998. Effect of calcium carbonate saturation of seawater on coral calcification. *Global and Planetary Change*, **18**, 37–46.

GEISTER, J. 1989. Qualitative aspects of coral growth and carbonate production in a Middle Jurassic reef. *Memoir Association of Australasian Palaeontologists*, **8**, 425–432.

GILI, E., MASSE, J. P. & SKELTON, P. W. 1995. Rudists as gregarious sediment dwellers, not reef builders, on Cretaceous carbonate platforms. *Palaeogeography, Palaeoclimatology, Palaeoecology*, **118**, 245–267.

GLASER, K. S. & DROXLER, A. W. 1991. High production and highstand shedding from deeply submerged carbonate banks, northern Nicaragua Rise. *Journal of Sedimentary Petrology*, **61**, 128–142.

GLYNN, P. W. 1997. Bioerosion and coral-reef growth: a dynamic balance. *In:* Birkeland, C. (ed.) *Life and Death of Coral Reefs.* Chapman & Hall, New York, 68–95.

GOLDHAMMER, R. K. & HARRIS, M. T. 1989. Eustatic controls on the stratigraphy and geometry of the Latemar buildup (Middle Triassic), the Dolomites of northern Italy. *In:* CREVELLO, P. D., WILSON, J. L., SARG, F. & READ, J. F. (eds) *Controls on Carbonate Platform and Basin Development.* SEPM, Tulsa, Special Publication, **44**, 323–338.

GOLONKA, J. & FORD, D. 1997. Pangean (Late Carboniferous to Middle Jurassic) paleoenvironment and lithofacies (abstract). *Gaea Heidelbergensis*, **3**, 143–144.

——, FORD, D. W., EDRICH, M., PAUKEN, R. J. WILDHARBER, J. & BOCHAROVA, N. Y. 1997. Zuni (Jurassic-Cretaceous) Paleoenvironment and Lithofacies (abstract). *CSPG–SEPM Joint Convention, Program with Abstracts.* Canadian Society of Petroleum Geologists, Calgary, 110.

GRADSTEIN, F. M. & OGG, J. G. 1996. Geological time scale for the Phanerozoic. *Episodes*, **19**, 3–4.

GREGOR, C. B. 1985. The mass–age distribution of Phanerozoic sediments. *In:* Snelling, N. J. (ed.) *The Chronology of the Geologic Record.* Geological Society of America, Boulder, Memoir, **10**, 284–289.

GRIGG, R. W. 1997. Paleooceanography of coral reefs in the Hawaiian-Emperor Chain: Revisited. *Coral Reefs*, **16**, S33-S38.

HALLAM, A. 1984. Pre-Quarternary changes of sea level. *Annual Reviews of Earth and Planetary Science*, **12**, 205–243.

HALLOCK, P. 1988. The role of nutrient availability in

bioerosion: Consequences to carbonate buildups. *Palaeogeography, Palaeoclimatology, Palaeoecology*, **63**, 275–291.

—— & SCHLAGER, W. 1986. Nutrient excess and the demise of coral reefs and carbonate platforms. *Palaios*, **1**, 389–398.

HALLOCK, P. A., HINE, A. C., VARGO, G. A., ELROD, J. A. & JAAP, W. C. 1988. Platforms of the Nicaraguan rise: Examples of the sensitivity of carbonate sedimentation to excess trophic resources. *Geology*, **16**, 1104–1107.

HAMMES, U. 1995. Initiation and development of small-scale sponge mud-mounds, Late Jurassic, southern Franconian Alb, Germany. *In:* MONTY, C. L. V., BOSENCE, D. W. J., BRIDGES, P. H. & PRATT, B. R. (eds) *Carbonate Mud-Mounds – Their Origin and Evolution.* International Association of Sedimentologists, Special Publications, **23**, Blackwell, Oxford, 335–357.

HANSEN, K. S. 1999. Development of a prograding carbonate wedge during sea level fall: Lower Pleistocene of Rhodes, Greece. *Sedimentology*, **46**, 559–576.

HARDIE, L. A. 1996. Secular variation in seawater chemistry: An explanation for the coupled secular variation in the mineralogies of marine limestones and potash evaporites over the past 600 m.y. *Geology*, **24**, 279–283.

HARRIS, M. T. 1994. A volume-based approach to reef productivity and submarine erosion rates – a case-study of a Middle Triassic reef margin (Latemar buildup, northern Italy). *Journal of Geology*, **102**, 603–610.

HIGHSMITH, R. C., RIGGS, A. C. & D'ANTONIO, C. M. 1980. Survival of hurricane generated coral *Acropora palmata* fragments and a disturbance model of reef calcification growth rates. *Oecologia*, **46**, 322–329.

HÖFLING, R. 1997. Eine erweiterte Riff-Typologie und ihre Anwendung auf kretazische Biokonstruktionen. *Bayerische Akademie der Wissenschaften mathematisch-naturwissenschaftliche Klasse Abhandlungen, Neue Folge*, **169**, 1–127.

HUBBARD, D. K., BURKE, R. B. & GILL, I. P. 1998. Where's the reef: the role of framework in the Holocene. *Carbonates & Evaporites*, **13**, 3–9.

——, MILLER, A. I. & SCATURO, D. 1990. Production and cycling of calcium carbonate in a shelf-edge reef system (U. S. Virgin Islands): Applications to the nature of reef systems in the fossil record. *Journal of Sedimentary Petrology*, **60**, 335–360.

HUGHES, T. P. 1999. Off-reef transport of coral fragments at Lizard Island, Australia. *Marine Geology*, **157**, 1–6.

HUTCHINGS, P. A. 1986. Biological destruction of coral reefs. *Coral Reefs*, **4**, 239–252.

INGELS, J. J. C. 1963. Geometry, paleontology, and petrography of Thornton Reef complex, Silurian of northeastern, Illinois. *AAPG Bulletin*, **47**, 404–440.

JAN, F. G. B. L. 1998. The role of high-energy events (hurricanes and/or tsunamis) in the sedimentation, diagenesis and karst initiation of tropical shallow-water carbonate platforms and atolls. *Sedimentary Geology*, **118**, 3–36.

JOHNSON, C. C., BARRON, E. J., KAUFFMAN, E. G., ARTHUR, M. A., FAWCETT, P. J. & YASUDA, M. K. 1996. Middle Cretaceous reef collapse linked to ocean heat transport. *Geology*, **24**, 376–380.

KAZMIERCZAK, J., ITTEKKOT, V. & DEGENS, E. T. 1985. Biocalcification through time: environmental challenge and cellular response. *Paläontologische Zeitschrift*, **59**, 15–33.

KIESSLING, W., FLÜGEL, E. & GOLONKA, J. 1999. Paleo Reef Maps: Evaluation of a comprehensive database on Phanerozoic reefs. *AAPG Bulletin*, **83**, 1552–1587.

KLEYPAS, J. A. 1997. Modeled estimates of global reef habitat and carbonate production since the last glacial maximum. *Paleoceanography*, **12**, 533–545.

——, BUDDEMEIER, R. W., ARCHER, D., GATTUSO, J.-P., LANGDON, C. & OPDYKE, B. N. 1999. Geochemical consequences of increased atmospheric carbon dioxide on coral reefs. *Science*, **284**, 118–120.

LAND, L. S. 1979. The fate of reef-derived sediment on the north Jamaican island slope. *Marine Geology*, **29**, 55–71.

LEINFELDER, R. R. 1994. Distribution of Jurassic reef types: a mirror of structural and environmental changes during breakup of Pangea. *In:* Embry, A. F., Beauchamp, B. & Glass, D. J. (eds) *Pangea: Global Environments and Resources.* Canadian Society of Petroleum Geologists, Calgary, Memoir, **17**, 677–700.

MACINTYRE, I. G., GRAUS, R. R., REINTHAL, P. N., LITTLER, M. M. & LITTLER, D. S. 1987. The Barrier Reef sediment apron: Tobacco Reef, Belize. *Coral Reefs*, **6**, 1–12.

MACKENZIE, F. T. & MORSE, J. W. 1992. Sedimentary carbonates through Phanerozoic time. *Geochimica et Cosmochimica Acta*, **56**, 3281–3295.

MARSAGLIA, K. M. & KLEIN, G. D. 1983. The paleogeography of Paleozoic and Mesozoic storm depositional systems. *Journal of Geology*, **91**, 117–142.

MARTIN, R. E. 1995. Cyclic and secular variation in microfossil biomineralization: clues to the biogeochemical evolution of Phanerozoic oceans. *Global and Planetary Change*, **11**, 1–23.

—— 1996. Secular increase in nutrient levels through the Phanerozoic: implications for productivity, biomass, and diversity of the marine biosphere. *Palaios*, **11**, 209–219.

MICHAILOVA, M. V. 1968. Biogermnie massivy v verchnejurskich otlozheniyach gornogo Krima i Severnogo Kavkaza. *In: Iskopaemie rifi i metodika ich izuchenya. Trudy tretei paleoekologo – litologicheskoi sessii.* Uraliskii Filial Akademii Nauk SSSR, Moscow, 196–209.

MILLIMAN, J. D. 1993. Production and accumulation of calcium carbonate in the ocean;: Budget of a non-steady state. *Global Biochemical Cycles*, **7**, 927–957.

—— & DROXLER, A. W. 1996. Neritic and pelagic carbonate sedimentation in the marine envrionment: ignorance is not bliss. *Geologische Rundschau*, **85**, 496–504.

MORROW, J. R., SCHINDLER, E. & WALLISER, O. H. 1996. Phanerozoic development of selected global environmental features. *In:* WALLISER, O. H. (ed.) *Global Events and Event Stratigraphy.* Springer, Berlin, 53–61.

NAJJAR, R. G. 1992. Marine biogeochemistry. *In:* TRENBERTH, K. E. (ed.) *Climate System Modeling.* Cambrige University, Cambridge, 241–280.

NEUMANN, C. & MACINTYRE, I. G. 1985. Reef response to rising sea level: keep up, catch up or give up. *Proceedings 5th International Coral Reef Symposium,* Tahiti, **3**, 105–110.

PERRY, C. T. 1998. Macroborers within coral framework at Discovery Bay, north Jamaica: species distribution and abundance, and effects on coral preservation. *Coral Reefs*, **17**, 277–287.

PHILIP, J., MASSE, J.-P. & CAMOIN, G. 1995. Tethyan carbonate platforms. *In:* NAIRN, A. E. M., RICOU, L.-E., VRIELYNCK, B. & DERCOURT, J. (eds) *The Ocean Basins and Margins. Volume 8: The Tethys Ocean.* Plenum, New York, 239–265.

POL'STER, L. A., CHERPELYUGIN, A. B., SHEREMET'YEV, Y. F. & SHERMET'YEVA, G. A. 1985. Formirovaniye krupnykh zon i ploshchadei neftegazonakopleniya v paleozoyskisk rifakh Prikaspiyskoy sineklizy. *In: Usloviya formirovaniya krupnykh uon neftegazonakopleniya.* Moscow, 163–168.

PSUCLIM 1999. Storm activity in ancient climates, 2, An analysis using climate simulations and sedimentary structures. *Journal of Geophysical Research – Atmospheres*, **D104**, 27295–27320.

RONOV, A. B., KHAIN, V. E., BALUKHOVSKY, A. N., AND SESLAVINSKY, K. B. 1980. Quantiative analysis of Phanerozoic sedimentation. *Sedimentary Geology*, **23**, 311–325.

RONOV, A., KHAIN, V. & SESLAVINSKI, A. 1984. *Atlas of Lithological Paleogeographical Maps of the World: Late Precambrian and Paleozoic of the Continents.* USSR Academy of Sciences, Leningrad.

——, —— & BALUKHOVSKI, A. 1989. *Atlas of Lithological Paleogeographical Maps of the World: Mesozoic and Cenozoic of the Continents.* USSR Academy of Sciences, Leningrad.

ROUGERIE, F. & FAGERSTROM, J. A. 1994. Cretaceous history of Pacific Basin guyot reefs: a reappraisal based on geothermal endo-upwelling. *Palaeogeography, Palaeoclimatology, Palaeoecology*, **112**, 239–260.

RUPPEL, S. C. & KERANS, C. 1987. Paleozoic buildups and associated facies, Llano Uplift, central Texas. *Austin Geological Society Guidebook*, **10**, 1–33.

SCATURO, D. M., STROBEL, J. S., KENDALL, C. G. S. C., WENDTE, J. C., BISWAS, G., BEZDEK, J. & CANNON, R. 1989. Judy Creek: a case study for a two-dimensional sediment deposition simulation. *In:* CREVELLO, P. D., WILSON, J. L., SARG, F. & READ, J. F. (eds), *Controls on Carbonate Platform and Basin Development.* SEPM, Tulsa, Special Publication, **44**, 63–76.

SCHÄFER, P. 1979. Fazielle Entwicklung und palökologische Zonierung zweier obertriadischer Riffstrukturen in den Nördlichen Kalkalpen ('Oberrhät'-Riff-Kalke, Salzburg). *Facies*, **1**, 3–245.

SCHLAGER, W. 1999. Scaling of sedimentation rates and drowning of reefs and carbonate platforms. *Geology*, **27**, 183–186.

——, REIJMER, J. J. G. & DROXLER, A. 1994. Highstand shedding of carbonate platforms. *Journal of Sedimentary Research*, **B64**, 270–281.

SCHUHMACHER, H. & PLEWKA, M. 1981. Mechanical resistance of reefbuilders through time. *Oecologia*, **49**, 279–282.

SCOFFIN, T. P. 1993. The geological effects of hurricanes on coral reefs and the interpretation of storm deposits. *Coral Reefs*, **12**, 203–221.

——, STEARN, C. W., BOUCHER, D., FRYDL, P., HAWKINS, C. M., HUNTER, I. G. & MACGEACHY, J. K. 1980. Calcium carbonate budget of a fringing reef on the west coast of Barbados 2. Erosion sediments and internal structure. *Bulletin of Marine Science*, **30**, 475–508.

SKELTON, P. W., GILI, E., ROSEN, B. R. & VALLDEPERAS, F. X. 1997. Corals and rudists in the late Cretaceous: A critique of the hypothesis of competitive displacement. *Boletín de la Real Sociedad Española de Historia Natural (Sección Geológica)*, **92**, 225–239.

SLOSS, L. L. 1963. Sequences in the cratonic interior of North America. *Geological Society of America Bulletin*, **74**, 93–113.

SPALDING, M. D. & GRENFELL, A. M. 1997. New estimates of global and regional coral reef areas. *Coral Reefs*, **16**, 225–230.

STANLEY, S. M. & HARDIE, L. A. 1998. Secular oscillations in the carbonate mineralogy of reef-building and sediment-producing organisms driven by tectonically forced shifts in seawater chemistry. *Palaeogeography, Palaeoclimatology, Palaeoecology*, **144**, 3–19.

STEWART, W. D. & DURKEE, E. F. 1985. Petroleum potential of the Papuan Basin. *Oil and Gas Journal*, **83**, 151–160.

STODDART, D. R. 1972. Catastrophic damage to coral reef communities by earthquake. *Nature*, **239**, 51–52.

SUN, S. Q. & ESTEBAN, M. 1994. Paleoclimatic controls on sedimentation, diagenesis, and reservoir quality: lessons from Miocene carbonates. *AAPG Bulletin*, **78**, 519–543.

TARDY, Y. 1989. The global water cycles and continental erosion during Phanerozoic time (570 my). *American Journal of Science*, **289**, 455–483.

TUDHOPE, A. W. & RISK, M. J. 1985. Rate of dissolution of carbonate sediments by microboring organisms, Davies reef, Australia. *Journal of Sedimentary Petrology*, **55**, 440–447.

ULMISHEK, G. F. 1988. Upper Devonian-Tournaisian facies and oil resources of the Russian Craton's eastern margin. *In:* MCMILLAN, N. J., EMBRY, A. F. & GLASS, D. J. (eds) *Devonian of the World. Volume I: Regional Syntheses.* Canadian Society of Petroleum Geologists, Calgary, Memoir, **14**, 527–549.

VAIL, P. R., MITCHUM, R. M. J., TODD, R. G., *et al.* 1977. Seismic stratigraphy and global changes of sea-level. *In:* Payton, C. E. (ed.) *Seismic Stratigraphy – Applications to Hydrocarbon Exploration.* AAPG, Memoir, **26**, 49–212.

VECSEI, A. & SANDERS, D. G. K. 1997. Sea-Level highstand and lowstand shedding related to shelf margin aggradation and emersion, Upper Eocene–Oligocene of Maiella carbonate platform, Italy. *Sedimentary Geology*, **112**, 219–234.

VEEVERS, J. J. 1990. Tectonic-climatic supercyle in the billion-year plate-tectonic eon; Permian Pangean icehouse alternates with Cretaceous dispersed-continents greenhouse. *Sedimentary Geology*, **68**, 1–16.

VEIZER, J., BUHL, D., DIENER, A., *et al.* 1997. Strontium isotope stratigraphy: potential resolution and event correlation. *Palaeogeography, Palaeoclimatology, Palaeoecology*, **132**, 65–77.

VINCELETTE, R. R. & SOEPARJADI, R. A. 1976. Oil bearing reefs in Salwati Basin of Irian Jaya Indonesia. *AAPG Bulletin*, **60**, 1448–1462.

VINOGRADOV, A. P. 1968. *Atlas of the Lithological-Paleogeographical Maps of the USSR. Vol.II: Devonian, Carboniferous, Permian*. Ministry of Geology of the USSR and Academy of Sciences of the USSR, Moscow.

WARE, J. R., SMITH, S. V. & REAKA-KUDLA, M. L. 1992. Coral reefs: sources or sinks of atmospheric CO_2? *Coral Reefs*, **11**, 127–130.

WEBBY, B. D. 1984. Ordovician reefs and climate: a review. *In:* BRUTON, D. L. (ed.) *Aspects of the Ordovician System.* Palaeontological Contributions from the University of Oslo, **295**, 89–100.

WEBSTER, J. M., DAVIES, P. J. & KONISHI, K. 1998. Model of fringing reef development in response to progressive sea level fall over the last 7000 years (Kikai-jima, Ryukyu Islands, Japan). *Coral Reefs*, **17**, 289–308.

WEIDLICH, O. & FAGERSTROM, J. A. 1999. Influence of sea-level changes on development, community structure, and quantitative composition of the upper Capitan-massive (Permian), Guadalupe Mountains. SEPM, Tulsa, Special Publications, **65**, 139–160.

WEISSERT, H. & MOHR, H. 1996. Late Jurassic climate and its impact on carbon cycling. *Palaeogeography, Palaeoclimatology, Palaeoecology*, **122**, 27–43.

WILKINSON, B. H. & WALKER, J. C. G. 1989. Phanerozoic cycling of sedimentary carbonate. *American Journal of Science*, **289**, 525–548.

WOLD, C. N. & HAY, W. W. 1990. Estimating ancient sediment fluxes. *American Journal of Science*, **290**, 1069–1089.

WOOD, R. 1993. Nutrients, predation and the history of reef-building. *Palaios*, **8**, 526–543.

WORSELY, T. R., MOORE, T. L., FRATICELLI, T. M. & SCOTESE, C. R. 1994. Phanerozoic CO_2 levels and global temperatures inferred from changing paleogeography. *In:* KLEIN, G. D. (ed.) *Pangea: Paleoclimate Tectonics and Sedimentation during Accretion Zenith and Breakup of a Supercontinent.* Geological Society of America, Boulder, Special Paper, **228**, 57–73.

ZANKL, H. 1977. Quantitative aspects of carbonate production in a Triassic reef complex. *Proceedings of the Third International Coral Reef Symposium* (Miami), **2**, 379–382.

Sedimentation rates and growth potential of tropical, cool-water and mud-mound carbonate systems

WOLFGANG SCHLAGER

Vrije Universiteit/Earth Sciences, De Boelelaan 1085, 1081 HV Amsterdam, The Netherlands

Abstract: Carbonate fixation in the ocean proceeds in three basic modes: abiotically, biotically induced (with organic trigger), and biotically controlled (where organisms determine timing, location and composition of the product). The three modes combine in a variety of ways to produce carbonate sediment. When viewed on the scale of formations and global facies belts, three benthic carbonate production systems, or 'factories', emerge: (1) the tropical shallow-water system, dominated by biotically controlled (mainly photo-autotrophic) and abiotic precipitates; (2) the cool-water system, dominated by biotically controlled (mainly heterotrophic) precipitates; and (3) the mud-mound system, dominated by abiotic and biotically induced (mainly microbial) precipitates.

The sedimentation rates of all three factories decrease as the time span of observation increases. This scaling trend is not just an artifact of ratio correlation; it is almost certainly related to the episodic, pulsating nature of sedimentation.

The growth potential of the three systems can be estimated by drawing the envelope of the maximum observed rates of aggradation. The tropical system shows the highest rates: 10^4 μm a^{-1} at 10^3 years, decreasing to 10^2 μm a^{-1} at 10^7 years. The maximum rates of the cool-water system are comparable to those of the tropical carbonates for intervals shorter than 2×10^5 years but they amount to only 25% of the tropical standard in the domain of 10^6–10^8 years. The high short-term rates of cool-water carbonates are probably caused by reworking and local trapping of sediment. The growth rates of the mud-mound system significantly exceed cool-water rates and rival the aggradation rates of tropical shallow-water carbonates. However, the mud-mound system exports less sediment than its tropical counterpart and is, therefore, somewhat less productive.

The sedimentation rate is an important characteristic of any sediment accumulation. For shallow-water carbonate systems this rate is particularly important because the material is produced within the depositional setting and the rate of sediment accumulation provides a first clue to the growth potential of a carbonate system, i.e. the maximum rate at which it can aggrade and produce sediment. The growth potential determines whether the system can keep up with a given relative rise of sea level or whether it will fall behind, leave accommodation unfilled and eventually drown.

A central point of this paper is the comparison of different carbonate production systems. Since World War II, carbonate sedimentologists have generated a wealth of data on carbonate deposits of shallow tropical seas. Tropical carbonates have become a standard of shallow-water carbonate sedimentation because of their large volume, conspicuous appearance and importance as hydrocarbon reservoirs. Work on other systems has lagged for some time but gained momentum in the past two decades. These more recent studies demonstrated conclusively that besides the photosynthetically driven system of the sunlit tropics there are other 'carbonate factories', i.e. other settings where biotic and abiotic factors combine in characteristic ways to produce significant carbonate accumulations. First, it was shown that the modern carbonate deposits beyond about 30° of latitude differ profoundly from their tropical counterparts and that these 'cool-water carbonates' can be traced through the geological record (e.g. Lees & Buller 1972; Nelson *et al.* 1988; James 1997). Secondly, work on caves and foreslopes of modern reefs as well as certain intervals of the distant geological past revealed yet another production system, primarily governed by microbial precipitation, that differs from both the shallow-water tropical and the cool-water factory (Monty *et al.* 1995; Reitner & Neuweiler 1995). It is termed the 'mud-mound system' in this report. Mud-mound accumulations are not particularly conspicuous in modern oceans but they are a dominant feature in certain parts of the geological past such as the late Palaeozoic.

Another topic recently brought into focus and relevant to the discussion of production systems is the scaling of sedimentation rates. When viewed over a wide range of time intervals, sedimentation rates decrease with increasing length

From: INSALACO, E., SKELTON, P. W. & PALMER, T. J. (eds) 2000. *Carbonate Platform Systems: components and interactions*. Geological Society, London, Special Publications, **178**, 217–227. 0305–8719/00/$15.00

of time (e.g. Sadler 1981). The mathematical problem of ratio correlation does not seem to influence this pattern in any fundamental way (Gardner *et al.* 1987; Schlager *et al.* 1998; Schlager 1999). Scaling of sedimentation rates is a real phenomenon and should be considered in any discussion of carbonate production and growth potential.

This paper compares accumulation rates of tropical, cool-water and mud-mound carbonates, determines their change with the changing length of the time interval considered and estimates the growth potential of these systems.

Modes of precipitation and carbonate factories

Precipitation of solid sedimentary material from the dissolved load of the ocean is partly biotic, partly abiotic. The ratio of abiotic to biotic products varies widely among the major sediment families. Evaporites are essentially abiotic in origin, opaline sediments entirely biotic. Marine carbonate sediments contain significant portions of abiotic and biotic precipitates and pathways of precipitation are particularly diverse. In their thorough review, Lowenstam & Weiner (1989) distinguished three basic modes of carbonate precipitation and their system is adopted here (Fig. 1).

(1) Abiotic precipitation is controlled by the laws of thermodynamics and further constraints imposed by reaction kinetics. Kinetics is important because marine carbonate precipitation proceeds at rather low temperatures (see Morse & Mackenzie 1990 for review). Examples of abiotic precipitation are marine cements of aragonite or magnesian calcite (Morse & Mackenzie 1990).

(2) Biotically induced precipitation implies that organisms trigger the process but have little influence on the pathway such that the final product resembles abiotic precipitates in most aspects. Examples are most microbial micrites, particularly those that precipitate after the death of the microbes (Reitner & Neuweiler 1995; Monty 1995).

(3) Biotically controlled precipitation stands for processes where organisms determine the location, the onset and termination of the process, as well as the composition and texture of the precipitate. The skeletons of molluscs, scleractinian corals and calcareous algae are examples of biotically controlled precipitates (Lowenstam & Weiner 1989). This category may be subdivided into precipitation by autotrophic and heterotrophic organisms. In the autotrophic category, photosynthetic organisms are the most important carbonate producers.

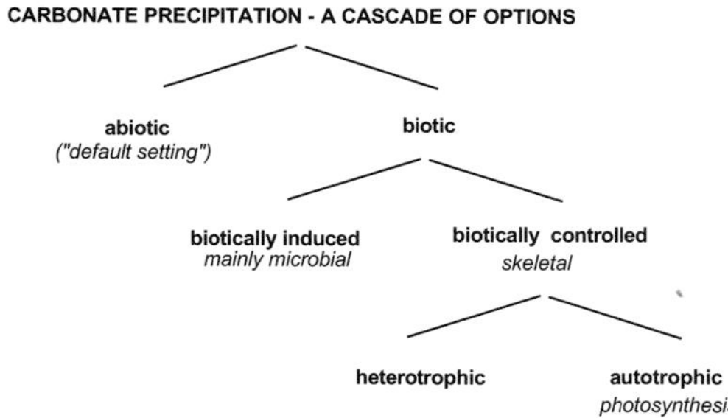

Fig. 1. The chemical modes of marine carbonate precipitation – a cascade of options. All these options are currently realized in the oceans and most probably also throughout the Phanerozoic. However, proportions of precipitates have varied considerably in time and space.

While the categories themselves are well identifiable and useful, the designation of certain products to one of the categories remains debatable. The controversy on the origin of ooids and whitings is a case in point. In most carbonate depositional environments material is produced in all three modes, albeit in varying proportions.

If one considers carbonate precipitation not at the molecular or microscopic scale but at scales from mappable sediment bodies ($\geq 10^2$ m) to global facies realms, different production systems or 'carbonate factories' may be distinguished. The boundaries between these systems are gradational and to some extent arbitrary. Accordingly, the number of basic types is a matter of debate with little hope of reconciling the views of lumpers and splitters. Personally, I believe that at the current state of knowledge three basic systems can be recognized and characterized in terms of depositional setting and modes of precipitation: the tropical, the cool-water and the mud-mound system (Fig. 2). A fourth system – the planktic production system of the open ocean – is fundamentally different as the sites of production and accumulation are separated by a large water column and the production process has little influence on the anatomy of the accumulation. It is not considered here.

Tropical factory

This system is developed in the tropical and subtropical surface waters of the ocean, approximately 30° N and S of the Equator. Production

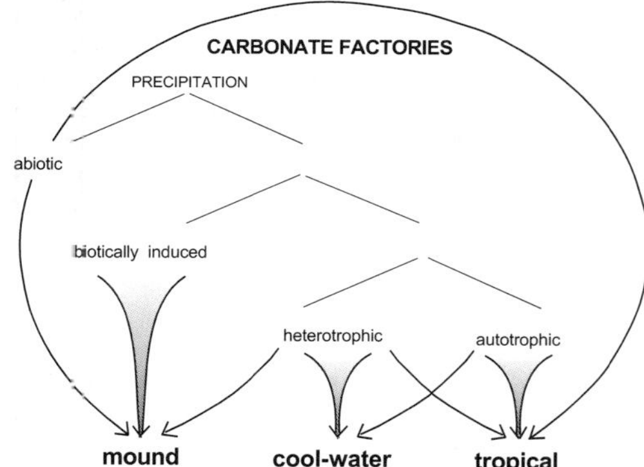

Fig. 2. Carbonate factories. Viewed at the scale of formations or environmental realms, the precipitation modes combine to different carbonate production systems or 'factories': tropical shallow-water, dominated by photo-autotrophic skeletal precipitates; cool-water, dominated by heterotrophic skeletal material; and mud-mound, dominated by microbial precipitates.

is largely biotically controlled. Characteristic are photosynthetic autotrophs, such as green algae, and animals with photosynthetic symbiotic algae such as hermatypic corals, certain foraminifers and certain molluscs. The other characteristic product is abiotic precipitates in the form of marine cements and, most probably, ooids (Morse & Mackenzie 1990). The origin of clay-size precipitates, the whitings, that contribute significantly to mud accumulations in the tropics is unsettled. A likely scenario is that biotically induced crystals provide the seeds for subsequent abiotic precipitation. Heterotrophs devoid of photo-symbionts are a major, but not diagnostic contributor. The tropical factory requires sunlit waters and high temperatures. In the present oceans, the high-latitude limit of the tropical factory is well described by the line where the mean temperature of the coldest month is about 20°C. Thus, in modern oceans the prime control on the limits of the tropical factory is temperature, and latitude is only a crude proxy indicator of its limits. The tropical factory may also pass into the cool-water factory downward in the water column, for instance at the boundary between the warm surface layer of the ocean and the thermocline. Furthermore, tropical–coolwater transitions may occur within shallow tropical waters where upwelling brings cool, nutrient-rich waters to the surface (Lees & Buller 1972; Pope & Read 1997, p. 423; Brandley & Krause 1997, p. 365).

Cool-water carbonate factory

In the surface waters of the ocean, the cool-water factory extends poleward from the limit of the tropical factory (at about 30°) to polar latitudes but it also occurs in the thermocline waters of the low latitudes (Lees & Buller 1972; James 1997). Carbonate precipitation is nearly always of the biotically controlled type. Heterotrophic organisms dominate; the contribution of photo-autotrophs in the form of red algae is highly variable but at times significant (Lees & Buller 1972; Henrich *et al.* 1995; James 1997). The sediment typically consists of skeletal hash of sand-to-granule size. Cool-water carbonates lack hermatypic coral reefs and ooids; carbonate mud and cement are scarce. The transition to the domain of the tropical carbonate factory normally extends over more than 1000 km (Schlanger 1981; Collins *et al.* 1997). The cool-water factory substitutes for the tropical factory in the tropics if environmental conditions are adverse (see above).

Mud-mound factory

Intensive work on ancient mud-mounds and micrite precipitates in the Recent has led to two conclusions that underline the independent position of this system and the fundamental differences from the tropical and cool-water factories (James & Bourque 1992; Monty 1995; Pratt 1995; Reitner & Neuweiler 1995). First, typical mud-mounds are formed by in-situ production rather than sweeping together of fine sediment; they are fundamentally different from hydrodynamic sediment accumulations such as lagoonal mudbanks or the contourite drifts of the deep sea. The micrite that constitutes the dominant component of the mounds was firm or hard during or soon after formation and does not represent lithified soft sediment. Second, the micrite is precipitated by a complex interplay of inorganic and organic chemical reactions with microbes and decaying organic tissue playing a central role (Reitner & Neuweiler 1995; Monty 1995). Consequently, I view mud-mounds as products of a third carbonate factory that differs fundamentally from its tropical and cool-water counterparts. The main product is micritic carbonate that can often be linked to microbially induced precipitation. However, other processes may also be operating. The term 'automicrite' for micrite precipitated in situ (Wolf 1965; Reitner & Neuweiler 1995) is useful in instances where the microbial origin is uncertain and the term 'microbialite' is not justified.

Important differences among the three factories

The three carbonate factories can be characterized in terms of dominant components and modes of precipitation but this is not to say that other components or precipitation modes are absent. The characteristic products of the tropical system are skeletons of photo-autotrophs (e.g. green algae) and abiotic precipitates (such as marine cements or ooids), yet tropical carbonate formations always include a significant proportion of skeletal heterotrophs or microbial micrite; locally, these may even dominate. Similarly, cool-water carbonates may contain significant, locally dominant contributions by photo-autotrophic red algae. In mud-mounds, the characteristic microbial micrite nearly always has admixtures of abiotic marine cement and skeletal heterotrophs. Thus, the products of the three factories overlap in composition. Furthermore, modern oceans show that the boundaries among factories are gradational in space. Finally, the geological record indicates that the balance among the factories has varied through time.

The notion of different carbonate production systems with different products and different environmental requirements is useful if one tries to capture the diversity of marine carbonate production. It should be emphasized that tropical, cool-water and mud-mound systems refer to sea-floor production only. If the entire ocean is considered, the planktic carbonate system would have to be added as a fourth factory as mentioned above.

The differences in the dominant mode of precipitation among the three systems imply significant variation in the depth range over which carbonate production can take place. The production window of the tropical system is the narrowest. It is restricted to the uppermost 50–100 m of the water column. Production of the cool-water and mud-mound systems is not, or only marginally, dependent on light. Production of both systems has been observed to extend to at least several hundred metres of water depth, far beyond the photic zone of the ocean. Because of the insensitivity to light levels, cool-water carbonates cannot be drowned, i.e. submerged in too much water. They must be terminated by other envinronmental change as James (1997) pointed out. The same holds for the mud-mound system.

Sedimentation rates and growth potential of the three factories

The main purpose of this study is to compare the sedimentation rates and estimated growth potential of the cool-water and mud-mound factories with the fairly well known rates of the tropical factory. The rates in Fig. 3 are set out against the length of the time interval of observation because sedimentation rates generally decrease as the time interval increases (Reineck 1960; Sadler 1981, 1994). It must be noted that regression analysis on direct plots of rate versus time are suspect because of the mathematical problem of ratio correlation. Rates in the geological record are determined by dividing observed thickness by observed difference in age. Accordingly, time appears on both axes – once as a single variable and once as part of the composite variable thickness/time. Under these circumstances, even random numbers may show strong correlation (e.g. Kenney 1982). Gardner *et al.* (1987) and Schlager *et al.*(1998) circumvented this problem by basing their analysis on the primary variables thickness and time. Both Gardner *et al.* (1987) and Schlager *et al.* (1998) concluded that the effect of ratio correlation on geological rates is minor and that the scaling trend has physical meaning.

In Fig. 3, sedimentation rates of all three carbonate systems decrease with increasing length of the observation span. Closer inspection shows that these trends are robust and not simply an artefact of ratio correlation. The carbonate data set of Schlager *et al.* (1998) in Fig. 3a serves as a representation of the tropical factory. On this

Fig. 3. Sedimentation rates of the three carbonate factories plotted against the length of the time interval of observation. Rates decrease with increasing length of time in all factories. Tropical rates are highest and best documented. Cool-water rates are lower overall (by a factor of 4); occasionally high short-term rates are caused by reworking and local trapping of sediment. Mud-mound rates approach tropical rates in the domain of 2×10^5– 10^7 years. Their short-term rates are poorly known. **(a)** Data from Schlager *et al.* (1998). **(b)** Data from McKenzie *et al.* (1980), Farrow *et al.* (1984), Collins (1988), Kamp *et al.* (1988), Scoffin (1988), Smith (1988), James & Bone (1989, 1991), Boreen & James (1997), Ferland & Roy (1997), Gillespie & Nelson (1997), Stemmerik (1997), Surlyk (1997). **(c)** Data from Schultz (1966), Lees *et al.* (1977), Phillips & Sevastopulos (1986), Lanier (1986), Strogen *et al.* (1990), Brachert & Dullo (1991), Somerville *et al.* (1992), Dronov (1993), Golubic *et al.* (1993), Wendt (1993), De Freitas & Dixon (1995), Kaufmann (1995), Lees & Miller (1995), Neuweiler (1995), Reitner & Neuweiler (1995), Warnke & Meischner (1995), Weller (1995), Boulvain & Herbosch (1996), Macintyre *et al.* (1996), Labiaux (1997), Wendt *et al.* (1997).

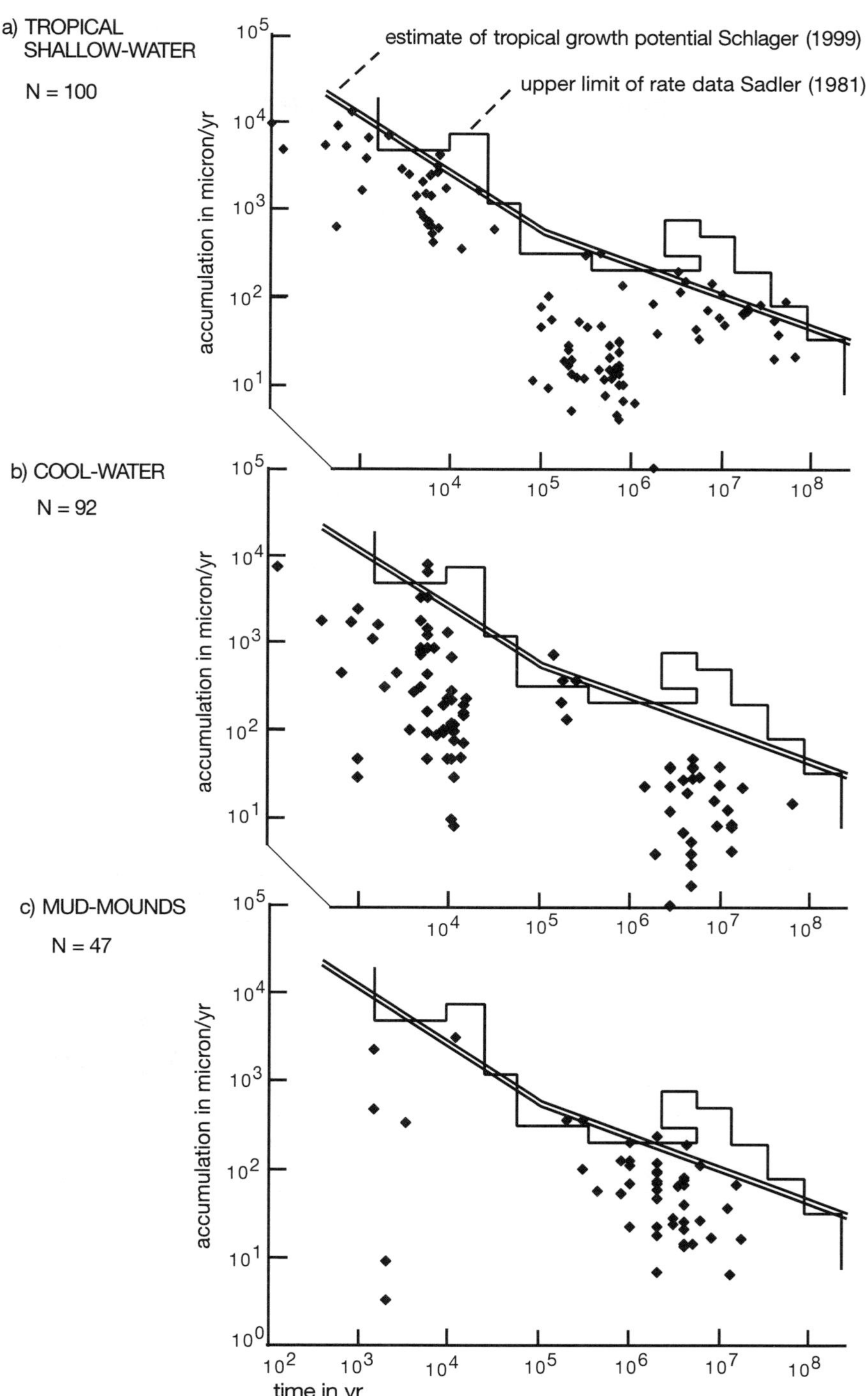
a) TROPICAL SHALLOW-WATER
N = 100
estimate of tropical growth potential Schlager (1999)
upper limit of rate data Sadler (1981)
accumulation in micron/yr
b) COOL-WATER
N = 92
accumulation in micron/yr
c) MUD-MOUNDS
N = 47
accumulation in micron/yr
time in yr

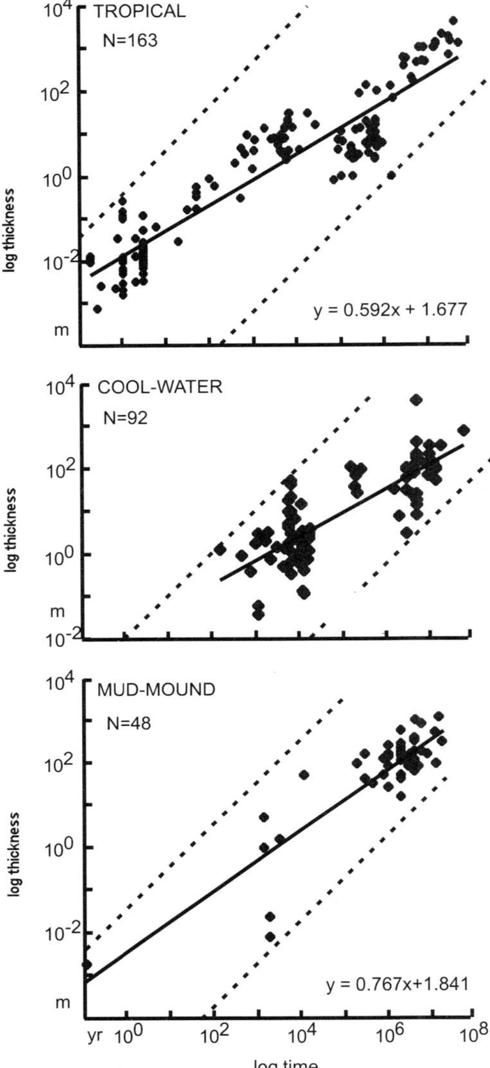

Fig. 4. Data of Fig. 3 plotted as thickness versus time. If sedimentation rates were independent of the length of time, the primary variables thickness and time would scatter around a regression line of slope + 1 (dotted lines). Observed slopes are significantly smaller, confirming that the decrease of sedimentation rates with time has physical meaning. Tropical data have also been tested by piecewise regression on the primary variables with the same result (Schlager *et al.* 1998).

data set the decrease of rates with increasing time span has been demonstrated by splitting the sample into three subsets (of time intervals 10^{-1}–10^2, 10^2–10^5 and 10^5–10^8 years) and determining the average sedimentation rates by regression on the primary variables thickness and time (Schlager *et al.* 1998). The data on the cool-water and mud-mound factories in Fig. 3b and c do not lend themselves well to regression analysis in segments because the data sets are small and spread rather unevenly over time. Figure 4 shows an alternative test to demonstrate that at least the first-order trend over the entire time range of 10^2–10^8 years has physical meaning. The test, proposed by Gardner *et al.* (1987), is again performed on the primary variables thickness and time. It relies on the fact that on a log–log plot, data where scaling is purely the result of ratio correlation will scatter around a regression line with a slope of +1. The regression equations in Fig. 4 have slopes of <1 (slopes are 0.592, 0.548 and 0.767 for tropical, cool-water and mud-mound systems respectively). This confirms that the decrease of sedimentation rates with increasing time interval is real for accumulations of all three factories.

As we cannot experiment with geological systems, the growth potential of the three factories, i.e. the maximum rate of aggradation they are capable of, must be estimated from the geological record. Schlager (1999) suggested that the growth potential scales similarly to mean sedimentation rate. This implies that the upper limits of the data in Fig. 3a–c can be viewed as (conservative) estimates of the respective growth potential of the systems. The estimated growth potentials differ considerably. In the geologically most relevant (and relatively well covered) time interval 10^6–10^7 years the upper limit of tropical rates decreases from about 250 to 100 $\mu m\ a^{-1}$. The growth potential of the cool-water factory (as indicated by the upper limit of observed rates) is about one-fourth that of the tropical system. Rates of the mud-mound system in the domain of 10^6–10^7 years clearly exceed those of the cool-water system and approach those of the tropical factory. I estimate them at 80–100% of the tropical rates.

The high growth potential of the mud-mound system is supported by case studies in the Recent and distant geological past. In Tahiti, Camoin *et al.* (1999) observed that microbial micrite formed a volumetrically important, often dominant component of the coral barrier reef in the early Holocene. This reef grew at rates of up to 8000 $\mu m\ a^{-1}$ sustained over 10^4 years. Micrite formed in dimly lit cavities between the coral framework in pore waters with elevated nutrient levels. Microbes seem to trigger precipitation but organic matter from metazoans is also important. A similar association of corals and automicrite has been described from Holocene reefs in the Indian Ocean (Montaggioni & Camoin 1993).

High production rates on the time scale of 10^5–10^6 years are indicated by slopes of the Triassic Sella platform in the southern Alps of Italy (Kenter 1990; Russo *et al.* 1998; Keim & Schlager 1999). The Sella platform is an atoll with preserved platform-basin transitions on all sides. The slopes consist of planar foresets of about 35-37° of primary dip. The planar shape and the dip angle suggest that the slopes were prograding near the angle of repose of sand and rubble; yet 25% or more of the clinoforms consist of layers of automicrite with abundant primary vugs and marine fibrous cements. Biostratigraphy and field mapping indicate that the slope prograded approximately 2800 m in 1 Ma (Neri *et al.* 1994; Keim & Schlager 1999). At 35° slope angle, this translates into a displacement rate of the sea floor perpendicular to the slope of 1606 m Ma^{-1}, sustained over 1 Ma – certainly a rapid sedimentation setting.

Discussion

Justification of the three factories

We have seen that the products of the three factories overlap in composition and the boundaries of the factory domains are gradational in space and time. Thus, one may ask if this three-fold subdivision of the carbonate production spectrum really is justified. On the other hand, one might focus on the considerable variation within each of the present categories and find reasons to distinguish many more carbonate factories. For the cool-water system, James (1997) cautiously suggested a subdivision that builds on the fact that the cool-water domain spans nearly 60° of latitude and a wide range of skeletal composition. Similar arguments apply to the other two factories. The tropical system shows continuous change of reef communities through time and considerable variation in space. For instance, in the Recent, some products of the tropical factory are dominated by organic reefs and skeletal sands while others consist mainly of abiotic or biotically induced precipitates such as ooids and aragonite mud from whitings. Products of the mud-mound system as characterized here are also diverse, including stromatolites built by photosynthetic cyanobacteria as well as deeper-water microbialites and cements.

Variation in geological time is considerable but a three-fold subdivision of carbonate production can be maintained at least for the Phanerozoic. The Phanerozoic provides good evidence for the almost continuous existence of a tropical shallow-water system with prominent photo-autotrophic carbonate producers (e.g. Kiessling *et al.* 1999). In addition, major accumulations by the mud-mound factory have been documented from most pre-Cenozoic epochs of the Phanerozoic (e.g. Monty 1995; Pratt 1995; Lees & Miller 1995; Reitner & Neuweiler 1995). The identification of cool-water carbonates is hampered by their similarity with tropical deposits but recent studies indicate a growing number of cool-water accumulations (e.g. James 1997; Surlyk 1997; Stemmerik 1997; Brandley & Krause 1997).

The three carbonate factories are easily identifiable in terms of facies anatomy of their accumulations. Each factory tends towards a characteristic morphology that can also be recognized in seismic evidence and sea-floor topography. Tropical carbonates tend towards a flat-topped rimmed platform; when the rate of sea-level rise approaches the growth potential, the flat top may give way to the empty-bucket shape with a raised rim and empty lagoon. Cool-water carbonates tend to form seaward-sloping shelves in equilibrium with wave action, which are similar to siliciclastic shelves. Products of the mud-mound factory develop their characteristic morphology of convex-upward mounds in settings below the zone of strong wave action. They become flat-topped (and change facies) when building into the zone of permanent wave action.

Further important differences among the three factories concern the variation of production with water depth. These differences relate to the dominant mode of precipitation in each factory. The production window of the tropical system is the narrowest. It is restricted to the uppermost 50–100 m of the water column because the photo-autotrophic carbonate producers, the key element of the tropical factory, are dependent on light. Production of the cool-water and mud-mound systems is not, or only marginally, dependent on light. Production of both systems has been observed to extend to at least several hundred metres of water depth, far beyond the euphotic zone of the ocean. Because of the insensitivity to light levels, cool-water carbonates cannot be drowned, i.e. submerged in too much water. They must be terminated by other environmental factors as James (1997) pointed out. The same holds true for the mud-mound system.

The three factories also differ in their sensitivity to nutrients and temperature. The tropical system is particularly well adapted to warm submarine deserts, such as the nutrient-depleted waters of the subtropical gyres. The other systems do better at higher nutrient levels and can adjust to a wide range of temperatures.

Data quality and coverage.

The amount of data is limited and rather unevenly distributed. All factories show a gap around 10^5 years that reflects the combined effect of limited resolution of the distant past and eustatic turmoil in the recent past. The stratigraphy of pre-Quaternary shallow-water carbonates is generally unable to resolve 10^5 years intervals with any certainty. In the Quaternary, on the other hand, eustatic sea-level oscillations continually shifted the loci of carbonate deposition and precluded the accumulation of shallow-water carbonates in a single spot for 10^5 years or longer. Apart from the 10^5 years 'thin spot', the tropical data are fairly well spread. With the cool-water data, the gap is bigger. The data really consist of two subpopulations: the Holocene record of 10^2–10^4 years, based mainly on radiocarbon ages, and the distant past of 10^6–10^7 years based on a combination of biostratigraphy, magnetic stratigraphy and isotope stratigraphy. The mud-mound factory, much like its cool-water counterpart, shows good coverage in the range of 10^6–10^7 years. The 10^5–10^6 year domain is fairly well covered owing to detailed conodont stratigraphy of the Carboniferous. Rates of even shorter time intervals are scarce because mud-mounds have not been in fashion during the late Cenozoic. I believe that the microbially dominated micrite precipitates of modern coral reefs (Camoin *et al.* 1999), in shallow-water stromatolites (Macintyre *et al.* 1996) and in the deep forereef (Brachert & Dullo 1991) represent variations of the Palaeozoic and Mesozoic mud-mound system. Some well constrained growth rates of these deposits have been included in the mud-mound data set of Figs 3 and 4.

Scaling of sedimentation rates

As mentioned above, data of all three factories show scaling, i.e. rates decrease as the time window increases. Furthermore, tests in Fig. 4 and in Schlager *et al.* (1998) confirm that this scaling of rates is real and not simply an effect of ratio correlation. The slopes of the regression lines on the primary variables, thickness and time (Fig. 4), are similar for the tropical and the mud-mound factories. In the cool-water data, the slope is considerably flatter (implying that the rate scaling is more pronounced). However, cool-water rates show the most scatter and the correlation coefficient is low, probably due to intensive reworking on time scales shorter than 2×10^5 years (see next section for further evaluation).

As rates of all three systems are scaled, they need to be normalized to the same time interval before they are compared. For instance, the use of sedimentation rates as an argument for assigning deposits of the distant past to the cool-water or tropical carbonate category carries little weight unless rate scaling is taken into account, a point already made by Wright & Burchette (1996). Furthermore, it should be noted that the diagnostic difference between the tropical and the cool-water system lies not so much in the average rates but in the maximum rates – the growth potential, discussed in the next section.

Growth potential of the three factories

The comparison of rates of the three factories in Figs 1–3 is a crude estimate but it does lead to some interesting conclusions. With regard to cool-water carbonates, these data shed light on some conflicting reports in the literature. Many authors (e.g. Wright & Burchette 1996; Henrich *et al.* 1995; James 1997) have concluded that accumulation rates of cool-water carbonates are significantly lower than those of tropical systems. However, there are also repeated statements that production and accumulation rates of cool-water carbonates may equal those of tropical carbonates (Sarnthein 1973; Henrich *et al.* 1995). Figures 1 and 2 confirm both observations but for different time domains. In the domain of 10^6–10^8 years, the upper limit of cool-water rates, approximately equal to the growth potential of these systems, is about one-fourth of that of tropical systems. The relationship is very different for time intervals shorter than 2×10^5 years where cool-water rates approximately equal tropical rates. These data strongly suggest that the high rates are the result of reworking and local trapping of sediment. The raw data that form the basis of these high rates show highly variable rates in vertical succession within a core, along with occasionally inverted age gradients, i.e. older on top of younger material (e.g. Farrow *et al.* 1984) or they display giant foresets indicative of large-scale lateral transport (e.g. Kamp *et al.* 1988).

If rates are used as a diagnostic criterion, for instance to differentiate between cool-water carbonates and tropical carbonates in the geologic record (e.g. James 1997), one has to compare rates measured over similar lengths of time. In view of the pervasive scaling of sedimentation rates it is not very meaningful to compare rates of Holocene tropical reefs that were measured in the thousand-year domain with rates of ancient deposits that were averaged over millions of years.

The high rates of the mud-mound factory are rather unexpected. Several studies have concluded that mud-mounds demand sediment starvation (Reitner & Neuweiler 1995, p. 63; Neuweiler 1995). A difference in scale offers a possible explanation for this discrepancy. The conclusions about sediment starvation were drawn from detailed analysis of microfacies and generally pertain to short time intervals, whereas the high rates reported here refer to observations in the million-year domain. If sedimentation is pulsating and episodic, there may be sufficient periods of quiescence for biofilms to grow and induce micrite precipitation while short intensive pulses of sedimentation guarantee relatively high long-term rates of accumulation.

In considering the above comparisons, it should be noted that the rate of aggradation shown here is not a reliable measure of a system's production by volume (or mass). Mud-mounds seem to export relatively little sediment (e.g. Lees & Miller 1995; Wright & Burchette 1996, p. 326); they feed only narrow aprons at the toe of the slope (typically 10^2 m wide). This contrasts sharply with the aprons of tropical platforms that are typically 10^3–10^4 m in width (Schlager & Chermak 1979; Mullins *et al.* 1984). On the other hand, the production of cool-water carbonate shelves may be significantly higher than their aggradation potential. The shelves are unrimmed and the sediments largely unlithified as James (1997) pointed out. We do not yet know how much sediment is dissolved or swept into the deep but it is probably more than the amount that accumulates on the shelf.

Conclusions

1. Viewed at the elementary, molecular level, marine carbonate precipitation may be subdivided into three modes: abiotic, biotically induced, and biotically controlled.
2. Viewed at the level of formations and global environmental belts, a subdivision of the carbonate production spectrum into three systems or 'factories' is useful: (1) tropical shallow-water system, dominated by abiotic precipitates and biotically controlled (mainly photo-autotrophic) skeletal particles; (2) cool-water system, dominated by biotically controlled, heterotrophic skeletal precipitates; (3) mud-mound system, dominated by biotically induced, mainly microbial micrite precipitates with significant abiotic cements.
3. Sedimentation rates of all three systems decrease with increasing length of the time interval. The maximum rates of the best documented factory, the tropical one, are 10^4 μm a^{-1} at 10^3 years and decreasing to 10^2 μm a^{-1} at 10^7 years.
4. The maximum observed sedimentation rates, an approximation of the system's growth potential, are different for the three systems. Tropical carbonates have the highest rates. The maximum rates of the cool-water system are comparable to those of the tropical carbonates for time intervals of up to 10^5 years but they amount to only 25% of the tropical standard in the domain of 10^6–10^7 years. The mound system rivals the maximum aggradation rates of the tropical system in the domain of 10^6–10^7 years. However, it exports less sediment; thus, its overall production rates may be somewhat less than those of the tropical shallow-water system.

Much of the research was supported by the VU Industrial Associates in Sedimentology. I thank P. Skelton and E. Insalaco for inviting me to the Geological Society's symposium on reefs and carbonate platforms, and D. Bosence (Imperial College), W. Piller (Graz) and C. Dullo (Kiel) for fruitful discussions. M. Tucker and V. P. Wright contributed important comments as journal reviewers.

References

BOREEN, T. D. & JAMES, N. P. 1995. Stratigraphic sedimentology of Tertiary cool-water limestones, SE Australia. *Journal of Sedimentary Research*, **B65**(1), 142–159.

BOULVAIN, F. & HERBOSCH, A. 1996. Anatomie des monticules micritiques du Frasnien belge et contexte eustatique. *Bulletin Societe geologique de France*, **167**, 391–398.

BRACHERT, T. C. & DULLO, W. C. 1991. Laminar micrite crusts and associated foreslope processes, Red Sea. *Journal of Sedimentary Petrology*, **61**(3), 354–363.

BRANDLEY, R. T. & KRAUSE, F. F. 1997. Upwelling, thermoclines and wave-sweeping on an equatorial carbonate ramp: Lower Carboniferous strata of western Canada. *In*: JAMES, N. P. & CLARKE, J. A. D. (eds) *Cool-water Carbonates*. Society for Sedimentary Geology, Special Publications, **56**, 365–390.

CAMOIN, G. F., GAUTRET, P. *et al.* 1999. Nature and environmental significance of microbialites in Quaternary reefs: the Tahiti paradox. *Sedimentary Geology*, **126**, 271–304.

COLLINS, L. B. 1988. Sediments and history of the Rottnest Shelf, southwest Australia: a swell-dominated, non-tropical carbonate margin. *Sedimentary Geology*, **60**, 15–49.

——, *ET AL.* 1997. Cool-water carbonates, James and Clarke.

DE FREITAS, T. A. & DIXON, O. A. 1995. Silurian microbial buildups of the Canadian Arctic. *In*: MONTY, C. L. V., BOSENCE, D. W. J., BRIDGES, P. H., &

PRATT, B. R. *Carbonate Mud-Mounds – their Origin and Evolution.* International Association of Sedimentologists, Special Publications, **23,** 151–169.

DRONOV, A. V. 1993. Middle Paleozoic Waulsortian-type mud mounds in Southern Fergana (Southern Tien-Shan, Commonwealth of Independent States): The shallow-water atoll model. *Facies,* **28,** 169–180.

FARROW, G. E., ALLEN, N. H., *et al.* 1984. Bioclastic carbonate sedimentation on a high-latitude, tide-dominated shelf: northeast Orkney Islands, Scotland. *Journal of Sedimentary Petrology,* **54**(2), 373–393.

FERLAND, M. A. & ROY, P. 1997. Southeastern Australia: a sea-level dependent, cool-water carbonate margin. *In*: James, N. P. & Clarke, J. A. D. (eds) *Cool-water Carbonates.* Society for Sedimentary Geology, Special Publications, **56,** 37–52.

GARDNER, T. W., JORGENSEN, D. W., *et al.* 1987. Geomorphic and tectonic process rates: effects of measured time interval. *Geology,* **15**, 259–261.

GILLESPIE, J. L. & NELSON, C. S. 1997. Mixed siliciclastic-skeletal carbonate facies on Wanganui shelf, New Zealand: a contribution to the temperate carbonate model. *In*: JAMES, N. P. & CLARKE, J. A. D. (eds) *Cool-water Carbonates.* Society for Sedimentary Geology, Special Publications, **56,** 127–140.

GOLUBIC, S., VIOLANTE, C., *et al.* 1993. Algal control and early diagenesis in Quaternary travertine formation (Rocchetta a Volturno, Central Apennines). *Boll. Soc. Paleont. Ital., Special Volume* **1**, 231–247.

HENRICH, R., FREIWALD, A., *et al.* 1995. Controls on modern carbonate sedimentation on warm-temperate to Arctic coasts, shelves and seamounts in the northern hemispere: implications for fossil counterparts. *Facies,* **32**, 71–108.

HOLDGATE, G. & GALLAGHER, S. 1997. Microfossil paleoenvironments and sequence stratigraphy of Tertiary cool-water carbonates, onshore Gippsland Basin, southeastern Australia. *In*: JAMES, N. P. & CLARKE, J. A. D. (eds) *Cool-water Carbonates.* Society for Sedimentary Geology, Special Publications, **56,** 205–220.

JAMES, N. P. 1997. The cool-water carbonate depositional realm. *In*: JAMES, N. P. & CLARKE, J. A. D. (eds) *Cool-water Carbonates.* Society for Sedimentary Geology, Special Publications, **56,** 1–20.

—— & BONE, Y. 1989. Petrogenesis of Cenozoic temperate water calcarenites, south Australia: a model for meteoric/shallow burial diagenesis of shallow water calcite sediments. *Journal of Sedimentary Research,* **59**, 191–203.

—— & —— 1991. Origin of a cool-water, Oligo-Miocene deep shelf limestone, Eucla Platform, southern Australia. *Sedimentology,* **38**, 323–341.

—— & Bourque, P. A. 1992. Reefs and mounds. *In*: WALKER, R. G. & JAMES, N. P.(eds) *Facies Models: Response to Sea Level Change.* Geological Association of Canada, St. Johns**,** 323–347.

KAMP, P. J. J., HARMSEN, F. J., *et al.* 1988. Barnacle-dominated limestone with giant cross-beds in a non-tropical, tide-swept, Pliocene forearc seaway, Hawke's Bay, New Zealand. *Sedimentary Geology,* **60**, 173–195.

KAUFMANN, B. 1995. Part IX Middle Devonian mud mounds of the Ma'der Basin in the eastern Anti-Atlas, Morocco. *Facies,* **32**, 49–57.

KEIM, L. & SCHLAGER, W. 1999. Automicrite facies on steep slopes (Triassic, Dolomites, Italy). *Facies,* **41**, 15–26

KENNEY, B. C. 1982. Beware of spurious self-correlation. *Water Resources Research,* **18**, 1041–1048.

KENTER, J. A. M. 1990. Carbonate platform flanks: slope angle and sediment fabric. *Sedimentology,* **37**, 777–794.

KIESSLING, W., FLÜGEL, E. & GOLONKA, J. 1999 Paleoreef maps: evaluation of a comprehensive database on Phanerozoic reefs. *AAPG Bulletin*, **83**, 1552–1587.

LABIAUX, S. 1997. Sponges in Waulsortian-type mud-mounds at Tralee Bay, Co. Kerry, southwest Ireland. *Facies,* **36**, 253–256.

LANIER, W. P. 1986. Approximate growth of early Proterozoic microstromatolites as deduced by biomass productivity. *Palaios,* **1**, 525–542.

LEES, A. & BULLER, A. T. 1972. Modern temperate-water and warm-water shelf carbonate sediments contrasted. *Marine Geology,* **13**, M67-M73.

—— & MILLER, J. 1995. Waulsortian banks. *In*: MONTY, C. L. V., BOSENCE, D. W. J., BRIDGES, P. H., & PRATT, B. R. (eds) *Carbonate Mud-Mounds – their Origin and Evolution.* International Association of Sedimentologists, Special Publications, **23,** 191–271.

LOWENSTAM, H. A. & WEINER, S. 1989. *On Biomineralization.* Oxford University, New York.

MACINTYRE, I. G., REID, R. P., *et al.* 1996. Growth history of stromatolites in a Holocene fringing reef, Stocking Island, Bahamas. *Journal of Sedimentary Research,* **66**(1), 231–242.

MCKENZIE, J., BERNOULLI, D. & SCHLANGER, S. O. 1980. Shallow-water carbonate sediments from the Emperor Seamounts: their diagenesis and paleogeographic significance. *Deep Sea Drilling Program Initial Reports,* **55**, 415–455.

MONTAGGIONI, L. F. & CAMOIN, G. F. 1993. Stromatolites associated with coralgal communities in Holocene high-energy reefs. *Geology,* **21**, 149–152.

MONTY, C. L. V. 1995. The rise and nature of carbonate mud-mounds: an introductory actualistic approach. *In*: MONTY, C. L. V., BOSENCE, D. W. J., BRIDGES, P. H., & PRATT, B. R. *Carbonate Mud-Mounds – their Origin and Evolution.* International Association of Sedimentologists, Special Publications, **23**.

——, BOSENCE, D. W. J., BRIDGES, P. H. & PRATT, B. R. 1995 *Carbonate Mud-Mounds – their Origin and Evolution.* International Association of Sedimentologists, Special Publications, **23**.

MORSE, J. W. & MACKENZIE, F. T. 1990. *Geochemistry of Sedimentary Carbonates.* Elsevier, Amsterdam.

MULLINS, H. T., HEATH, K. C., *et al.* 1984. Anatomy of modern deep-ocean carbonate slope: Northern Little Bahama Bank. *Sedimentology,* **31**, 141–168.

NELSON, C. S., KEANE, S. L., *et al.* 1988. Non-tropical

carbonate deposits on the modern New Zealand shelf. *Sedimentary Geology,* **60**, 71–94.

NERI, C., MASTANDREA, A., *et al.* 1994. New biostratigraphic data on the S. Cassiano Formation around Sella platform (Dolomites, Italy). *Palaeopelagos,* **4**, 13–21.

NEUWEILER, F. 1995. *Dynamische Sedimentationsvorgaenge, Diagenese und Biofazies unterkretazischer Plattformraender (Apt/Alb; Soba-Region, Prov. Cantabria, N-Spanien).* Berliner Geowissenschafliche Abhandlungen, **E17**.

PHILLIPS, W. E. A. & SEVASTOPULOS, G. D. 1986. The stratigraphic and structural setting of Irish mineral deposits. *In*: ANDREW, C. J., CROWE, R. W. A., FENNEL, W. M. & PYNE, J. F. (eds) *Geology and Genesis of Mineral Deposits in Ireland.* Irish Association for Economic Geology, 1–30.

POPE, M. C. & READ, J. F. 1997. High-resolution stratigraphy of the Lexington Limestone (late Middle Ordovician), Kentucky, U. S. A.: a cool-water carbonate-clastic ramp in a tectonically active foreland basin. *In*: James, N. P. & Clarke, J. A. D. (eds) *Cool-water Carbonates.* Society for Sedimentary Geology, Special Publications, **56,** 411–430.

PRATT, B. R. 1995. The origin, biota and evolution of deep-water mud-mounds. *In*: MONTY, C. L. V., BOSENCE, D. W. J., BRIDGES, P. H., & PRATT, B. R. *Carbonate Mud-Mounds – their Origin and Evolution.* International Association of Sedimentologists, Special Publications, **23,** 49–126.

REINECK, H. E. 1960. Über Zeitlücken in rezenten Flachsee-Sedimenten. *Geol. Rundschau,* **49**(1), 149–161.

REITNER, J. & NEUWEILER, F. 1995. Mud mounds: a polygenetic spectrum of fine-grained carbonate buildups. *Facies,* **32**, 1–70.

RUSSO, F., NERI, C., *et al.* 1997. The mud mound nature of the Cassian platforms of the Dolomites. *Facies,* **36**, 25–36.

SADLER, P. M. 1981. Sediment accumulation and the completeness of stratigraphic sections. *Journal of Geology,* **89**, 569–584.

—— 1994. The expected duration of upward-shallowing peridtidal carbonate cycles and their terminal hiatuses. *Geological Society of America Bulletin,* **106**, 791–802.

SARNTHEIN, M. 1973. Quantitative Daten über benthische Karbonatsedimentation in mittleren Breiten. *Veröffentlichungen der Universität Innsbruck,* **86**, 267–279.

SCHLAGER, W. 1999. Scaling of sedimentation rates and drowning of reefs and carbonate platforms. *Geology,* **27**(2), 183–186.

—— & CHERMAK, A. 1979. Sediment facies of platform-basin transition, Tongue of the Ocean, Bahamas. *In*: DOYLE, L. J. & PILKEY, O. H. *Geology of Continental Slopes.* SEPM, Special Publications, **27,** 193–207.

——, MARSAL, D., *et al.* 1998. Sedimentation rates, observation span, and the problem of spurious correlation. *Mathematical Geology,* **30**, 547–556.

SCHLANGER, S. O. 1981. Shallow-water limestones in oceanic basins as tectonic and paleoceanographic indicators. *In*: WARME, J. E., DOUGLAS R. G. & WINTERER E. L. *The Deep Sea Drilling Project: A decade of progress***.** SEPM, Special Publications, **32,** 209–226.

SCHULTZ, R. W. 1966. Lower Carboniferous cherty ironstones at Tynagh, Ireland. *Economic Geology,* **61**, 311–342.

SCOFFIN, T. P. 1988. The environments of production and deposition of calcareous sediments on the shelf west of Scotland. *Sedimentary Geology,* **60**, 107–124.

SMITH, A. M. 1988. Preliminary steps toward formation of a generalized budget for cold-water carbonates. *Sedimentary Geology,* **60**, 323–331.

SOMERVILLE, I. D., STROGEN, P., *et al.* 1992. Mid-Dinantian Waulsortian buildups in the Dublin Basin, Ireland. *Sedimentary Geology,* **79**, 91–116.

STEMMERIK, L. 1997. Permian (Artinskian-Kazanian) cool-water carbonates in north Greenland, Svalbard and the western Barents Sea. *In*: JAMES, N. P. & CLARKE, J. A. D. (eds) *Cool-water Carbonates.* Society for Sedimentary Geology, Special Publications, **56**, 349–364.

STROGEN, P., JONES, G. L., *et al.* 1990. Stratigraphy and sedimentology of Lower Carboniferous (Dinantian) boreholes from West Co. Meath, Ireland. *Geological Journal,* **25**, 103–137.

SURLYK, F. 1997. A cool-water carbonate ramp with bryozoan mounds: Late Cretaceous–Danian of the Danish Basin. *In*: JAMES, N. P. & CLARKE, J. A. D. (eds) *Cool-water Carbonates.* Society for Sedimentary Geology, Special Publications, **56,** 293–308.

WARNKE, K. & MEISCHNER, D. 1995. Part VII Origin and depositional environment of Lower Carboniferous mud mounds of northwestern Ireland. *Facies,* **32**, 42.

WELLER, H. 1995. Part VIII The Devonian mud mound of Ruebeland in the Harz mountains Germany. *Facies,* **32**, 43–49.

WENDT, J. 1993. Steep-sided carbonate mud mounds in the Middle Devonian of the eastern Anti-Atlas, Morocco. *Geological Magazine,* **130**(1), 69–83.

——, BELKA, Z., *et al.* 1997. The world's most spectacular carbonate mud mounds (Middle Devonian, Algerian Sahara). *Journal of Sedimentary Research,* **A67**(3), 424–436.

WOLF, K. H. 1965. Gradational sedimentary products of calcareous algae. *Sedimentology*, **5**, 1–37.

WRIGHT, V. P. & BURCHETTE, T. P. 1996. Shallow-water carbonate environments. *In*: READING, H. G. (ed.) *Sedimentary Environments: Processes, Facies, Stratigraphy*. Blackwell, Oxford, 325–394.

Index

Numbers in *italic* refer to figures; numbers in **bold** refer to tables.

abrasion 92, 100
Acanthaster 118, 120, 121, 125
accelerated evolution 123
accommodation space 1, 83, 85, 113, 115, 142
accretion rates 71, 82
Acropora 78, 80, 82, 84, 124
Acropora formosa 37
Aka 39, 42
anoxic facies 158–159, 165, 171
antecedent topography 140–141
arborescent coralline algae 98
Archaean
  Ghaap Group 52–55
  oolitic fabrics 60–63
  sea-water composition 65–66
  stromatolitic fabrics 55–59
  taphonomic evolution 63–65
Astraeopora myriophthalma 82

back-reef environment 37, 43
barnacles 34, 37
barrier reefs 37, *76*, 82
Belize, carbonate platforms 135–145
benthic communities 177
Bermuda 37
biochemical degradation 63–64
bioconstructors 9
biodeposits 28, 30, *see also* faeces contamination
bioeroders 9, 120, *see also* macroborers, microborers
bioerosion 1, 27, 33, 91–92, 100, 120, 121, 125, 201
biogeochemical interactions 52
biogeomorphology 9, 17–18
bioherms 74, *see also* coral reefs
biokarst 9
biomass production 29
bioprotectors 9
biostromes 72, *see also* coral carpets
bivalves 34, 37, 42, *see also* rudist bivalves
bleaching *see* coral bleaching
Bristol Channel 18
bryozoans 34, 42, 98

calcification 119–120, 129
caprinids 181
carbonate factories 218–225
carbonate platform development
  Cretaceous 6, 177
  Precambrian 51
  Quaternary 140–145
carbonate platform size, Phanerozoic 6, 205–206
carbonate precipitation 218
carbonate production systems 217, *see also* global carbonate production; reefal carbonate production
carbonate ramp 165, 170
Cassian Formation 39
catch-up 82, 85

cement 63, 66, 67, 92, 98, 218–219
Cima delle Murelle Formation 178–180
cirripedes 39
climatic change 85, 117–118, 126–129
Clionidae 34, 37, 44
conceptacles 90, 92
congregations, hippuritid 113–115, 184
conodont biostratigraphy 159–162
Conophyton 54
constratal growth 109, 113
cool-water carbonate factory 6, 219–225
coral bleaching 118, 120–129
coral carpets 71–85
coral communities, definitions 74–75
coral evolution 123
coral reefs 1, 71–72, *74*, *76*, 78, 82
  extinction 117–129
  framework-building 71, 77, 82–85
    climatic change 85
    oceanographic change 84
coralline red algae
  diversity 90, 97, 100–102
  early diagenesis 92
  growth forms 90, 97–98, 100–102
  taphonomy 91–92, 98–102
  taxonomic concepts 90
corallivores 119–121
Corculum 29
Crassostrea virginica 29
Cretaceous, carbonate platform development 177
cuspate microbialites 57
cyanobacteria 39, 51, 54, 57, 63–64, 66–67
Cymopolia 102
Cyphastrea 80

Devonian, Timan-Pechora 149–151, *see also* Frasnian
diagenesis 1, 92, 100
dinoflagellate blooms 118–120
disarticulation 91, 98–100
Discovery Bay 37
Distefanella 182–185
dolomitization 57–59, *60*, *61*, 63, 65, 66–68

Echinophyllia aspera 81
Echinopora 80
El Niño-Southern Oscillation 117–129
  effects on coral 118–121, 125–126
  effects on other taxa 121
elevator growth strategy 21, 30, *see also* rudist bivalves
encrustation 91, 97–98, 100
encrusting reefs 10
ENSO *see* El Niño-Southern Oscillation
Entobia **35**, 43
epithallial sloughing 92
epithallus 91, 92
Eunice mutilata 37

eustatic sea-level change 202, 204
eutrophication 118
extinctions 44, 117, 124–129, 169, 199

faeces contamination 26, 28, 30, 113
Favia 80
Favites 80
fore-reef environment 37, 39, 43
fragmentation 92, 100
Fragum 29
framework building 75, 77, 82–85
Frasnian
  New York, biostratigraphy 169–172
  Timan-Pechora
    anoxic facies 158–159
    biostratigraphy 159–165
    reef complexes 157–158, 165–168
    stratigraphy 151–157
fringing reef *76*, 82

Galápagos Islands 118, 120, 121, 123, 125
Gardineroseris 125–126
Gastrochaenolites **35**, 43
Ghaap Group 52–55
global carbonate production 206–209
global warming 117, 123–124
Glovers Reef 135–145, *136*
Goniastrea 80
goniatite biostratigraphy 162–165
Gorjanovicia 25–28, 182
Gornji Grad, Paleogene succession 92, *93*
Gornji Grad limestones
  coralline algae 97–104
  facies 93–97, 100–102
Great Barrier Reef 37, 83
greenhouse conditions 1, 118, 123
growth fabric 1–2, 75, 84
  rudist bivalves 24, 27–28, 110–112
growth potential, carbonate factories 220–222, 224–225
Gulf of Aqaba 72, 78
Gulf of Suez 72, 78, 81

Halimeda 102
herbivory 92
Hilbre Island 10
Hippuritids *22*, 29
  congregations 113–115, 184
  lithosomes 109–115, 182–184
Holocene, carbonate platform development 140–145
hydrocarbons, Timan-Pechora area 150–151

icehouse conditions 1, 128
ichnotaxa 34, 42
Indo-Pacific region 38

Jamaica 37

karstification 140–141
keep-up 82, 85, 141–142

La Niña 118–119
lagoonal environment 27, 28, 30, 37, 39, 43
lamellae 98
Leptastrea 80
Lighthouse Reef 135–145, *136*
Lithophaga bisulcata 37, 43
Lithophaga laevigata 38
Lithoporella 98, 102
lithosomes 109–115, 181–189
Lithothamnion 98, 102
Litophyton 81, 82
Lobophytum 82

macroborers 33, **36**, 92
  Cretaceous 40–42, 45
  environmental controls 45
  Holocene 35–38
  Jurassic 39–40, 44–45
  Neogene 43–44
  Palaeogene 42–43, 44
  Pleistocene 43–44
  Triassic 38–39, 44
Maeandropolydora **35**, 43
mäerl 92
Maiella carbonate platform margin 177–178
  facies associations 178–181
  lithosome types 181–187
mangrove growth 143
marine extinction events 117–129
Marine Nature Conservation Review 12
micrite 219–220, 222
micritization 92
microbial degradation 63–65, 67, *see also* sulphate reduction
microbial ecosystems 52, 67–68
microbial mats 54, 56–58
microbialites 51, *see also* stromatolites
microborers 33, **35**, 92
Millepora 78, 124
modern bivalves, carbonate production 29–30
modern reef environments 4, 35
Mont-Saint-Michel bay 10
Moorea Atoll 38
mud-mound carbonate factory 1, 219–225
mussel beds 29–30
Mycedium elephantotus 81
Mytilus edulis 29, 30

necrolysis 91
Neogoniolithon 98, 101
New York, biostratigraphy 169–172
non-framework communities 82, 83
nutrient level 204, 223
nutrient pulses 118–119

oceanographic change 84–85
oligotrophic environment 39, 42
ooids 51, 60–63, 67, 219
oolitic fabrics 60–63, 67
organic diagenesis 66
organism-environment feedback 2, 29–31, 71, 82, 84
oyster beds 29

Pacific Ocean, coral mortality 118–123
patch reefs 39, *76*, 77, 82
Pavona 80
Phanerozoic reefs, carbonate production 191–211
Phascolosoma perlucens 37
photosymbiosis 29
Plagioptychus paradoxus 110
Platygyra lamellina 82
polychaete worms 2, 34–35, 37
Polystrata 98, 102
Porites 78, 82, 84, 124
Porites lobata 37
Precambrian platformal carbonates 51
protuberances 98
Pyrenees, hippuritid congregations 109

Quaternary, carbonate platform development 140–145

Radiolitidae *22*, 24, 182–183, 184–187
Red Sea 38, 71–72, *73*, 78, 83
  ecological zonation 77–82
  sedimentary evolution 72
reef calcification 120, 129
reef communities 33
reefal carbonate production
  Phanerozoic
    calculation 194–195
    controls 199–204
    database 191–194
    fluctuation 195–199
reefs
  definitions 1, 72–75
  encrusting 10
  South Wales 14–15
  see also coral reefs
rhodoliths 90
rudist bivalves
  Upper Cretaceous *22*, 109
    carbonate production 23–29
      comparison with modern bivalves 29
    lithosomes 109–115, 181–189
    skeletal growth rates 24–26

Sabellaria alveolata
  characteristics 10, *11*
  distribution 12
  effects on environment 17–18
  exposure block trials 13–14
  growth rates 14, 15–16
  habitat 10, *11*
  research 12
Safaga Bay 72, 75, 77, 82
Sarcophyton 81, 82
Sauvagesia 24, 182
sclerochronology 21
scleractinian coral 38–39, 43, 77–78, 123
sea-level change 83, 84, 115, 118–119, 141–143, 158–159
  eustatic 202, 204
  Frasnian
    New York 169–172
    Timan-Pechora 158–159, 165–168
sea warming 118, 128
sea-water composition 29
  Archaean 65–66
sediment baffling 29, 30
sedimentation rates 217, 220–223, 224
Siderastrea savignyana 82
silicification 59
Sinai 38
sipunculans 34, 37
skeletal growth rates, rudist bivalves 24–26
solar radiation 118, 120–121
South Africa, Campbellrand subgroup 52–54
South Wales, carbonate shore platforms 1, 10, 13
sponges 33, 37, 42, 110, 112
Spongites 98, 102
Sporolithon 98
storms, effect on reefs 30, 118–119, 204
stromatolites 51, 54–68
stromatolitic fabrics 55–60
Stylophora 78, 81, 82
sulphate reduction 58, 63–66
superstratal frameworks 81–82
Syphonophycus transvaalensis 56, 58–59

tabulae 25, 113
taphofacies, Gornji Grad Limestones 100, *101*
taphonomic diagenesis 52, 63–65
taphonomic feedback 1, 6, 113, 115
taphonomy, coralline algae 91–92, 98–102
Tethyan environments 21, 29, 39, 92, 109
thallus 91, 92, 98
Timan-Pechora
  biostratigraphy 159–165
  Devonian 149–151
  Frasnian reef development 151–158, 165–168
Torreites 26
trace fossils 34, *see also* ichnotaxa
Tridacna gigas 29
tropical carbonate factory 219–225
Trypanites **35**, 43
Turbinaria mesenterina 81
Turneffe Islands 135–145, *136*

ultraviolet radiation 118, 120–121
upwelling 124

Vaccinites 110, 112, 115, 182–184
Vaccinites cornuvaccinum 26–28
Vaccinites ultimus 25–27

wave energy 142–145, 201
wave-resistant reefs 10

Yvaniella alpani 25–27

zooxanthellae 118, 123–124, 128